Current Research in Marine Biotechnology

Current Research in Marine Biotechnology

Editor: Gilbert King

RCALLISTO
REFERENCE

www.callistoreference.com

Callisto Reference,
118-35 Queens Blvd., Suite 400,
Forest Hills, NY 11375, USA

Visit us on the World Wide Web at:
www.callistoreference.com

ISBN: 978-1-64116-834-2 (Hardback)

Cataloging-in-Publication Data

Current research in marine biotechnology / edited by Gilbert King.
      p. cm.
Includes bibliographical references and index.
ISBN 978-1-64116-834-2
 1. Marine biotechnology--Research. 2. Marine biology--Research. I. King, Gilbert.
TP248.27.M37 C87 2023
660.6--dc23

Table of Contents

Preface

The world is advancing at a fast pace like never before. Therefore, the need is to keep up with the latest developments. This book was an idea that came to fruition when the specialists in the area realized the need to coordinate together and document essential themes in the subject. That's when I was requested to be the editor. Editing this book has been an honour as it brings together diverse authors researching on different streams of the field. The book collates essential materials contributed by veterans in the area which can be utilized by students and researchers alike.

Marine biotechnology refers to the use of biotechnology tools on marine resources. It utilizes resources from aquatic living organisms to create new commercially viable applications or products. Marine biotechnology has demonstrated its potential in a wide range of applications from biomedicine to the environment. Commercial applications of marine biotechnology include enhancing the production of marine organisms, bio-prospecting, creation of new products, as well as developing biosensors and diagnostics. The goal of current marine biotechnology research is to develop new methods and tools that will improve the economically and environmentally viable industrial production of compounds derived from marine sources. This book unfolds the current research in marine biotechnology, which will be crucial for the progress of this field in the future. Most of the topics introduced herein cover new techniques and the applications of this area of biotechnology. The book aims to serve as a resource guide for students and experts alike and contribute to the growth of the discipline.

Each chapter is a sole-standing publication that reflects each author's interpretation. Thus, the book displays a multi-facetted picture of our current understanding of application, resources and aspects of the field. I would like to thank the contributors of this book and my family for their endless support.E

Editor

Marine Extremophiles: A Source of Hydrolases for Biotechnological Applications

Gabriel Zamith Leal Dalmaso [1,2]**, Davis Ferreira** [3] **and Alane Beatriz Vermelho** [1,*]

[1]   BIOINOVAR—Biotechnology laboratories: Biocatalysis, Bioproducts and Bioenergy, Institute of Microbiology Paulo de Góes, Federal University of Rio de Janeiro, Av. Carlos Chagas Filho, 373, 21941-902 Rio de Janeiro, Brazil; E-Mail: gdalmaso@gmail.com

[2]   Graduate Program in Plant Biotechnology, Health and Science Centre, Federal University of Rio de Janeiro, Av. Carlos Chagas Filho, 373, 21941-902 Rio de Janeiro, Brazil

[3]   BIOINOVAR—Biotechnology Laboratories: Virus-Cell Interaction, Institute of Microbiology Paulo de Góes, Federal University of Rio de Janeiro, Av. Carlos Chagas Filho, 373, 21941-902 Rio de Janeiro, Brazil; E-Mail: davisf@micro.ufrj.br

*   Author to whom correspondence should be addressed; E-Mail: abvermelho@micro.ufrj.br

---

Abstract: The marine environment covers almost three quarters of the planet and is where evolution took its first steps. Extremophile microorganisms are found in several extreme marine environments, such as hydrothermal vents, hot springs, salty lakes and deep-sea floors. The ability of these microorganisms to support extremes of temperature, salinity and pressure demonstrates their great potential for biotechnological processes. Hydrolases including amylases, cellulases, peptidases and lipases from hyperthermophiles, psychrophiles, halophiles and piezophiles have been investigated for these reasons. Extremozymes are adapted to work in harsh physical-chemical conditions and their use in various industrial applications such as the biofuel, pharmaceutical, fine chemicals and food industries has increased. The understanding of the specific factors that confer the ability to withstand extreme habitats on such enzymes has become a priority for their biotechnological use. The most studied marine extremophiles are prokaryotes and in this review, we present the most studied archaea and bacteria extremophiles and their hydrolases, and discuss their use for industrial applications.

Keyword: enzymes; marine extremophiles; hydrolases

1. Extremophiles from Marine Habitats: A Source of Bioproducts

The oceans cover more than 70% of the surface of the planet Earth and contain a vast biological diversity, accounting for more than 95% of the whole biosphere. Marine habitats can be divided into coastal and open ocean environments of various natures, supporting marine life with a wide heterogeneity of microorganisms. They represent the largest reservoir of biodiversity on the planet, and have a great potential for the development of new natural products including enzymes. However, in the prokaryotic group only 1%–10% of the species has been described [1].

Extreme environments combine a range of physical parameters such as pressure, temperature, pH, salinity, oxidative stress, radiation, chemicals (oxygen, H_2S, CH_4) and metals (Fe, Cu, Mo, Zn, Cd, Pb and others). Some of these chemicals are toxic, however extremophiles microorganisms are able to adapt due to their highly flexible metabolism, allowing them to survive and thrive in these extreme conditions [2]. Over the last few decades, extremophiles have attracted the attention of research centers in the search for new bioactive substances, such as enzymes and biocides to be used in major sectors of the world economy, including the agricultural, chemical, food, textile, pharmaceutical, bioenergy and cosmetic industries.

The global market for industrial enzymes was nearly US$ 4.8 billion in 2013, and it is expected to reach US$ 7.1 billion by 2018, with a compound annual growth rate (CAGR) of 8% over the five-year period, according to BCC Research [3]. Besides their economic value, microbial enzymes are applied in technologies employing eco-friendly processes [4]. Microorganisms are easily cultivated in bioreactors with controllable growth conditions such as pH, temperature, aeration, medium composition and other parameters, leading to high reproducibility. In contrast, enzymes isolated from plant and animals present a series of limitations like soil composition, light incidence, seed homogeneity, pathogen control and other issues that make the reproducibility of these processes more difficult [5].

Competition for space and nutrients in the marine environment constitutes a selective force leading to evolution and generating multiple enzyme systems to adapt to the different environments. Marine microorganisms can be found in extreme conditions such as hypersaline habitats, high pressures and extreme temperature. Many marine extremophiles are capable of overcoming such extreme conditions and are a source of enzymes with special characteristics. Therefore, these microorganisms are of great interest for industrial processes, mainly in biocatalysis [6]. This vast variation in marine habitats has led to the development of new hydrolases with novel specificities and properties including tolerance to extreme conditions used in industrial processes [7,8]. Metagenomic studies have revealed that extremophile prokaryotes from marine habitats are a source of novel genes and consequently a source of new bioproducts, including enzymes and other active metabolites. Therefore, it is important to study and understand these microorganisms in order to be able to use the biochemical, ecological, evolutionary and industrial potential of these marine microbes [9,10].

Hydrolases from extremophiles have advantages when compared to chemical biocatalysts. Their catalyses are clean processes, ecologically friendly, highly specific and take place in mild reaction conditions. These hydrolases can also be active in the presence of organic solvents, an important feature for the preparation of single-isomer chiral drugs. Various applications have been described for these hydrolases. A metagenome library was created from the brine: seawater interface of the Urania hypersaline basin. One esterase presented high enantioselectivity toward an ester of the important chiral synthon solketal, an important drug intermediate. The esterase hydrolyzed solketal acetate, producing building blocks for the synthesis of pharmaceuticals, such as anti-AIDS drugs [11]. Lipases from psychrophiles can be used for the synthesis of a wide range of nitrogenized compounds that are used for the production of pharmaceuticals such as amines and amides. This procedure was studied in *Candida Antarctica*, a patented eukaryote [12].

Other hydrolases with medical applications are peptidases. They are useful catalysts for inorganic synthesis and have many industrial applications in the pharmaceutical field, such as anti-inflammatory and digestive agents. Peptidases can be isolated from marine extremophiles [10], such as from *Thermatoga maritime*, which is a hyperthermophilic isolate from a marine geothermal area near Vulcano, Italy, and has a homomultimeric peptidase (669 kDa) based on 31 kDa subunits, named Maritimacin [13]. This enzyme was found to have a structural and gene sequence similarity to bacteriocin from the mesophilic bacterium *Brevibacterium linens* which inhibits the growth of certain Gram-positive bacteria [14]. The thermophilic *Pyrococcus horikoshii* has an intracellular peptidase (PH1704) with remarkable stability. Recently this enzyme, a cysteine peptidase, was shown to be the first allosteric enzyme that has negative cooperativity with chloride ions (Cl^-). The discovery of new allosteric sites is very important for pharmaceutical developments [15]. Peptidases from halophiles have been used in peptide synthesis and an example is the extracellular peptidase from *Halobacterium halobium* which was exploited for efficient peptide synthesis in Water-N'-N'-dimethylformamide [16].

Given the promising potential for the biotechnological applications of these organisms, this work presents the recent advance in the knowledge of hydrolases from marine extremophiles and their biotechnological potential. Peptidases, lipases, amylases, cellulases and other cell wall-degrading hydrolases in halophiles, thermophiles, psychrophiles and piezophiles will be the main focus of this review.

2. Extremophiles

Extremophile organisms are classified as living organisms able to survive and proliferate in environments with extreme physical (temperature, pressure, radiation) and geochemical parameters (salinity, pH, redox potential). Polyextremophile microorganisms are those that can survive in more than one of these extreme conditions. The vast majority of extremophile organisms belong to the prokaryotes, and are therefore, microorganisms belonging to the Archaea and Bacteria domains [17,18]. A phylogenetic tree showing the microorganisms of different genera and their extremophilic characteristics is presented in Figure 1.

Extremophile microorganisms are classified according to the extreme environments in which they grow and the major types are summarized in Table 1. Different structural and metabolic characteristics are acquired by these organisms so that they can survive in these environments [19,20]. Because of the ability to withstand extreme situations, possible industrial applications of extremophiles have been widely investigated [21–25].

One of the most well known applications of an extremophile is the DNA polymerase of the extremophile *Thermus aquaticus* (Taq polymerase) [26] which is widely used in the polymerase chain reaction (PCR). The stability and enzymatic activity of extremophiles and their extremozymes are useful alternatives to conventional biotechnological processes [27].

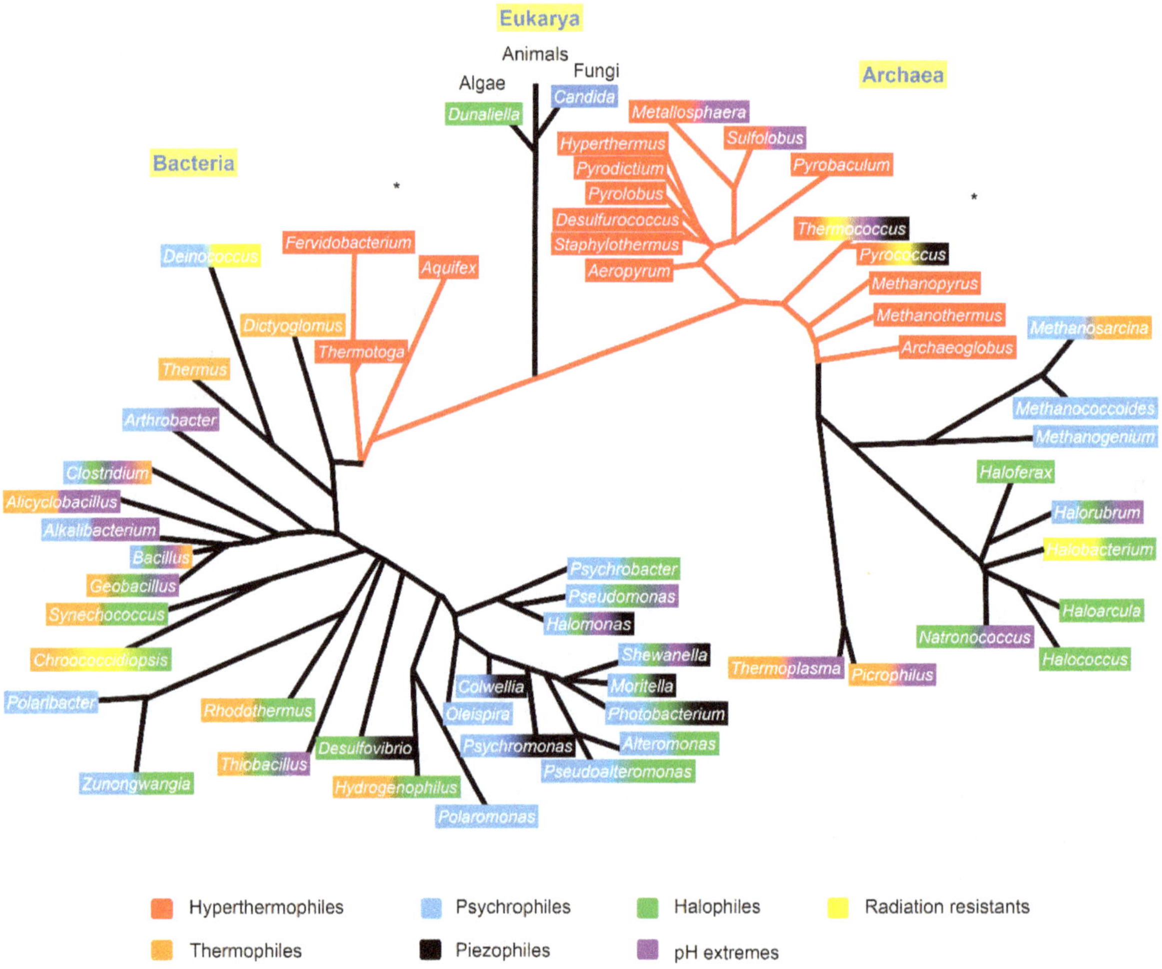

Figure 1. Phylogenetic tree showing the extremophiles and the resistant characteristics that appear in at least one species of each genera, identified with the color code. The phylogenetic tree was based on Woese *et al.* [28], Lang *et al.* [29] and Dereeper *et al.* [30]. The * indicates the phylogenetic branch that were according to Lang *et al.* [29].

Table 1. Extremophile microorganisms and their environments (adapted from Horikoshi and Bull [17]).

Extremophile Microorganism	Favorable Environment to Growth
Acidophile	Optimum pH for growth—Below 3
Alkaliphile	Optimum pH for growth—Above 10
Halophile	Requires at least 1M salt for growth
Hyperthermophile	Optimum growth at temperatures above 80 °C
Thermophile	Grows at temperatures between 60 °C and 85 °C
Eurypsychrophile (psychrotolerant)	Grows at temperatures above 25 °C, but also grow bellow 15 °C
Stenopsychrophile (psychrophile)	Grows at temperatures between 10 °C and 20 °C
Piezophile	Grows under high pressure—Above 400 atm (40 MPa)
Endolithic	Grows inside rocks
Hipolith	Grows on rocks and cold deserts
Oligotroph	Able to grow in environments of scarce nutrients
Radioresistant	Tolerance to high doses of radiation
Metallotolerant	Tolerance to high levels of heavy metals
Toxitolerant	Tolerates high concentrations of toxic agents (eg. Organic solvents)
Xerophile	Grows in low water availability, resistant to desiccation

2.1. Thermophiles

The thermophile marine microorganisms include several groups such as the phototrophic bacteria (cyanobacteria, green and purple bacteria), bacteria domains (*Actinobacteria* sp., *Bacillus* sp., *Clostridium* sp., *Desulfotomaculum* sp., *Thermus* sp., *Thiobacillus* sp., fermenting bacteria, spirochetes and numerous other genres) and the archaea domains (*Pyrococcus* sp., *Sulfolobus* sp., *Thermococcus* sp., *Thermoplasma* sp. and methanogenic) [31–33]. The maximum temperature that hyperthermophile organisms have been observed to tolerate is around 120 °C [34,35].

Thermophiles have several mechanisms to support extreme temperatures. It is believed that the thermostability of cellular components such as ATP, amino acids, and peptides may exceed 250 °C, suggesting that the maximum temperature for life goes beyond the temperatures that have been observed until now [36,37]. The proteins of organisms adapted to extreme temperatures generally have similar three-dimensional structures of mesophilic organisms but the amino acid content is different from ordinary proteins and the number of charged residues on their surfaces is much greater than nonadapted organisms. In addition, such proteins often have shorter loops, thus preventing the occurrence of nonspecific interactions due to their increased flexibility at high temperatures [38,39].

Extreme thermophile bacteria produce thermostable proteins that can be readily crystallized to obtain stable enzymes for structural and functional studies. Proteins from hyper/thermophiles require sufficient structural rigidity to resist unfolding. This is an important feature to characterize antidrug targets. A classical instance is the bacteria *Thermus thermophilus* that was originally isolated from a thermal vent within a hot spring in Izu, Japan, and is frequently used in genetic manipulation studies. The DNA gyrase from this extremophile has been used as an antidrug target model. DNA gyrase is a type IIA topoisomerase that introduces negative supercoils into closed circular bacterial DNA using ATP hydrolysis. It is an important antibacterial target that is sensitive to the widely-used fluoroquinolone drugs [40,41].

The thermal hypothesis determines that a G:C pair and its contents are related to thermostability. This is observed for several thermophilic bacteria. *Geobacillus thermoleovorans* CCB US3UF5 is a thermophilic bacterium that was isolated from a hot spring in Malaysia and is a source for thermostable enzymes. The bacteria contains a circular chromosome of 3,596,620 bp with a mean G:C content of 52.3% [42]. However, a study reported a comparative analyses of G:C composition and optimal growth temperature with 100 prokaryote genomes (Archaea and Bacteria domains) that failed to demonstrate this correlation (G:C/thermostability). Moreover the authors related that the G:C content of structural RNA (16S and 23S) is strongly correlated with optimal temperature and it is higher at high temperatures [43]. An increased number of disulfide bonds improve stability within thermophilic proteins and play a role in preventing the alteration of the quaternary structure [44].

Ether lipids are always present in thermophile archaea without exception, but mesophilic archaea also have ether lipids. The presence of isoprenoid chains in archaea membranes is associated with two properties to maintain the thermostability of the lipid membrane: A high permeability barrier and a liquid crystalline state. Bacterial membranes only keep these states at the transition phase of temperature [45].

Accordingly, enzymes adapted to higher temperatures bring advantages to industrial processes, promoting faster reactions, high solubility of the substrate, a lower risk for contamination of the system, and also lowering the solution viscosity and increasing the miscibility of the solvent [46]. Several thermophiles and thermostable enzymes are used in the biorefinery industry, first and second generation biofuels, paper and bleaching industries [47–49].

2.2. Psychrophiles

Earth is primarily a cold marine planet: 90% of the water in the oceans has temperatures of 5 °C and 20% of the terrestrial region of the Earth is permafrost (frozen soils), glaciers and ice sheets, polar sea ice and snow covered regions [50]. These regions have cold-adapted microorganisms which have a restricted temperature range for growth. Stenopsychrophiles (formerly true psychrophiles) have an upper temperature limit of 20 °C for growth. However, the majority of isolates are eurypsychrophile (formerly psychrotolerant) and have a broader temperature range, tolerating warmer environments [51]. Psychrophile microorganisms are often found in other extremely cold environments such as deep oceans, caves, land surfaces and even in the upper atmosphere [52]. They have been described performing DNA synthesis at −20 °C and active metabolism at −25 °C [53,54].

Various stenopsychrophiles isolated from Antarctic have been studied; for example the genera Arthrobacter, Colwellia, Gelidibacter, Glaciecola, Halobacillus, Halomonas, Hyphomonas, Marinobacter, Planococcus, Pseudoalteromonas, Pseudomonas, Psychrobacter, Psychroflexus, Psychroserpens, Shewanella and Sphingomona. Methanogens, members of Archaea, are the only group known to have individual species able to grow in a very wide temperature range from subzero to 122 °C [55,56].

The adaptations and mechanisms related to life in icy environments include responses to cold shock and RNA chaperones. Cold shock proteins (CSPs) act as cold-adaptive proteins in psychrophiles. They are small proteins that bind to RNA to preserve its single-stranded conformation and contain a nucleic-acid-binding domain, known as the cold shock domain (CSD), and also they have additional roles besides serving as RNA chaperones [57,58]. Small RNA-binding proteins (RBPs) can facilitate cold adaptation but together with the CSPs they have other functions in bacteria. RNA helicases are regulated during cold growth and are capable of unwinding secondary structures in an ATP-dependent manner in some psychrophiles [58]. The presence of dihydrouridine can enhance tRNA flexibility and is elevated in some psychrophilic bacteria and archaea [59].

Other factors involved in cold adaptation are the production of secondary cold active metabolites, enzymes that are activated and induced by cold, antifreeze proteins, and the production of pigments and membrane fluidity [60–62]. From the structural point of view, the proteins of these organisms have a higher content of α-helix relative to the β-sheets, which is considered to be an important factor to maintain flexibility even at low temperatures [63]. Also the cytoplasmic membranes of these microorganisms contain a higher proportion of unsaturated fatty acids (52%) compared to mesophilic (37%) and thermophilic (10%) organisms, favoring the maintenance of the semi-fluid state of membranes [64]. Marine psychrophiles participate in biogeochemical cycling, polar food web and produce a wide variety of enzymes including amylases, cellulases, peptidases, lipases, xylanases and other classes of enzymes [50,65]. High rates of catalysis at low temperatures are generally achieved by the flexible structure and low stability of cold-active enzymes. The most common adaptive feature of cold-active enzymes is a reaction rate (k_{cat}) that is largely independent of temperature [56,66].

Furthermore, enzymes adapted to lower temperatures allow efficient production at low cost, save energy, and are important in thermal protection, resulting in improvements in the quality of various products [67]. Enzymes suited to low temperatures are used in the food, cosmetics, pharmaceutical, and biofuels industries; also they are applied to substances for molecular biology, nanotechnology, in the manufacturing of household detergents, in the cleaning of animal wastes, with peptidases for cleaning contact lenses and pectinases to extract and clear fruit juices [67,68]. A clear example of the benefit generated by the industrial application of cold adapted microorganisms is the use in the hydrolysis of lactose at low temperatures in the process of milk storage. This bioprocess has enabled the production of milk for patients with milk intolerance and is now being patented for use on a large scale by Nutrilab NV (Bekkevoort, Belgium) [69].

2.3. Halophiles

Halophiles are microorganisms that require salt (NaCl) for growth, and they can be found in lakes, oceans, salt pans or salt marshes. Moreover about 25% of the available land on Earth is in the form of saline deposits [70]. According to the optimal salt concentration for growth, they are classified in three categories: (i) extreme halophile—Grows in an environment with 3.4–5.1 M (20% to 30%) NaCl; (ii) moderate halophile—Grows in an environment with 0.85–3.4 M (3% to 25%) NaCl; and (iii) slightly halophile—Grows in an environment with 0.2–0.85 M (1% to 5%) NaCl [31]. Halotolerant microorganisms do not show an absolute requirement for salt to grow but grow well in high salt concentrations [71].

Members of the family *Halobacteriaceae* have been isolated from different habitats including alkaline and salt lakes, marine salterns, the Dead Sea and saline soils. Deep hypersaline anoxic basins (DHABs) or deep-sea hypersaline anoxic lakes (DHALs) are extreme habitats that have been discovered on the sea floor in different oceanic regions, such as the Gulf of Mexico, the Red Sea and the Eastern Mediterranean Sea. DHABs are composed by dissolution of evaporitic deposits, entrapped in the sea floor, forming a very stable brine and sharply stratified in water columns, a chemocline. The *brines* enclosed in these basins are characterized by hypersalinity, 5–10 times higher than seawater, a lack of oxygen and highly reducing conditions, high pressure (around 350 atm–35 MPa) and absence of light. These physicochemical parameters make the DHABs one of the most extreme environments of the planet and they have also ensured that those habitats have been maintained isolated for thousands of years [11,72,73].

Since the discovery of the first Mediterranean DHAB named *Tyro* in 1983, six others have been unveiled: *l'Atalante*, *Bannock*, *Discovery*, *Medee*, *Thetis*, and *Urania* [74], and these habitats are a source of anaerobic halophilic microorganisms. Prokaryotes from the Bacteria and Archaea domains belonging to new taxonomic lineages were discovered in high abundance in DHABs by 16S rRNA libraries and fluorescent *in situ* hybridization (FISH). Microbiologically DHABs of the Eastern Mediterranean are the most studied. *Halorhabdus utahensis* constitutes 33% of the total archaeal community and tolerates up to 0.8 M $MgCl_2$ [75]. In the Mediterranean Ridge, a DHAL named *Kryos* has been identified. This lake is filled with $MgCl_2$-rich, athalassohaline brine (salinity > 470 practical salinity units). Two groups of halophilic euryarchaeal divisions (MSBL1 and HC1) account for ~85% of the rRNA-containing archaeal clones analyzed in the 2.27–3.03 M $MgCl_2$ layer [74]. An earlier study assumed that 2.3 M of $MgCl_2$ was the upper limit of concentration for life to survive [76], despite the fact that halophilic archaea have been identified in deep hypersaline anoxic basins composed of saturating concentrations of $MgCl_2$ [73,77]. Antunes *et al.* [78] recently published a long list of microorganisms in a microbiological review of these unique deep-sea anoxic environments.

In addition, some halophiles are thermostable and tolerant to a wide range of pH. The metabolic diversity of halophiles is widely spread, comprising the anoxic phototrophic, aerobic heterotrophic, fermenter, denitrifying, sulfate reducers and methanogenic organisms [79].

Halophiles have developed different adaptive strategies to support the osmotic pressure induced by the high NaCl concentrations in the environments they inhabit. Some extremely halophilic bacteria accumulate inorganic ions (K^+, Na^+, Cl^-) in the cytoplasm, which is a type of "salt-in" strategy to balance the osmotic pressure of the environment, and they have also developed specific proteins that are stable and active in the presence of salts [80–82]. Also, the halophile organisms contain enzymes that maintain their activity at high salt concentrations, alkaline pH and high temperatures [71]. Most proteins and enzymes denature when suspended in high salt concentrations. Halophilic proteins bind significant amounts of salt and water. This characteristic is dependent on the number of acidic amino acids on the surface of the protein.

The function of electrostatic interactions in the stability and folding of halophilic proteins has been investigated and is an important determinant of haloadaptation. The intracellular K^+ ions of haloarchaea have been found to be extremely high, near to 5 M [83]. Based on the comparative analyses of halophile and non-halophile proteomes, the amino acid composition of halophilic enzymes is in general characterized by an abundant content of acidic amino acid, a high proportion of aspartic and glutamic acids, a low frequency of lysine, and a high occurrence of amino acids with a low hydrophobic character. Structural analyses between halophilic and mesophilic proteins reveal that the major differences are concentrated on the surface of the protein. These characteristics allow cooperation with electrostatic interactions and the presence of a higher number of salt bridges [84]. The stability of the enzymes depends on the negative charge on the surface of the protein due to acidic amino acids, the hydrophobic groups in the presence of high salt concentrations and the hydration of the protein surface due to carboxylic groups present in aspartic and glutamic acids. In addition, negative surface charges are thought to be important for the solvation of halophilic proteins, to prevent denaturation, aggregation and precipitation [85,86].

Moderate halophiles use other haloadaptations based on biosynthesis and/or accumulation in the cytoplasm of high amounts of specific organic osmolytes, which function as osmoprotectants, providing osmotic balance and maintaining low intracellular salt concentrations without interfering in the normal metabolism of the cell. The osmolyte could be obtained by direct uptake from the environment [85]. These solutes can act as stabilizers for biological structures and allow the cells to adapt not only to salts but also to heat, desiccation, cold or even freezing conditions [87]. Many halophilic bacteria accumulate ectoine or hydroxyectoine as the predominant compatible solutes. Other types of osmolyte include glycine, betaine and other neutral glycerols [71].

One of the adaptation mechanism developed is the lipid composition. Structural adaptations have been observed in the S-layers of halophiles. The extreme halophile contains sulfated glucuronic acid residues and a higher degree of glycosylation, leading to an increased density in surface charges. This characteristic demonstrates an adaptation in response to the higher salt concentrations experienced by *Halobacterium salinarum*. Moreover, in *Haloarchaea*, some S-layer glycoproteins are enriched in acidic residues [88].

The halophiles have been used in biodegradation of organic pollutants, in desalinization of wastewater, in nanotechnology, in production of biopolymers and as osmoprotectors [89,90]. The halotolerance of hydrolases derived from halophilic bacteria can be exploited wherever enzymatic transformations are required to function under physical and chemical conditions, such as in the presence of organic solvents and extremes in temperature and salt content. Many halophiles can secrete extracellular hydrolytic enzymes, such as amylases, lipases, peptidases, xylanases and cellulases that are thermostable and adaptable to a wide range of pH [80].

2.4. Piezophiles

High hydrostatic pressure is one of the physical parameters in deep-ocean environments and it plays a selective role in the distribution of life on the planet. The oceans, which have an average depth of 3800 m and an average pressure of 380 atmosphere (atm) or 38 MPa, make up ~95% of the biosphere. In the deepest parts of the oceans, pressures of 700 to 1100 atm (70 to 110 MPa) prevent the growth of most microorganisms. Moreover, the temperature in the deep-sea is typically within the 1–3 °C range. However, there are hydrothermal vent habitats where high pressures and high temperatures are found, and in these regions marine microbes might be exposed to temperatures and pressures ranging from 1–300 °C and 1–1100 atm (0.1–110 MPa), respectively [91].

The effects of high hydrostatic pressure on microbial metabolisms occurs in the cellular structures and cellular processes such as cell division and motility [92]. Microorganisms called piezophiles (previously named barophile), such as deep-sea bacteria or archaea, live in high pressure environments and are of interest to various sectors of biotechnology [93]. The Mariana Trench is the deepest part of the ocean found on the planet; it has a maximum depth of 11 km and a pressure of 1100 atm (110 MPa). This extreme habitat harbors organisms that can grow in standard pressure and temperature and strict piezophiles, like *Moritella yayanosii* and *Shewanella benthica*, that have pressure growth conditions of between 700 and 800 atm (70 ~80 MPa), but not less than 500 atm (50 MPa) [18,94].

Microorganisms which possess optimal growth rates at pressures above atmospheric pressure are classified as piezophilic; and as piezotolerant those that grow at high pressure, as well as at atmospheric pressure but they do not have optimal growth rates at pressures above one [91]. Bacterial piezophiles are mainly psychrophiles belonging to five genera of γ-proteobacteria, *Photobacterium*, *Shewanella*, *Colwellia*, *Psychromonas* and *Moritella*, while piezophilic Archaea are mostly (hyper)thermophiles from *Thermococcales* [95].

The physiological adaptations required for growth under these extreme conditions are substantial and involve a combination of modifications of gene structure and regulation. The adaptation mechanisms of piezophiles are under investigation. Whether piezophilic adaptation requires the modification of a few genes, or metabolic pathways, or a more profound reorganization of the genome has not yet been fully elucidated [95]. As with psychrophiles, piezophiles contain lipids with highly unsaturated fatty acids [96]. Other adaptation mechanisms against the high pressures include reduction of cell division, modification of membrane and transport proteins and accumulation of osmolytes, which stabilize the proteins [94,97,98]. The occurrence of elongated helices in the 16S rRNA genes to increase adaptation to growth at elevated pressure has also been described [99].

2.5. Polyextremophiles

Microorganisms in their natural habitats are thought to experience stress during their life cycle. Many extremophiles inhabit environments with more than one extreme parameter, for example, extremophiles that thrive in the depth of the oceans or close to hot springs. In the first situation, if the extremophiles are found in ocean mud, they could be piezophiles or psychrophiles, but if they were found close to a hydrothermal vent, they could be piezophiles, thermophiles or acidophiles, due to the minerals released in the chimney, or even, if they were found in DHABs, they could be piezophiles, psychrophiles or halophiles.

In an effort to provide a comprehensive look at the extremes of temperature and pH, Capece *et al.* [100] tabulated over 200 extremophile species found in the literature. They are called thermoacidophiles, thermoalkaliphiles, psychroacidophiles and psychroalkaliphiles. Since membrane fluidity decreases at low temperatures, the lower permeability to protons (H$^+$) becomes an advantage for acidophiles and alkaliphiles. The pH homeostasis is controlled by H$^+$ movements across the membrane. The presence of psychroacidophiles and psychroalkaliphiles has not yet been found in nature; only the presence of psychrotolerant alkaliphiles, such as *Alkalibacterium psychrotolerans* has been observed [100,101].

However, the presence of thermoacidophiles is well documented. High temperatures increase the permeability to H$^+$ resulting in a lethal cytoplasmic acidification [102]. Modifications in RNA codon thermostability and a neutral surface charge in proteins prevent the occurrence of an acid hydrolysis. The existence of thermoacidophiles over 100 °C has not yet been observed [100,103]. The survival of thermoalkaliphiles is thought to be related more to the buffering capacity to maintain a stable intracellular cytoplasm than the maintenance of a bioenergetic gradient [100,102,104].

High salt concentrations allied to temperature extremes have been observed in psychrohalophiles and thermohalophiles. Many environments in polar sea ice are cold brine solutions, thus some degree of halophily is typical in most psychrophiles. Indeed, cold adaptation and salt adaptation have common approaches [100]. Thermohalophiles are rare. Their uncharged proteins became extremely unstable in hypersaline solutions, and are denatured by solvents at high temperatures and by decreasing the electrostatic interactions required to maintain the native folding [105,106]. *Thermococcus waiotapuensis* is the most hyper-thermohalophilic organism discovered to date, however, the basis of its biochemical stability is still unclear [107].

The correlation of temperature and pressure is also another parameter involved in extremophilic survivability. Increases in volume are favored at high temperatures but disfavored at high pressures. Thermopiezophile protein adaptations are synergistic, comprising a small surface charge with a strongly hydrophobic core. Other biochemical responses are the induction of both heat-shock and cold-shock response pathways reinforcing the synergistic reaction [100,108]. The cold-shock response is extremely important in psychropiezophiles, since they do not benefit from the synergistic temperature and pressure approach. They also incorporate monosaturated and polysaturated fatty acids to prevent membrane crystallization, which is caused by both extremes of temperature, a characteristic already incorporated in piezophiles [100,109].

Halo-acidophiles and halo-alkaliphiles are both found in nature. For halo-acidophiles, high extracellular concentrations allow a more favorable efflux of H^+. Halo-alkaliphiles are more commonly found, because the monovalent cations of salts are essential for pH homeostasis and energetic coupling [110,111]. On other hand, for halo-alkaliphiles, the less favorable influx of H^+, caused by high salt concentrations, leads to a lethal alkylation of the cytoplasm, and becomes a limitation to growth. Halo-alkaliphiles are able to maintain a gradient of about 1.0–1.5 units [102,112]. Finally, high pressures promote negative volume changes by the dissociation of acids and the protonation of amine groups in proteins. This leads to an acidification of the solution, creating an environment where piezo-acidophiles can be found. On the other hand, the alkalinification of the environment associated to high-pressures creates a propitious site for piezo-alkaliphiles. However, both of these organisms still need to be further investigated for systematic identification and characterization [100,113].

Polyextremophilic enzymes have been applied in the food, detergent, chemical, pulp and paper industries. A thermo-alkali-stable enzyme from *Bacillus halodurans* TSEV1 has applicability in pre bleaching of paper pulp and recently has been expressed in *Pichia pastoris* for the production of oligosaccharides [114,115]. Another strain of *B. halodurans* PPKS-2 produced an alkaliphilic, halotolerant, detergent and thermostable mannanase. This strain grows in agro wastes and can be applied for mannanase production on an industrial scale for detergent and pulp and paper bleaching [116]. The Antarctic cold-adapted halophilic Archeon *Halorubrum lacusprofundi* produces a recombinant polyextremophilic enzyme that is active in cold temperatures, high salinity and is stable in aqueous-organic mixed solvents. This enzyme is suitable for applications in synthetic chemistry [117].

3. Biocatalysis: Bioengineering and Other Strategies

The application of hydrolases in industrial processes sometimes fails due to the lack of robustness, stability and undesirable properties [5]. Recent experimental advances, associated to novel bioinformatic tools and protein engineering, has allowed the development of more efficient hydrolases for industrial purposes [46,118,119]. Since the introduction of site-directed mutagenesis and other integrated modern techniques such as synthetic biology and system biology, it is now possible to modify the characteristics of these enzymes, such as enhancing their stability and specificity [120,121].

Anaerobic, extremophilic and marine bacteria are a source of enzymes with superior chances of success in biotechnological processes. A great deal of laboratory effort has been concentrated on their production and characterization. Furthermore, the design of novel enzymes as well as molecular approaches such as enzyme evolution and metagenomic approaches can be used to identify and develop novel biocatalysts from uncultured bacteria—A treasure-trove of unknown proteins. Some characteristics of extremophiles are important for obtaining recombinants with specific properties. The advantage of psychrophile enzymes is their reduced energy consumption. Thus, psychrophiles could be used in the production of thermolabile proteins for genetic engineering [60].

Stabilization of hydrolytic enzymes is of interest due their potential applications in the medical, chemical and pharmaceutical industries. Methods to stabilize proteins include protein engineering or chemical modifications [122,123]. In the past two decades, protein engineering has become a powerful means to alter or improve enzymatic catalysis [124,125], and is divided into two methods: (1) directed evolution—A random mutagenesis is applied to a protein; and (2) rational protein design—Knowledge of the structure and function of the protein is necessary in order to change its characteristics. These methods have been successfully applied to increase protein activity, selectivity or thermostability [126]. The protein engineering technique is helping to improve the production of chemicals, such as optically pure tertiary alcohols and drugs, like the profen family (NSAIDs) [127,128]. Directed evolution has been reported to be laborious and costly, however it does provide the means for selecting mutants with improved properties [122,123]. Allying chemical modifications to recombinant DNA technology can generate improvements in enzyme stability and efficiency, preventing changes in the active-site and in targeting enzyme surface groups [129,130].

Mutations could be induced by physical and chemical methods such as UV-irradiation, γ-rays, fast neutron irradiation, nitrosoguanidine (NTG), diethyl sulfate and nitrous acid which have been applied to breed lipase-producing microorganisms. With these simple methods, the lipase yield can be improved by 1 to 10-fold; however, the low positive mutation rate, the long periodicity and the laborious screening has limited its widespread use [131].

Enhancing the stability of extremophilic enzymes is very beneficial because this could maintain the high activity at low/high physical-chemical parameters for prolonged periods of time, a characteristic that can be exploited in many industrial processes. However, basic issues involved in the stabilization mechanisms need to be analysed before any modifications can be made. These issues include the type and size of the protein, the structure and size of the modifying reagent, the chemical reactions involved in the modification procedures and the conditions of such modifications [122,123].

Siddiqui et al. [132] demonstrated that chemical modifications could be useful in providing details of structure, function and stability of proteins, turning them into potential guides for future target studies, including an attempt to convert mesophilic enzymes into cold-adapted ones. Tadeo et al. [84] demonstrated that it is possible to decrease the salt dependence of a halophilic protein to the level of a mesophilic one, and engineering the protein in an inverse form, from mesophilic to an obligate halophilic form, suggesting that the halophilicity is related exclusively to surface residues.

On the other hand, searching for enzymes that already exist in nature could be faster and more straightforward than the engineering routes. Metagenomic techniques could be a powerful tool for making such discoveries, by trying to find novel genes and enzymes from uncultured microorganisms [133]. Metagenomic analysis can be either sequence driven, for example, accessing the 16S rRNA, recA or other conserved sequences, or function driven, expressing features such as a specific enzyme activity or antibiotic production [134]. The 16S rRNA metagenomic studies provide an enormous diversity of free-living marine microorganisms from different environments and include the discovery of new enzymes of microbial origin [135]. A list of marine enzymes discovered by metagenomics can be found in Kennedy *et al.* [136].

Some advantages can be applied to functional metagenomics over the sequence methods. The functional approach recognizes the genes by their function, rather than their sequences, avoiding incorrect annotations or similar sequences of gene products with multiple functions. When you search for novel functions or a complete new class of genes, this could be crucial [136]. Currently platform databases, such as KEGG (Kyoto Encyclopedia of Genes and Genomes) and GenomeNet, are helping in the interpretation of a large amount of data generated by metagenomic output analyses. A large number of bioinformatic tools are now available for Data Mining of metagenomic sequences, based on a series of features [136,137]. Combining different approaches, like activity-based mining with tailoring of robust and selective enzymes, would give a biotechnological process a chance to take over from chemical catalysis, transforming industrial chemistry into a more eco-friendly process [138].

Currently all these techniques are powerful tools for biocatalysis. Miyake *et al.* [139] constructed a recombinant protein expression system using the psychrophilic bacterium *Shewanella* sp., isolated from Antarctic seawater. This bacterium grows at temperatures close to 0 °C. The enzyme produces β-lactamase reaching 91 mg/liter of culture at 4 °C and 139 mg/liter of culture at 18 °C. In another study, mesophilic enzymes were expressed in psychrophilic microorganisms for efficient 3-hydroxypropionaldehyde production from glycerol, using *Shewanella livingstonensis* Ac10 as a selected host [140]. A heat-sensitive esterase and two chaperones (Cpn60 and Cpn10) from a psychrophilic bacterium, *Oleispira antarctica* RB-8 T were simultaneously expressed in *E. coli*. The resulting enzyme from the recombinant strains was more active (180 fold) than the *E. coli* strain grown at 37 °C and was also active at low temperatures [141].

Other more suitable techniques can be applied in the search for novel enzymes from rare microorganisms, such as single cell genomics, metatranscriptomics and metaproteomics. The single cell genomics allow the study of the entire biochemical process of a single uncultured cell; the metatranscriptomics can access only the transcriptionally active genes in their specific population; and the metaproteomics can analyze the enzymes directly involved in a particular biochemical pathway [136].

Systems Biology is another integrated strategy used in industrial biotechnology. A quantitative analysis of biological systems is carried out using a mathematical model via computer simulation. Cellular metabolisms are analyzed and optimized for application in the development of strains and bioprocesses [121]. "Omics" technologies have been employed in studies such as: Transcriptome (genome-wide study of mRNA expression levels), proteome (analysis of structure and function of proteins and their interactions), metabolome (measurement of all metabolites to access the complete metabolic response to a stimulus), fluxome analysis (metabolic flux) and also in advanced mathematical modeling tools, such as genome-scale metabolic modeling [5,121].

Synthetic Biology is a field that has been growing in the last few years and is in many ways related to genetic engineering. It is an integrated and interdisciplinary theme including bioinformatics, microbiology, molecular biology, systems biology and biology in order to design and construct new biologic functions and systems not found in nature or improving certain functions through the creation of a synthetic genome [120].

4. Hydrolases and Their Biotechnological Potential

Currently more than 500 products are produced using enzymes and about 150 industrial processes benefit from the use of enzymes or catalysts from microorganisms. Moreover, more than 3000 enzymes are known and approximately 65% are hydrolases used in the detergent, textile, pulp, paper and starch industries and almost 25% of these are used for food processing [142,143]. Studies show that the diversity of extremophile microorganisms may be higher than we think [31,144]. However, the characterization and use of such a diversity of enzymes becomes complicated due to the difficulty of isolating and growing these microorganisms [142].

By definition, hydrolases are enzymes that catalyze reactions with the substrate through the hydrolysis of chemical bonds. Hydrolases are enzymes classified as Class 3 (EC 3) by the Nomenclature Committee of the International Union of Biochemistry and Molecular Biology (NC-IUBMB) that keeps an updated list of all the enzymes described in a database, available at ExplorEnz [145].

5. Amylases from Marine Extremophiles

Starch comprises an abundant source of available energy and is present in almost all higher plants. Starch is a polymer composed of glucose molecules that form a so-called straight-chain amylose via the α 1–4 linked type. The association of α 1–4 with the α 1–6 type branches creates amylopectin, the largest part of the starch molecule [146].

Amylases are enzymes, which hydrolyze the starch molecules to glucose monomers and can be classified according to the specificity of the substrate in which they operate, as showed in Table 2 [5,147]. Figure 2 illustrates the function of the three major amylolytic enzymes.

The amylolytic enzymes are one of the most interesting enzymes for industrial processes. The α-amylase of different species of the genus *Bacillus* are the amylases that are the most applied in biotechnological processes, because of their thermophilic properties and high conversion rates [148]. The research for extremozymes and their special characteristics for industrial processes has been expanded [149], and over the years several stable amylases from thermophiles, psychrophiles, alkaliphiles, acidophiles and halophiles have been reported [150–152]. They are found in different genera of marine extremophile archaea and bacteria from the surface to deep sea locations, and include *Desulfurococcus* sp., *Pseudoalteromonas* sp., *Pyrococcus* sp., *Rhodothermus* sp. and *Thermococcus* sp. [153–156]. *Geobacillus* sp. is an isolate from a geothermal vent, and has a remarkable alpha-amylase stability between 80 °C to 140 °C [157].

Table 2. Classification of Amylases.

		Enzyme	Classification	Cleavage	Product
Amylases	**Endoamylase**	α-amylases	EC 3.2.1.1	Internal α-1,4	Dextrins
	Exoamylases	β-amylase or maltase	EC 3.2.1.2	Outer regions of α-1,4	Maltose
		Glucoamylase	EC3.2.1.3		β-cyclodextrin and glucose
		α-glucosidase	EC 3.2.1.20		
	Debranching	Pullulanases	EC 3.2.1.41	α-1,6 linkages	Maltotriose
		Isoamylases	EC 3.2.1.68	Pullulan	Malto-oligosaccharides
		Dextrinases	EC 3.2.1.142	α-1,6 linkages	Maltose

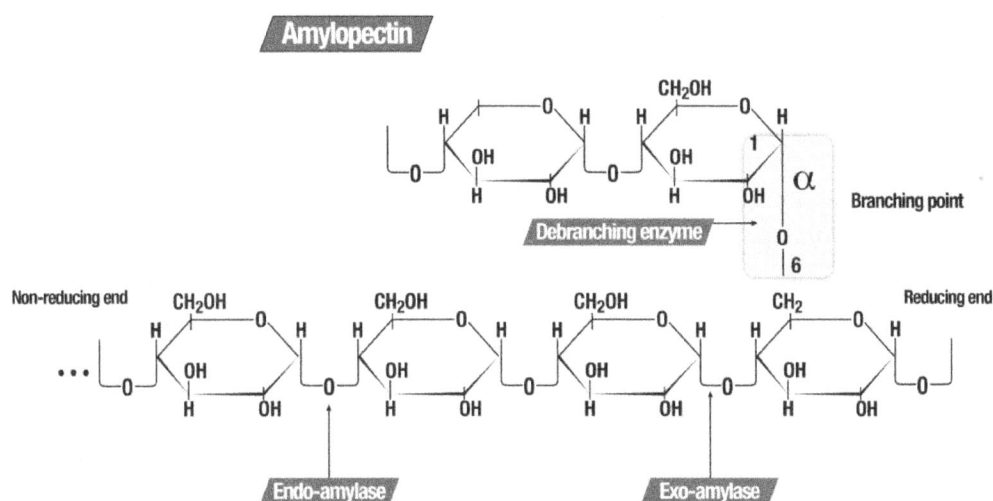

Figure 2. The three major amylolytic enzymes [5].

Several industrial processes can be performed with the use of amylases. The endoamylases, which have optimum activity for temperatures between 70 °C and 95 °C have been applied in the production of ethanol from corn starch, rice or as a pretreatment for forming sugars (saccharification) for fermentation [158,159]. Amylases are also used in the food industry for the production of glucose syrups, fructose and maltose, reduced viscosity of syrups, reducing the turbidity of juices and also for alcohol fermentation. They are also involved in the textile, papermaking, detergent, chemical, pharmaceutical and petroleum industries [5,160,161].

Thermostable amylases are important for starch hydrolysis under high temperatures, accelerating the reactions and reducing possible contaminations [148]. Many companies sell different thermostable and broad-range pH operating enzymes. This is the case of Fuelzyme®—Verenium Corporation (San Diego, CA, USA), an alpha-amylase which was originated from *Thermococcus* sp. isolated from a deep-sea hydrothermal vent. Fuelzyme® is applied to mash liquefaction during ethanol production, releasing dextrins and oligosaccharides with better solubility and with low molecular weight. It operates in a pH range of 4.0–6.5 and temperatures above 110 °C [162]. However, Fuelzyme® and Spezyme® (DuPont-Genencor Science) are only used for biofuel production. The use of a blend of these commercially available amylases and others from *Bacillus* sp. has been suggested to make industrial starch processing more efficient, and suitable for downstream applications [163].

Pullulanases are commonly associated to glucoamylases for the saccharification of starch, and there is a growing demand in industry for them [164]. Marine sources have been naturally appointed as providers of such enzymes [165]. Thermostable pullulanases from type II (amylopullulanases) are being used in the process of starch liquefaction and saccharification combined [166,167]. The hyperthermophilic archaea *Staphylothermus marinus*, an isolate from the deep-sea associated with geothermal activity or hydrothermal vents, presents an optimum growth at 98 °C. A new thermostable amylopullulanase of the glycosyl hydrolase family from *S. marinus*, has recently been described with degradation activity towards pullulan and cyclodextrin at 105 °C [168]. One of the most thermostable and thermo-active pullulanases type II was described for *Pyrococcus furiosus* and *Thermococcus litoralis*, with activity ranging from 130 °C to 140 °C in the presence of 5 mM Ca^{++} [169].

On the other hand, cold adapted amylases also have their uses in the detergents, textile and food industries, due to their considerable energy savings and reduction of bacterial contamination [150]. The most studied cold active alpha-amylase originates from *Alteromonas haloplanctis*, which is synthesized at 0–2 °C and has been successfully expressed in *Escherichia coli* [170]. The alpha-amylase from *Pseudoalteromonas haloplanktis* (AHA) (former *Alteromonas haloplanctis*) showed 80% of initial activity at 4.5 M of NaCl at 10 °C [171]. Structural features in AHA have been studied through site-directed mutagenesis and chemical modification, including a modification performed that provides support for the importance of arginine residues, instead of lysine, which enhances the enzyme thermostability to cold adaptation; however, it decreases their activity [132]. Recently, *Zunongwangia profunda* was isolated from the deep-sea and presented a cold adapted and salt tolerant alpha-amylase, one of the very few alpha-amylases that can tolerate both cold and salt conditions [172]. An alpha-amylase was found in an isolate from the genus *Bacillus* on a marine salt farm, with a hyperthermostable enzyme acting at 110 °C as optimum operating temperature [173]. The *Halomonas* sp. strain AAD21 was found to produce a halo and thermostable alpha-amylase [174]. The *Holoarcula* sp. stain S-1, produced an extracellular organic solvent-tolerant alpha-amylase that was stable and active in benzene, toluene and chloroform, with a maximal activity at 50 °C in 4.3 M of NaCl and pH 7.0 [175]. Extreme halophiles and piezophiles are not common sources of amylases. Some other examples of extremophile producers of amylases are summarized in Table 3.

Table 3. Extreme amylases from marine extremophiles.

Microorganism	Domain	Natural Isolation Site	Metabolism	Enzyme	Type	Reference
Pyrococcus furiosus (recombinant)	Archaea	Thermal marine sediments	Hyperthermophile	Amylase Endoamylase	α-amylase	[176]
Fervidobacterium pennavorans V5 (recombinant)	Bacteria	Hot springs, Azores islands	Hyperthermophile	Amylase debranching	Pullulanase type I	[177]
Pseudoalteromonas haloplanktis	Bacteria	Antarctica	Psycrophille	Amylase Endoamylas	α-amylase	[171]
Halothermothrix orenii	Bacteria	Tunisian salt lake	Halophile/poliextremophile	Amylase Endoamylase	α amylase	[178]
Haloferax mediterranei	Archaea	Saltern, Spain	Extreme halophile	Amylase Endoamylase	a-amylase	[179]

6. Cell Wall-Degrading Hydrolases from Marine Extremophiles

Along with starch, the cell wall is another element present in the structure of all plants and which also maintains an energy reserve, although rarely used and of difficult access. This element consists of three major polymers: Cellulose, hemicellulose and lignin. Cellulose is the most abundant macromolecule on Earth and one of the major constituents of plants, formed by β-1,4-linked glucose molecules. Due to its compactness and crystalline disposition in nature, cellulose is very resistant to hydrolysis and degradation. Hemicelluloses are non-cellulosic polysaccharides composed of complex carbohydrate polymers, where xylan and glucomannan are the main components. Lignin, with hemicellulose and pectin, fills the spaces between the cellulose fibers acting as a bonding material between the cell wall components [180].

Cellulases are complex hydrolases capable of degrading insoluble cellulose polymers, present in plants, fungi and bacteria. Cellulases are among the enzymes that are the most produced for industrial purposes and it is expected that within a few years, their production will increase further, due to their use in biofuel conversion [181]. Regarding the hydrolysis of xylan, a wide variety of enzymes become necessary, and they differ in their specificity and mechanism of action [5,182]. Table 4 summarizes the classification of cellulases and xylanases. Figure 3 illustrates the dynamics of polysaccharide catalysis.

Cellulolytic microorganisms have developed complex forms of cellulolytic systems which actively hydrolyze the cellulose fibrils, and are capable of producing cellulolytic enzymes with additional functions. Furthermore, they are organized in the form of multiprotein complexes such as cellulosomes and xylanossomas [180,183,184]. The use of different vegetables such as sugarcane, corn, beets, among others for the bioenergy industry generates residues that could be reused. Biotechnological research has been stimulated to develop technologies for second-generation ethanol production from plant biomass containing lignocellulose [184,185]. Extremophile microorganisms, especially thermophilic and alkaliphiles, are widely used in lignocellulolytic processes. The thermophilic and psychrophilic cellulases have also been used in different industrial processes in the food and fermented beverages, textile, pulp and paper and animal feed industries [47,186,187]. Glycoside hydrolases are commonly found in marine thermophile-microorganisms, like *Pyrococcus* sp., *Thermococcus* sp., and *Thermotoga* sp. [156,188,189].

Beta-glycosidase from the hyperthermophilic bacteria *Thermotoga maritima*, used in transglycosylation reactions, is thermostable and resistant to a large number of proteolytic denaturants found in nature and the presence of alcohol or organic compounds stimulates its activities [190]. Thermostable xylanases are being largely used in the paper bleaching industry [191]. Xylosidases and xylanases are expressed in *Thermotoga neapolitana* during the bio-production of hydrogen, using many different carbohydrates as feedstock [192]. Recombinant xylanase is being improved to achieve more extremophilic characteristics, like *XynB* from *T. maritima*, which has been successfully expressed in *E. coli*, exhibiting thermo and alkaline stability, an attractive characteristic for bleaching kraft pulp in the paper industry [193]. Aiming for crystalline cellulose hydrolysis, a variant designed from a beta-1,4-endoglucanase (EGPh) of *P. horikoshii* exhibited stronger activity than the wild type EGPh [194].

Table 4. Classification of cell wall-degrading hydrolases.

		Enzyme	Classification	Cleavage	Product
Cellulases	Endoglucanases	Endo-β-1,4-glucanase	EC 3.2.1.4	Intramolecular bonds of β-1,4-glycosidic	New chain ends
	Exoglucanases	β-glucosidase	EC 3.2.1.21	Ends of the cellulose	Glucose or soluble cellulose
		Exo-β-1,4-glucan cellobiohydrolase	EC 3.2.1.91	Glycosidic terminals	Cellobiose
Xylanases		β-1,4-endoxylanase	EC 3.2.1.8	Internal glycosidic linkages along heteroxylan main skeleton	Polymerization degree of the substrate
		α-D-xylosidase	EC 3.2.1.177	Small xylo-oligosaccharides and xylobiose	Xylose

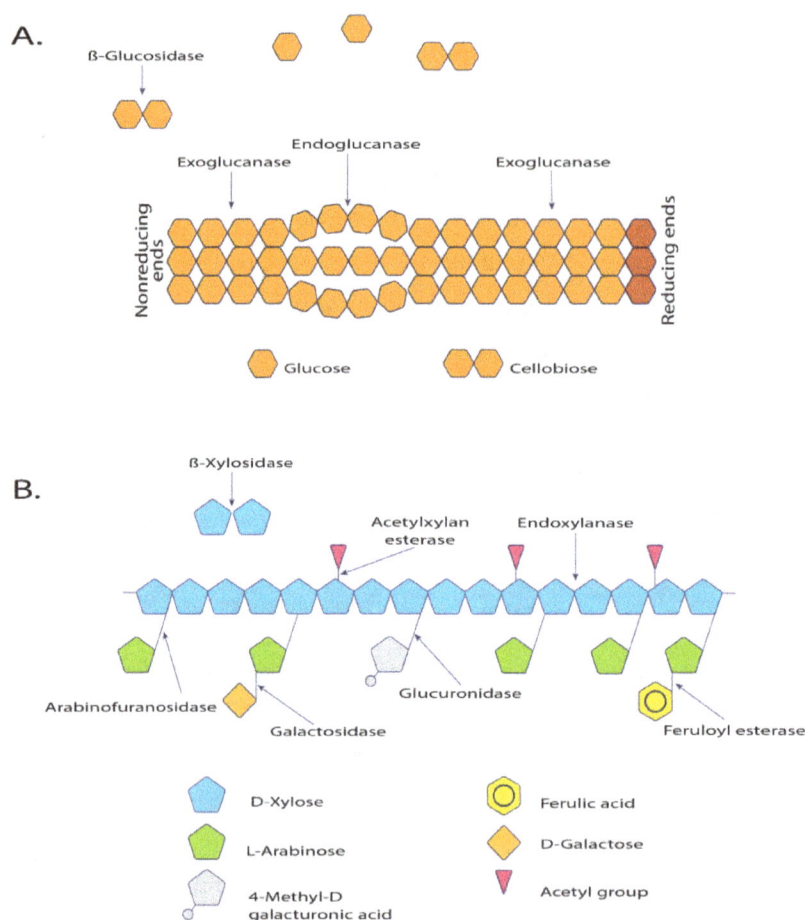

Figure 3. (**A**) Enzymes involved in the hydrolysis of cellulose and (**B**) in xylan hydrolysis [183]. Exo-β-1,4-glucanase and cellobiohydrolase hydrolyze the glycosidic terminals releasing cellobiose units. The β-glucosidases act directly on cellulose, hydrolyzing it to glucose. The 1,4-β-glucosidase is essential to complete the hydrolysis process of the cellulose [5]. The α-D-xylosidases are exoglycosidases that act in the non-reducing end, hydrolyzing small xylo-oligosaccharides and xylobioses, releasing xylose.

Psychrophiles are a source of these enzymes. The first cold-active and alkali-stable β-glucosidase was isolated from *Martelella mediterranea*. This enzyme retains 80% of its activity at pH 11.0 and 50% at 4 °C [195]. A deep-sea mud *Exiguobacterium oxidotolerans* also presents a cold active beta-glycosidase, maintaining 61% of its maximum activity at 10 °C and a pH range from 6.6 to 9.0 [196]. The Antarctic bacterium *P. haloplanktis CelG* gene was purified and expressed in *E. coli* for kinetic and structural optimization purposes [197]. Psychrophilic xylanases, like *TAH3a* from *P. haloplanktis* and *MSY-2* from *Flavobacterium* sp. have also been used in the baking industry [198,199]. Ideal enzymes for treating milk should work well at 4–8 °C and pH 6.7–6.8, and this is still highly desired in industrial applications. *Arthrobacter* sp. is an Antarctic isolate that produces a β-D-galactosidase capable of working at 4–8 °C, and it retains 15% of its maximum activity at 0 °C [200].

The slightly halophilic bacterium *Alteromonas macleodii* has been isolated from the Pacific Ocean, Mediterranean Sea, English Channel, Black Sea and Gulf of Thailand. The activities for beta-D-galactosidase, alpha-D-galactosidase and beta-D-glucosidase were found using a sample isolated from the Black Sea [201].

7. Marine Extremophilic Proteases

Peptidases or proteases are proteolytic enzymes that catalyze the hydrolysis of peptide bonds on proteins or peptides. Peptidases are classified in subclass 3.4 (peptide-hydrolases). The rating uses three criteria: I—Chemical mechanism of catalysis. In this system, peptidases are classified according to their catalytic type in serine, cysteine, threonine, aspartic, glutamic, asparagine, metallo and unknown catalytic (S, C, T, A, G, N, M, and U, respectively) and P for peptidases with protein nucleophiles of mixed catalytic types. II—Type of reaction they catalyze. These are subdivided into exo-peptidases (EC 3.4.11-19), when enzymes cleave terminal amino acids, and endo-peptidases (EC 3.4.21-99), when they hydrolyze peptide bonds in the middle of the polypeptide chain. III—Molecular structure and homology. This is the most modern classification concept, based on amino acid sequences and three-dimensional structures. The peptidases are classified in families and clans. In the families, the classification of peptidases is grouped by amino acid sequence comparisons: 251 families of peptidases can be found in MEROPS Release 9. Currently the number of peptidases described and indexed in the MEROPS database exceeds more than 4000 enzymes [202]. Figure 4 summarizes the classification of peptidases.

Peptidases are enrolled in different biological processes such as regulation, localization, modulation and activities of protein interactions [203]. An important function of peptidases, is to perform posttranslational processing events, leading to the activation or inactivation of proteins, including enzymes [204]. Peptidases are also a potential drug target for microbial diseases, taking part in pathogenesis, inactivating the host immune defense mediators, in the processing of host or parasite proteins and in the digestion of host proteins. These features make peptidases very valuable to the pharmaceutical industries [5]. Other innumerous applications of peptidases are in the detergent, cosmetic, chemical and food industries [5,142].

Most peptidases from extremophilic archaea belong to the serine peptidases family, although, other families are also represented. Hyperthermophilic peptidases include serine peptidases, cysteine peptidases and the threonine-dependent proteasomes. *Pyrococus* sp. are archaeon hyperthermophiles, strictly anaerobes and obligate heterotrophs and are found in diverse habitats such as thermal marine sediments and shallow hydrothermal vents [205]. *P. furiosus* was isolated from deep-sea vents and volcanic marine mud of Italy [206]. An intracellular peptidase from *P. horikoshii* is an oligomeric cysteine peptidase [15] similar to the *P. furiosus* intracellular peptidase I [207]. In this genus, pyrolysin, a serine peptidase, metalopeptidases and ATP-dependent peptidases such as Lon A and subunits of proteasome have been characterized [208]. *Thermococcus litoralis* is a hyperthermophile archaea that is found around deep-sea hydrothermal vents as well as near shallow submarine thermal springs and oil wells. A proline dipeptidase named prolidase cleaves dipeptides having proline at the C-terminus and a nonpolar residue (Met, Leu, Val, Phe, Ala) at the amino terminus [209].

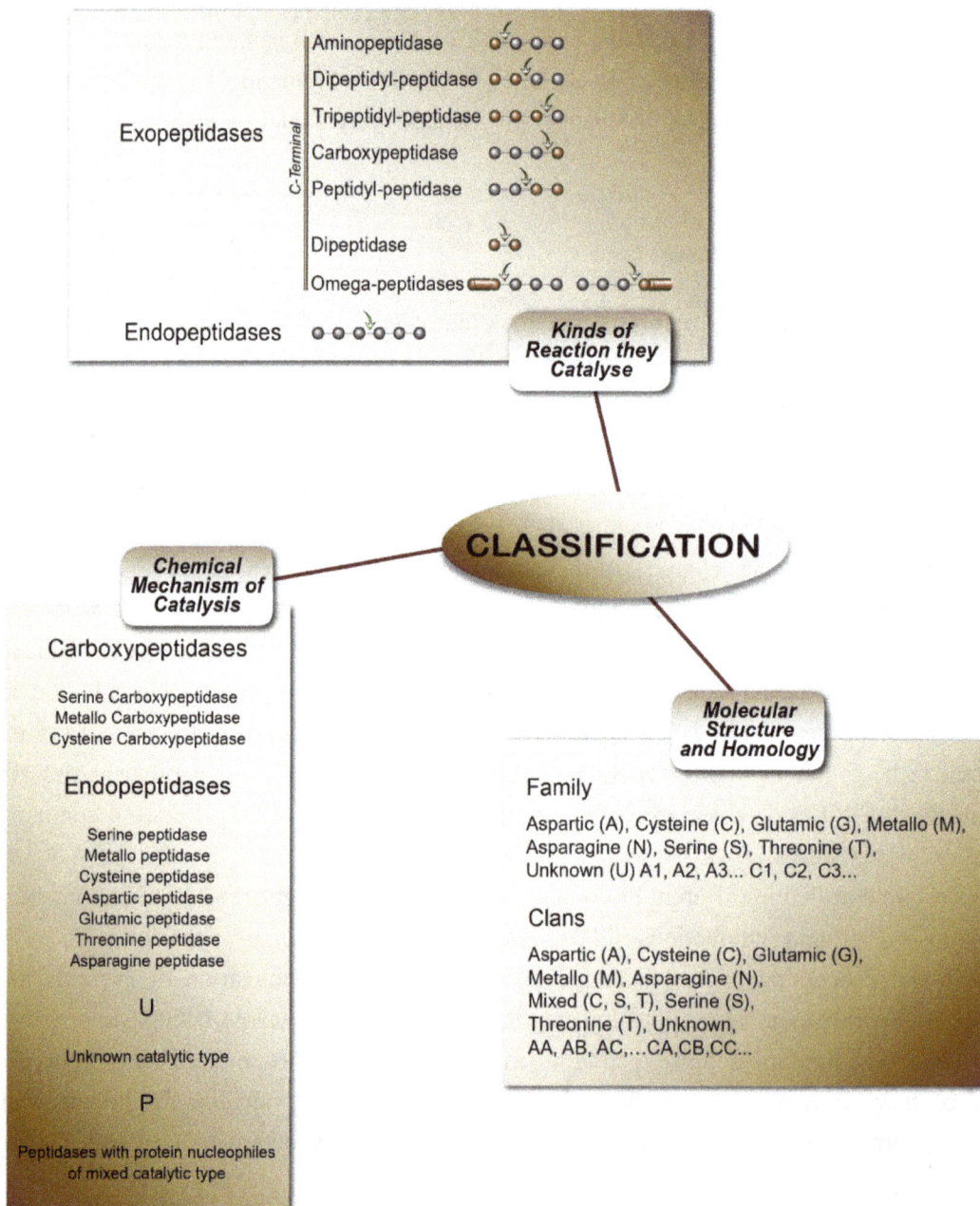

Figure 4. The classifications of peptidases [5].

Psychrophiles are a source of peptidases and other hydrolases. The potential of the enzymes produced by them has been reviewed [65,210,211]. The diversity of extremophile microorganisms is a valuable resource in the search for new proteolytic enzymes, mainly those active at low temperatures derived from psychrophiles [69,186,212]. *S. livingstonensis* is a psychrophilic, gram-negative, isolate from Antarctic coastal areas that produces a serine alkaline peptidase [213]. *Colwellia psychrerythraea is* an obligate psychrophile, Gram-negative bacteria that can be found in cold marine environments including in Arctic and Antarctic sea-ice. The bacteria produce a peptidase of the family M1 aminopeptidase. The enzyme with a molecular mass of 71 kDa displayed an optimum temperature at 19 °C [214].

Natrialba magadii is an extremophile archaea that lives in alkaline hypersaline conditions (pH 9.5, 3.5 M NaCl). A Lon peptidase from this halophile was cloned and sequenced by Sastre *et al.* [215]. The ATP-dependent Lon peptidase is universally distributed in bacteria, eukaryotic organelles and archaea. In this species, like other archaea, there is a Lon B type serine peptidase, with a molecular mass of 85 kDa. *Haloferax volcanii*, a haloarchaeon isolated from the Dead Sea, also presents a Lon Peptidase. In a study by Cerletti *et al.* [216] of this species, the group demonstrated that a suboptimal cellular level of the Lon protein affected its growth rate, cell shape, cell pigmentation, lipid composition and sensitivity to various antibiotics. An extracellular serine peptidase with 45 kDa was characterized in the same species by Gimenez *et al.* [217]. *Natrialba asiatica*, which was isolated from a beach in Japan, *and H. mediterranei* isolated from seawater evaporation ponds near Alicante, Spain produced a halolysin, a serine peptidase [218]. Other extracellular and intracellular peptidase classes including metalopeptidases, and 20S proteasomes/threonine peptidases are also found in the group (for review see De Castro *et al.* [219]). Table 5 summarizes other peptidases found in marine extremophiles.

Halophilic microorganisms present some advantages in fermentation processes that occur in the presence of salt. The high salt tolerance of extreme halophiles enables their cultivation under non-sterile and thus cost-reducing conditions [220]. Some peptidases such as bromelain, papain, and pepsin have been used as biocatalysts of protein hydrolysis in fish sauce fermentation. These peptidases are not stable in high salt concentration. In contrast, peptidases from extreme halophiles require salt to carry out their activities. In addition, halophilic peptidases have a wide application in processing activities for the food, leather and detergent industries [220,221].

The *Deinococcus* genus has been isolated from a variety of habitats including Antarctica, water and activated sludge, hot springs and aquifers [222–224]. The fresh water radioactive site isolate, *Deinococcus aquiradiocola*, proved to be a proteolytic, capable of hydrolyzing gelatin and casein [225]. *Deinococcus indicus*, an arsenic-resistant bacterium isolated from an aquifer, has been positively characterized for casein and gelatin hydrolysis [226]. Some species have been isolated from marine habitats such as *Deinococcus geothermalis* from deep-ocean subsurfaces [227,228]. Pietrow *et al.* [229] described the production of the thermo-alkali-stable peptidase from *D. geothermalis*. The peptidase presented stability at 60 °C and pH 9.0; these properties are appropriate for the detergent industry.

Table 5. Peptidases from marine extremophiles.

Peptidase	Extremophile	Habitats	Reference
Serine peptidase	*Fervidobacterium pennavorans*	Hot spring	[230]
	Sulfolobus solfataricus	Volcanic hot spring	[231]
	Thermococcus litoralis	Deep-sea hydrothermal vent Shallow submarine thermal springs and oil wells	[232]
	Thermoanaerobacter yonseiensis	Geothermal hot stream at Sileri on Java island, Indonesia	[34,233]
Cysteine peptidase	*Aciduliprofundum boonei*	*Marine* hydrothermal vent	[234]
	Haloferax volcanii	Dead Sea, the Great Salt Lake, and oceanic environments with high NaCl	[235]
Metallocarboxypeptidase	*Caldithrix abyssi*	Deep-sea hydrothermal chimneys	[234]

8. Lipases and Esterases from Marine Extremophiles

Lipases and esterases are hydrolases that catalyze the cleavage and formation of ester bonds (EC 3.1) and belong to Carboxylic Ester Hydrolases (EC 3.1.1) (Table 6). Both types of enzymes belong to the family of serine hydrolases and share structural and functional characteristics, such as an α/β-hydrolase fold in their core structure, and have a characteristic catalytic triad (serine, aspartic/ glutamic and histidine). Lipases are mainly active against water-insoluble substrates, such as triglycerides composed of long-chain fatty acids (C ≥ 8). Esterases preferentially hydrolyze "simple" esters and usually only triglycerides composed of fatty acids shorter than C 8. Based on amino acid sequences, esterases and lipases have been grouped into eight families. Enzymes in Family 1 are called true lipases and are further classified into six subfamilies. Enzymes belonging to Families 2–8 are esterases. The genome analysis of bacteria has shown putative lipases/esterases genes that are not included in any family [236,237].

Table 6. Lipases and esterases according the Nomenclature Committee of the International Union of Biochemistry and Molecular Biology (NC-IUBMB) [145].

Lipases	Systematic Name	E.C.	Reaction Catalysed
Carboxylesterase (other names: Esterases, serina esterases, *etc.*)	Carboxylesterase	3.1.1.1	carboxylic ester + H_2O = alcohol + carboxylate
Triacylglycerol lipase (other names: Lipase; triglyceride lipase, *etc.*)	Triacylglycerol acylhydrolase	3.1.1.3	triacylglycerol + H_2O = diacylglycerol + carboxylate

The biotechnological applications of lipases have a broad spectrum of use including in the food, dairy, pharmaceutical, cosmetic, agrochemical, biosurfactant, detergent and paper industries [238–241]. Biodiesel production has been one of the major sectors stimulating the search for lipases [242,243]. Biodiesel is composed of methyl-esterified fatty acids derived from transesterification of triglycerides by enzymatic action, providing a number of advantages such as a reduction of the operational process in the manufacture and separation of glycerol by-products. For the environment, the advantages are countless, such as the reduction of particles emissions, low toxicity and high biodegradability [5,244].

The genus *Pseudomonas* is the most exploited for lipases [245]. The industrial use of lipases and esterases from extremophile microorganisms has been increasing steadily in recent years [246], mainly with the use of thermostable enzymes from thermophiles and psychrophiles [247–249].

The *Thermococcus* genus has been isolated from black smokers (hydrothermal vents) or freshwater springs, with 1%–3% salt (NaCl) concentration, other marine hydrothermal areas and oil reservoirs. *Thermococcus sibiricus* is a hyperthermophile from a high-temperature Samotlor oil reservoir (Western Siberia). Genomic analysis revealed that *T. sibiricus* contains 15 genes that encode for lipases/esterases. Four of these putative enzymes contain signal peptides suggesting that they are extracellular enzymes [250].

A thermostable esterase gene from the aquatic hyperthermophilic *Archaeoglobus fulgidus* DSM 4304 was cloned in *E. coli*. The esterase was tested with various acyl chains of ρ-nitrophenol and the highest activity was found for ρ-nitrophenyl butyrate at 80 °C [251]. The archaeon *Pyrobaculum calidifontis* VA1, also a hyperthermophilic isolated from a hot spring in the Philippines, presented a thermostable carboxylesterase, with an optimum operation at 90 °C, and organic tolerance, supporting concentrations up to 80% [252]. Other thermostable esterases have been described in *S. solfataricus* P1. Arylesterase, the least understood among the esterases, showed high stability against denaturing agents at an optimum temperature and pH of 94 °C and 7.0, respectively [253]. Esterases have been purified and studied from the extremely *Sulfolobus acidocaldarius* [254] and *Sulfolobus shibatae* [255]. A thermostable esterase gene from *P. furiosus* has been cloned in *E. coli*, purified and studied [256].

Lipases from psychrophile microorganisms are some of the most widely used classes of enzymes in biotechnology applications, organic chemistry, detergent industry, for bioremediation purposes and in the food industry [5,12,240]. A cold-adapted lipase (M37) from *Photobacterium lipolyticum*, previously isolated from an intertidal flat of the Yellow Sea in Korea, maintains activity between 5 °C to 25 °C, and was expressed in *E. coli* at 18 °C [257]. The lipase M37 has been tested as an alternative to *C. antarctica* lipase B (Novozym435®—Novozymes Corporation (Bagsvaerd, Denmark), in the production of biodiesel. Novozym435 is unstable in a medium containing high concentrations of methanol, required for efficient biodiesel production [258,259].

C. antarctica was discovered to be a lipase producer in the 1980's and it was found that two types of lipases (A and B) were being produced. Since then, innumerous applications have been described, such as the production of non-steroidal anti-inflammatory drugs (NSAIDs), biofuel and organic compounds [260–262]. Lipase B is commonly used for desymmetrization, ester synthesis and production of peracids [130,263]. Due to such abilities, a great effort has been made to improve the stability of lipase B through chemical modifications and directed evolution [130,261,264].

Several Antarctic isolates from Terra Nova Bay, including *Pseudoalteromonas* sp., *Psychrobacter* sp. and *Vibrio* sp. have exhibited cold adapted lipases and esterases. The enzymes of six of them work better at a 4 to 15 °C range, with pH changing from 5 to 9.0 and the NaCl concentrations between 1% and 7.0% [265]. Until now, only a few esterases from psychrophilic microorganisms have been cloned and characterized [156]. The cold-adapted esterase from *Pseudoalteromonas arctica* shows optimum activity at 25 °C, but retains more than 50% of its activity at 0 °C [266]. In addition, after the complete genome of *P. haloplanktis* was sequenced, a new cold-active lipase (Lip1) that configures a new lipolytic family was found [237]. Other psychrophilic cold-adapted lipases and esterases are summarized in Table 7.

Table 7. Lipases and esterases from psychrophiles.

Prokaryote	Habitat	Lipase/Esterase	Reference
Desulfotalea psychrophila (sulfate-reducing bacteria)	Marine sediments that are permanently cold	Esterase	[267]
Pseudoalteromonas haloplanktis	Marine Antarctic	Lipase	[237]
Psychrobacter sp. wp37	Deep-sea sediments	Lipase	[268]
Colwellia psychrerythraea 34H (recombinant in *E. coli*)	Cold marine environments including Arctic and Antarctic sea ice	Lipase	[269]

One halotolerant strain identified as *Salinivibrio* sp. strain SA-2 was isolated from a hypersaline brackish water with 14% salinity (Garmsar-Iran) and presented a thermostable lipase, retaining 90% of its activity at 80 °C in a pH range of 7.5–8 for 30 min, tolerating a range of 0 to 3.0 M NaCl concentrations [270]. The Haloarcula marismortui is an extreme halophilic archaeon isolated from the Dead Sea, a hypersaline lake, producing different enzymes such as alkaline phosphatase, lipases, alcohol dehydrogenases and other bio-products [271]. An esterase from *H. marismortui* was expressed in *E. coli* and has been characterized as a salt-dependent esterase, with maximum activity at 3.0 M of KCl and no activity in the absence of this salt [272]. A lipase was isolated from *Natronococcus* sp., an extremely halophilic archaeon [273]. According to Ferrer *et al.* [11], five esterase genes were found in the Urania deep-sea hypersaline anoxic basins (Eastern Mediterranean), through a screening in the metagenome expression library. After expression in *E. coli* at least two were able to function in an extreme condition, with high-pressure and saline affinity, and one possesses an adaptive structure, showing halotolerance and conferring a wide range of catalytic activities. These esterases may possibly take part in the production of intermediate pharmaceuticals, synthesizing optically pure biological active substances.

9. Conclusions and Future Perspectives

The study of marine extremophiles has been significantly reinforced using the modern techniques of bioengineering and molecular biology. Such interactions have favored a better understanding of these microorganisms and their biotechnological applications. The evolutionary adaptation to extreme conditions has facilitated the selection of a more tolerant metabolism, of enzymes and other products with special features not found in any other prokaryotes. Good examples are hydrolases with the ability to operate in temperatures as low as those of the microorganisms isolated from Antarctica to those active at high temperatures such as the hyperthermophilic microorganisms isolated from thermal fractures; enzymes that act in high concentrations of salts found in anoxic basin environments, or high pressure, are found in several polyextremophiles. These properties make these unique microorganisms a source of genes that may be mapped by bio-prospecting metagenomic studies and used for cloning and expression in other organisms to obtain enzymes with properties suitable for industrial bioprocesses. However, although the studies related to the marine extremophiles cited here have increased greatly over the past decade, there are still other extremozymes in promising habitats such as deep sea floors, and also in the Red Sea that have so far attracted few studies and are open for exploration. Marine extremophiles are a group of prokaryotic organisms that are being intensively studied worldwide.

Although major advances have been made recently, our knowledge of the physiology, metabolism, enzymology and genetics of this fascinating group of microorganisms is still limited. Therefore, an important growth is expected to take place in this sector, pursuing a better understanding of the application of these hydrolases from marine extremophiles.

Acknowledgments

The authors wish to acknowledge the Brazilian National Council for Scientific and Technological Development (MCT-CNPq), Carlos Chagas Filho Foundation for Research Support in the State of Rio de Janeiro (FAPERJ) and Coordination for the Improvement of Higher Education Personnel (CAPES) for the financial support. This manuscript was reviewed by a professional science editor and by a native English-speaking copy editor to improve readability.

Author Contributions

The authors contributed equally to this work.

References

1. Bull, A.T.; Ward, A.C.; Goodfellow, M. Search and discovery strategies for biotechnology: The paradigm shift. *Microbiol. Mol. Biol. Rev.* **2000**, *64*, 573–606.

2. Nath, I.V.A.; Bharathi, P.A.L. Diversity in transcripts and translational pattern of stress proteins in marine extremophiles. *Extremophiles* **2011**, *15*, 129–153.

3. BCC Research. *Global Markets for Enzymes in Industrial Applications*; BIO030H; BCC Research: Wellesley, MA, USA, 2014; p. 146.

4. Díaz-Tenaa, E.; Rodríguez-Ezquerroa, A.; Marcaidea, L.N.L.L.; Bustinduyb, L.G.; Sáenzb, A.E. Use of Extremophiles Microorganisms for Metal Removal. *Procedia Eng.* **2013**, *63*, 67–74.

5. Vermelho, A.B.; Noronha, E.F.; Filho, E.X.; Ferrara, M.A.; Bon, E.P.S. Diversity and biotechnological applications of prokaryotic enzymes. In *The Prokaryotes*; Rosenberg, E., DeeLong, E.F., Lory, S., Stackebrandt, E., Thompson, F., Eds.; Springer-Verlag Heidelberg: Berlin, Germany, 2013; pp. 213–240.

6. Chandrasekaran, M.; Kumar, S.R. Marine microbial enzymes. In *Biotechnology*; Werner, H., Roken, S., Eds.; EOLSS: Paris, France, 2010; Volume 9, pp. 47–79.

7. Fulzele, R.; Desa, E.; Yadav, A.; Shouche, Y.; Bhadekar, R. Characterization of novel extracellular protease produced by marine bacterial isolate from the Indian Ocean. *Braz. J. Microbiol.* **2011**, *42*, 1364–1373.

8. Samuel, P.; Raja, A.; Prabakaran, P. Investigation and application of marine derived microbial enzymes: Status and prospects. *Int. J. Ocean. Mar. Ecol. Syst.* **2012**, *1*, 1–10.

9. Russo, R.; Giordano, D.; Riccio, A.; di Prisco, G.; Verde, C. Cold-adapted bacteria and the globin case study in the Antarctic bacterium *Pseudoalteromonas haloplanktis* TAC125. *Mar. Genomics* **2010**, *3*, 125–131.

10. Trincone, A. Potential biocatalysts originating from sea environments. *J. Mol. Catal. B-Enzym.* **2010**, *66*, 241–256.

11. Ferrer, M.; Golyshina, O.V.; Chernikova, T.N.; Khachane, A.N.; Martins Dos Santos, V.A.; Yakimov, M.M.; Timmis, K.N.; Golyshin, P.N. Microbial enzymes mined from the Urania deep-sea hypersaline anoxic basin. *Chem. Biol.* **2005**, *12*, 895–904.

12. Joseph, B.; Ramteke, P.W.; Thomas, G. Cold active microbial lipases: Some hot issues and recent developments. *Biotechnol. Adv.* **2008**, *26*, 457–470.

13. Hicks, P.M.; Rinker, K.D.; Baker, J.R.; Kelly, R.M. Homomultimeric protease in the hyperthermophilic bacterium *Thermotoga maritima* has structural and amino acid sequence homology to bacteriocins in mesophilic bacteria. *FEBS Lett.* **1998**, *440*, 393–398.

14. Valdes-Stauber, N.; Scherer, S. Nucleotide sequence and taxonomical distribution of the bacteriocin gene lin cloned from *Brevibacterium linens* M18. *Appl. Environ. Microbiol.* **1996**, *62*, 1283–1286.

15. Zhan, D.; Sun, J.; Feng, Y.; Han, W. Theoretical study on the allosteric regulation of an oligomeric protease from *Pyrococcus horikoshii* by Cl-Ion. *Molecules* **2014**, *19*, 1828–1842.

16. Kim, J.; Dordick, J.S. Unusual salt and solvent dependence of a protease from an extreme halophile. *Biotechnol. Bioeng.* **1997**, *55*, 471–479.

17. Horikoshi, K.; Bull, A.T. Prologue: Definition, categories, distribution, origin and evolution, pioneering studies, and emerging fields of extremophiles In *Extremophiles Handbook*; Horikoshi, K., Antranikaian, G., Bull, A.T., Robb, F.T., Stetter, K.O., Eds.; Springer: Tokyo, Japan, 2011; pp. 4–15.

18. Stan-Latter, H. Physico-chemical boundaries of life. In *Adaption of Microbial Life to Environmental Extremes*; Stan-Latter, H., Fendrihan, S., Eds.; Springer-Verlag: New York, NY, USA, 2012; pp. 1–14.

19. Horikoshi, K.; Antranikaian, G.; Bull, A.T.; Robb, F.T.; Stetter, K.O. *Extremophiles Handbook*; Springer: Tokyo, Japan, 2011; p. 1247.

20. Seckbach, J.; Oren, A.; Stan-Latter, H. *Polyextremophiles: Life Under Multiple Forms of Stress*; Springer: Dordrecht, The Netherlands, 2013; p. 634.

21. Arora, R.; Bell, E.M. Biotechnological applications of extremophiles: Promise and prospects. In *Life at Extremes: Environments, Organisms and Strategies for Survival*; Bell, E.M., Ed.; CAB International: Dunbeg, UK, 2012; pp. 498–521.

22. Gabani, P.; Singh, O.V. Radiation-resistant extremophiles and their potential in biotechnology and therapeutics. *Appl. Microbiol. Biotechnol.* **2013**, *97*, 993–1004.

23. Karan, R.; Capes, M.D.; Dassarma, S. Function and biotechnology of extremophilic enzymes in low water activity. *Aquat. Biosyst.* **2012**, *8*, 4.

24. Morozkina, E.V.; Slutskaia, E.S.; Fedorova, T.V.; Tugai, T.I.; Golubeva, L.I.; Koroleva, O.V. Extremophilic microorganisms: Biochemical adaptation and biotechnological application (review). *Prikl. Biokhim. Mikrobiol.* **2010**, *46*, 5–20.

25. Singh, O.V. *Extremophiles: Sustainable Resources and Biotechnological Implications*; Wiley-Blackwell: Hoboken, NJ, USA, 2012; p. 456.

26. Gelfand, D.H.; Stoffel, S.; Lawyer, F.C.; Saiki, R.K. Purified Thermostable Enzyme. US 4,889,818, 26 December 1989.

27. Rampelotto, P.H. Biotechnological applications of extremophiles. *Curr. Biotechnol.* **2013**, *2*, 273–274.

28. Woese, C.R.; Kandler, O.; Wheelis, M.L. Towards a natural system of organisms: Proposal for the domains Archaea, Bacteria, and Eucarya. *Proc. Natl. Acad. Sci. USA* **1990**, *87*, 4576–4579.

29. Lang, J.M.; Darling, A.E.; Eisen, J.A. Phylogeny of bacterial and archaeal genomes using conserved genes: Supertrees and supermatrices. *PLoS ONE* **2013**, *8*, e62510.

30. Dereeper, A.; Guignon, V.; Blanc, G.; Audic, S.; Buffet, S.; Chevenet, F.; Dufayard, J.F.; Guindon, S.; Lefort, V.; Lescot, M.; *et al.* Phylogeny.fr: Robust phylogenetic analysis for the non-specialist. *Nucleic Acids Res.* **2008**, *36*, W465–W469.

31. Pikuta, E.V.; Hoover, R.B.; Tang, J. Microbial extremophiles at the limits of life. *Crit. Rev. Microbiol.* **2007**, *33*, 183–209.

32. Rothschild, L.J.; Mancinelli, R.L. Life in extreme environments. *Nature* **2001**, *409*, 1092–1101.

33. Stetter, K.O. History of discovery of hyperthermophiles. In *Extremophiles Handbook*; Horikoshi, K., Antranikaian, G., Bull, A.T., Robb, F.T., Stetter, K.O., Eds.; Springer: Tokyo, Japan, 2011; pp. 404–426.

34. Canganella, F.; Wiegel, J. Anaerobic thermophiles. *Life* **2014**, *4*, 77–104.

35. Takai, K.; Nakamura, K.; Toki, T.; Tsunogai, U.; Miyazaki, M.; Miyazaki, J.; Hirayama, H.; Nakagawa, S.; Nunoura, T.; Horikoshi, K. Cell proliferation at 122 degrees C and isotopically heavy CH4 production by a hyperthermophilic methanogen under high-pressure cultivation. *Proc. Natl. Acad. Sci. USA* **2008**, *105*, 10949–10954.

36. Fushida, S.; Mizuno, Y.; Masuda, H.; Toki, T.; Makita, H. Concentrations and distributions of amino acids in black and white smoker fluids at temperatures over 200 °C. *Org. Geochem.* **2014**, *66*, 98–106.

37. White, R.H. Hydrolytic stability of biomolecules at high temperatures and its implication for life at 250 degrees C. *Nature* **1984**, *310*, 430–432.

38. Colletier, J.P.; Aleksandrov, A.; Coquelle, N.; Mraihi, S.; Mendoza-Barbera, E.; Field, M.; Madern, D. Sampling the conformational energy landscape of a hyperthermophilic protein by engineering key substitutions. *Mol. Biol. Evol.* **2012**, *29*, 1683–1694.

39. Tehei, M.; Madern, D.; Franzetti, B.; Zaccai, G. Neutron scattering reveals the dynamic basis of protein adaptation to extreme temperature. *J. Biol. Chem.* **2005**, *280*, 40974–40979.

40. Aung, H.L.; Samaranayaka, C.U.; Enright, R.; Beggs, K.T.; Monk, B.C. Characterisation of the DNA gyrase from the thermophilic eubacterium *Thermus thermophilus*. *Protein. Expr. Purif.* **2015**, *107*, 62–67.

41. Hidalgo, A.; Berenguer, J. Biotechnological applications of *Thermus thermophilus* as host. *Curr. Biotechnol.* **2013**, *2*, 304–312.

42. Sakaff, M.K.L.M.; Rahman, A.Y.A.; Saito, J.A.; Hou, S.B.; Alam, M. Complete genome sequence of the thermophilic bacterium *Geobacillus thermoleovorans* CCB_US3_UF5. *J. Bacteriol.* **2012**, *194*, 1239–1239.

43. Hurst, L.D.; Merchant, A.R. High guanine-cytosine content is not an adaptation to high temperature: A comparative analysis amongst prokaryotes. *Proc. Biol. Sci.* **2001**, *268*, 493–497.

44. Reed, C.J.; Lewis, H.; Trejo, E.; Winston, V.; Evilia, C. Protein adaptations in archaeal extremophiles. *Archaea* **2013**, *2013*, 373275.

45. Koga, Y. Thermal adaptation of the archaeal and bacterial lipid membranes. *Archaea* **2012**, *2012*, 789652.

46. Liszka, M.J.; Clark, M.E.; Schneider, E.; Clark, D.S. Nature *versus* nurture: Developing enzymes that function under extreme conditions. *Annu. Rev. Chem. Biomol. Eng.* **2012**, *3*, 77–102.

47. Klippel, B.; Antranikian, G. Lignocellulose converting enzymes from thermophiles. In *Extremophiles Handbook*; Horikoshi, K., Antranikaian, G., Bull, A.T., Robb, F.T., Stetter, K.O., Eds.; Springer: Tokyo, Japan, 2011; pp. 444–474.

48. Turner, P.; Mamo, G.; Karlsson, E.N. Potential and utilization of thermophiles and thermostable enzymes in biorefining. *Microb. Cell. Fact.* **2007**, *6*, 9.

49. Yeoman, C.J.; Han, Y.; Dodd, D.; Schroeder, C.M.; Mackie, R.I.; Cann, I.K. Thermostable enzymes as biocatalysts in the biofuel industry. *Adv. Appl. Microbiol.* **2010**, *70*, 1–55.

50. Junge, K.; Christner, B.; Staley, J.T. Diversity of psychrophilic bacteria from sea ice—and glacial ice communities. In *Extremophiles Handbook*; Horikoshi, K., Antranikaian, G., Bull, A.T., Robb, F.T., Stetter, K.O., Eds.; Springer: Tokyo, Japan, 2011; pp. 794–815.

51. Cavicchioli, R. Cold-adapted archaea. *Nat. Rev. Microbiol.* **2006**, *4*, 331–343.

52. Margesin, R.; Miteva, V. Diversity and ecology of psychrophilic microorganisms. *Res. Microbiol.* **2011**, *162*, 346–361.

53. Bakermans, C.; Bergholz, P.W.; Rodrigues, D.; Vishnevetskaya, T.A.; Ayala-del-Río, H.L.; Tiedje, J. Genomic and expression analyses of cold-adapted microorganisms. In *Polar Microbiology: Life in a Deep Freeze*; Miller, R.V., Whyte, L.G., Eds.; ASM Press: Washington, DC, USA, 2011; pp. 126–155.

54. Mykytczuk, N.C.; Foote, S.J.; Omelon, C.R.; Southam, G.; Greer, C.W.; Whyte, L.G. Bacterial growth at −15 °C; molecular insights from the permafrost bacterium *Planococcus halocryophilus* Or1. *ISME J.* **2013**, *7*, 1211–1226.

55. Kurosawa, N.; Sato, S.; Kawarabayasi, Y.; Imura, S.; Naganuma, T. Archaeal and bacterial community structures in the anoxic sediment of Antarctic meromictic lake Nurume-Ike. *Polar Sci.* **2010**, *4*, 421–429.

56. Siddiqui, K.S.; Williams, T.J.; Wilkins, D.; Yau, S.; Allen, M.A.; Brown, M.V.; Lauro, F.M.; Cavicchioli, R. Psychrophiles. *Annu. Rev. Earth Planet. Sci.* **2013**, *41*, 87–115.

57. Jones, P.G.; Inouye, M. The cold-shock response--a hot topic. *Mol. Microbiol.* **1994**, *11*, 811–818.

58. Lim, J.; Thomas, T.; Cavicchioli, R. Low temperature regulated DEAD-box RNA helicase from the Antarctic archaeon, *Methanococcoides burtonii*. *J. Mol. Biol.* **2000**, *297*, 553–567.

59. Noon, K.R.; Guymon, R.; Crain, P.F.; McCloskey, J.A.; Thomm, M.; Lim, J.; Cavicchioli, R. Influence of temperature on tRNA modification in Archaea: *Methanococcoides burtonii* (optimum growth temperature [T_{opt}], 23 °C) and *Stetteria hydrogenophila* (T_{opt}, 95 °C). *J. Bacteriol.* **2003**, *185*, 5483–5490.

60. Casanueva, A.; Tuffin, M.; Cary, C.; Cowan, D.A. Molecular adaptations to psychrophily: The impact of "omic" technologies. *Trends Microbiol.* **2010**, *18*, 374–381.

61. De Maayer, P.; Anderson, D.; Cary, C.; Cowan, D.A. Some like it cold: Understanding the survival strategies of psychrophiles. *EMBO Rep.* **2014**, *15*, 508–517.

62. Lorv, J.S.; Rose, D.R.; Glick, B.R. Bacterial ice crystal controlling proteins. *Scientifica* **2014**, *2014*, 976895.

63. Madigan, M.T.; Martinko, J.M.; Dunlap, P.V.; Clarck, D.P. *Brock Biology of Microorganisms*, 14th ed.; Benjamin Cummings: San Francisco, NC, USA, 2014; p. 1136.

64. Deming, J.W. Extremophiles: Cold environments. In *The Desk Encyclopedia of Microbiology*; Schaechter, M., Ed.; Academic Press: Oxford, UK, 2009; pp. 147–157.

65. Joshi, S.; Satyanarayana, T. Biotechnology of cold-active proteases. *Biology* **2013**, *2*, 755–783.

66. Siddiqui, K.S.; Cavicchioli, R. Cold-adapted enzymes. *Annu. Rev. Biochem.* **2006**, *75*, 403–433.

67. Fendriham, S.; Negoiţă, T.G. Psychrophilic microorganisms as important source for biotechnological processes. In *Adaption of Microbial Life to Environmental Extremes*; Stan-Latter, H., Fendrihan, S., Eds.; Springer-Verlag: New York, NY, USA, 2012; pp. 133–172.

68. Huston, A.L. Biotechnological aspects of cold adapted enzymes. In *Psychrophiles: From Biodiversity to Biotechnology*; Margesin, R., Schinner, F., Marx, J.-C., Gerday, C., Eds.; Springer: Heidelberg-Berlin, Germany, 2008; pp. 347–364.

69. Feller, G. Psychrophilic enzymes: From folding to function and biotechnology. *Scientifica* **2013**, *2013*, 512840.

70. Jackson, R.B.; Carpenter, S.R.; Dahm, C.N.; McKnight, D.M.; Naiman, R.J.; Postel, S.L.; Running, S.W. Water in a changing world. *Ecol. Appl.* **2001**, *11*, 1027–1045.

71. Siglioccolo, A.; Paiardini, A.; Piscitelli, M.; Pascarella, S. Structural adaptation of extreme halophilic proteins through decrease of conserved hydrophobic contact surface. *BMC Struct. Biol.* **2011**, *11*, 50.

72. Javor, B. Deep sea hypersaline basins. In *Hypersaline Environments*; Springer Heidelberg, Berlin, Germany, 1989; pp. 176–188.

73. Van der Wielen, P.W.; Bolhuis, H.; Borin, S.; Daffonchio, D.; Corselli, C.; Giuliano, L.; D'Auria, G.; de Lange, G.J.; Huebner, A.; Varnavas, S.P.; *et al.* The enigma of prokaryotic life in deep hypersaline anoxic basins. *Science* **2005**, *307*, 121–123.

74. Yakimov, M.; La Cono, V.; Ferrer, M.; Golyshin, P.; Giuliano, L. Metagenomics of deep hypersaline anoxic basins. In *Encyclopedia of Metagenomics*; Nelson, K.E., Ed.; Springer: New York, NY, USA, 2014; pp. 1–9.

75. Mapelli, F.; Borin, S.; Daffonchio, D. Microbial diversity in deep hypersaline anoxic basins. In *Adaption of Microbial Life to Environmental Extremes*; Stan-Lotter, H., Fendrihan, S., Eds.; Springer: Vienna, Austria, 2012; pp. 21–36.

76. Hallsworth, J.E.; Yakimov, M.M.; Golyshin, P.N.; Gillion, J.L.; D'Auria, G.; de Lima Alves, F.; La Cono, V.; Genovese, M.; McKew, B.A.; Hayes, S.L.; *et al.* Limits of life in $MgCl_2$-containing environments: Chaotropicity defines the window. *Environ. Microbiol.* **2007**, *9*, 801–813.

77. McGenity, T.J.; Oren, A. Hypersaline environments. In *Life at Extremes: Environments, Organisms and Strategies for Survival*; Bell, E.M., Ed.; CAB International: Dunbeg, UK, 2012; pp. 402–437.

78. Antunes, A.; Ngugi, D.K.; Stingl, U. Microbiology of the Red Sea (and other) deep-sea anoxic brine lakes. *Environ. Microbiol. Rep.* **2011**, *3*, 416–433.

79. Oren, A. Microbial life at high salt concentrations: Phylogenetic and metabolic diversity. *Saline Syst.* **2008**, *4*, 2.

80. Moreno, M.L.; Pérez, D.; García, M.T.; Mellado, E. Halophilic bacteria as a source of novel hydrolytic enzymes. *Life* **2013**, *3*, 38–51.

81. Ginzburg, M.; Sachs, L.; Ginzburg, B.Z. Ion metabolism in a *Halobacterium*. I. Influence of age of culture on intracellular concentrations. *J. Gen. Physiol.* **1970**, *55*, 187–207.

82. Lanyi, J.K.; Silverman, M.P. The state of binding of intracellular K + in *Halobacterium cutirubrum. Can. J. Microbiol.* **1972**, *18*, 993–995.

83. Roberts, M.F. Organic compatible solutes of halotolerant and halophilic microorganisms. *Saline Syst.* **2005**, *1*, 5.

84. Tadeo, X.; Lopez-Mendez, B.; Trigueros, T.; Lain, A.; Castano, D.; Millet, O. Structural basis for the aminoacid composition of proteins from halophilic archaea. *PLoS Biol.* **2009**, *7*, e1000257.

85. DasSarma, S.; DasSarma, P. Halophiles. In *Els*; John Wiley & Sons: Chichester, UK, 2012; pp. 1–11.

86. Tokunaga, H.; Arakawa, T.; Tokunaga, M. Engineering of halophilic enzymes: Two acidic amino acid residues at the carboxy-terminal region confer halophilic characteristics to *Halomonas* and *Pseudomonas* nucleoside diphosphate kinases. *Protein Sci.* **2008**, *17*, 1603–1610.

87. Delgado-Garcia, M.; Valdivia-Urdiales, B.; Aguilar-Gonzalez, C.N.; Contreras-Esquivel, J.C.; Rodriguez-Herrera, R. Halophilic hydrolases as a new tool for the biotechnological industries. *J. Sci. Food Agric.* **2012**, *92*, 2575–2580.

88. Eichler, J. Facing extremes: Archaeal surface-layer (glyco)proteins. *Microbiology* **2003**, *149*, 3347–3351.

89. Enache, M.; Popescu, G.; Itoh, T.; Kamekura, M. Halophilic microorganisms from man-made and natural hypersaline environments: Physiology, ecology, and biotechnological potential. In *Adaption of Microbial Life to Environmental Extremes*; Stan-Latter, H., Fendrihan, S., Eds.; Springer-Verlag: New York, NY, USA, 2012; pp. 173–197.

90. Le Borgne, S.; Paniagua, D.; Vazquez-Duhalt, R. Biodegradation of organic pollutants by halophilic bacteria and archaea. *J. Mol. Microbiol. Biotechnol.* **2008**, *15*, 74–92.

91. Abe, F.; Horikoshi, K. The biotechnological potential of piezophiles. *Trends Biotechnol.* **2001**, *19*, 102–108.

92. Bartlett, D.H.; Bidle, K.A. Membrane-based adaptations of deep-sea piezophiles. In *Enigmatic Microorganisms and Life in Extreme Environments*; Seckbach, J., Ed.; Springer: Dordrecht, The Netherlands, 1999; pp. 503–512.

93. Mota, M.J.; Lopes, R.P.; Delgadillo, I.; Saraiva, J.A. Microorganisms under high pressure-adaptation, growth and biotechnological potential. *Biotechnol. Adv.* **2013**, *31*, 1426–1434.

94. Kato, C. High pressure and prokaryotes. In *Extremophiles Handbook*; Horikoshi, K., Antranikaian, G., Bull, A.T., Robb, F.T., Stetter, K.O., Eds.; Springer: Tokyo, Japan, 2011; pp. 658–668.

95. Oger, P.; Cario, A. [The high pressure life of piezophiles]. *Biol. Aujourdhui* **2014**, *208*, 193–206.

96. Bartlett, D.H. Pressure effects on *in vivo* microbial processes. *Biochim. Biophys. Acta* **2002**, *1595*, 367–381.

97. Lauro, F.M.; Bartlett, D.H. Prokaryotic lifestyles in deep sea habitats. *Extremophiles* **2008**, *12*, 15–25.

98. Simonato, F.; Campanaro, S.; Lauro, F.M.; Vezzi, A.; D'Angelo, M.; Vitulo, N.; Valle, G.; Bartlett, D.H. Piezophilic adaptation: A genomic point of view. *J. Biotechnol.* **2006**, *126*, 11–25.

99. Lauro, F.M.; Chastain, R.A.; Blankenship, L.E.; Yayanos, A.A.; Bartlett, D.H. The unique 16S rRNA genes of piezophiles reflect both phylogeny and adaptation. *Appl. Environ. Microbiol.* **2007**, *73*, 838–845.

100. Capece, M.C.; Clark, E.; Saleh, J.K.; Halford, D.; Heinl, N.; Hoskins, S.; Rothschild, L.J. Polyextremophiles and the constraints for terrestrial habitability. In *Polyextremophiles: Life under Multiple Forms of Stress*; Seckbach, J., Oren, A., Stan-Latter, H., Eds.; Springer: Dordrecht, The Netherlands, 2013; Volume 27, pp. 3–60.

101. Yumoto, I.; Hirota, K.; Nodasaka, Y.; Yokota, Y.; Hoshino, T.; Nakajima, K. *Alkalibacterium psychrotolerans* sp. nov., a psychrotolerant obligate alkaliphile that reduces an indigo dye. *Int. J. Syst. Evol. Microbiol.* **2004**, *54*, 2379–2383.

102. Padan, E.; Bibi, E.; Ito, M.; Krulwich, T.A. Alkaline pH homeostasis in bacteria: New insights. *Biochim. Biophys. Acta* **2005**, *1717*, 67–88.

103. Baker-Austin, C.; Dopson, M. Life in acid: pH homeostasis in acidophiles. *Trends Microbiol.* **2007**, *15*, 165–171.

104. Wiegel, J. Anaerobic alkaliphiles and alkaliphilic poly-extremophiles. In *Extremophiles Handbook*; Horikoshi, K., Antranikaian, G., Bull, A.T., Robb, F.T., Stetter, K.O., Eds.; Springer: Tokyo, Japan, 2011; pp. 81–97.

105. Fukuchi, S.; Yoshimune, K.; Wakayama, M.; Moriguchi, M.; Nishikawa, K. Unique amino acid composition of proteins in halophilic bacteria. *J. Mol. Biol.* **2003**, *327*, 347–357.

106. Zaccai, G. Molecular adaptations to life in high salt: Lessons from *Haloarcula marismortui*. In *Origins and Evolution of Life: An Astrobiological Perspective*; Gargaud, M., López-García, P., Martin, H., Eds.; Cambridge University Press: New York, NY, USA, 2011; pp. 375–388.

107. Gonzalez, J.M.; Sheckells, D.; Viebahn, M.; Krupatkina, D.; Borges, K.M.; Robb, F.T. *Thermococcus waiotapuensis* sp. nov., an extremely thermophilic archaeon isolated from a freshwater hot spring. *Arch. Microbiol.* **1999**, *172*, 95–101.

108. Vanlint, D.; Michiels, C.W.; Aertsen, A. Piezophysiology of the model bacterium *Escherichia coli*. In *Extremophiles Handbook*; Horikoshi, K., Antranikaian, G., Bull, A.T., Robb, F.T., Stetter, K.O., Eds.; Springer: Tokyo, Japan, 2011; pp. 669–686.

109. Kato, C.; Li, L.; Nogi, Y.; Nakamura, Y.; Tamaoka, J.; Horikoshi, K. Extremely barophilic bacteria isolated from the Mariana Trench, Challenger Deep, at a depth of 11,000 meters. *Appl. Environ. Microbiol.* **1998**, *64*, 1510–1513.

110. Blum, J.S.; Bindi, A.B.; Buzzelli, J.; Stolz, J.F.; Oremland, R.S. *Bacillus arsenicoselenatis*, sp. nov., and *Bacillus selenitireducens*, sp. nov.: Two haloalkaliphiles from Mono Lake, California that respire oxyanions of selenium and arsenic. *Arch. Microbiol.* **1998**, *171*, 19–30.

111. Mesbah, N.M.; Hedrick, D.B.; Peacock, A.D.; Rohde, M.; Wiegel, J. *Natranaerobius thermophilus* gen. nov., sp. nov., a halophilic, alkalithermophilic bacterium from soda lakes of the Wadi An Natrun, Egypt, and proposal of *Natranaerobiaceae* fam. nov. and *Natranaerobiales* ord. nov. *Int. J. Syst. Evol. Microbiol.* **2007**, *57*, 2507–2512.

112. Mesbah, N.M.; Wiegel, J. Halophiles exposed concomitantly to multiple stressors: Adaptive mechanisms of halophilic alkalithermophiles. In *Halophiles and Hypersaline Environments*; Ventosa, A., Oren, A., Ma, Y., Eds.; Springer: Berlin, Germany, 2011; pp. 249–273.

113. Abe, F.; Kato, C.; Horikoshi, K. Pressure-regulated metabolism in microorganisms. *Trends Microbiol.* **1999**, *7*, 447–453.

114. Kumar, V.; Satyanarayana, T. Thermo-alkali-stable xylanase of a novel polyextremophilic *Bacillus halodurans* TSEV1 and its application in biobleaching. *Int. Biodeter. Biodegr.* **2012**, *75*, 138–145.

115. Kumar, V.; Satyanarayana, T. Generation of xylooligosaccharides from microwave irradiated agroresidues using recombinant thermo-alkali-stable endoxylanase of the polyextremophilic bacterium *Bacillus halodurans* expressed in *Pichia pastoris*. *Bioresour. Technol.* **2015**, *179*, 382–389.

116. Vijayalaxmi, S.; Prakash, P.; Jayalakshmi, S.K.; Mulimani, V.H.; Sreeramulu, K. Production of extremely alkaliphilic, halotolerent, detergent, and thermostable mannanase by the free and immobilized cells of *Bacillus halodurans* PPKS-2. Purification and characterization. *Appl. Biochem. Biotechnol.* **2013**, *171*, 382–395.

117. Karan, R.; Capes, M.D.; DasSarma, P.; DasSarma, S. Cloning, overexpression, purification, and characterization of a polyextremophilic beta-galactosidase from the Antarctic haloarchaeon *Halorubrum lacusprofundi*. *BMC Biotechnol.* **2013**, *13*, 3.

118. Bommarius, A.S.; Blum, J.K.; Abrahamson, M.J. Status of protein engineering for biocatalysts: How to design an industrially useful biocatalyst. *Curr. Opin. Chem. Biol.* **2011**, *15*, 194–200.

119. Kumar, A.; Singh, S. Directed evolution: Tailoring biocatalysts for industrial applications. *Crit. Rev. Biotechnol.* **2013**, *33*, 365–378.

120. Chen, Z.; Wilmanns, M.; Zeng, A.P. Structural synthetic biotechnology: From molecular structure to predictable design for industrial strain development. *Trends Biotechnol.* **2010**, *28*, 534–542.

121. Otero, J.M.; Nielsen, J. Industrial systems biology. *Biotechnol. Bioeng.* **2010**, *105*, 439–460.

122. Venkatesh, R.; Sundaram, P.V. Upward shift of thermotolerance of cold water fish and mammalian trypsins upon chemical modification. *Ann. N. Y. Acad. Sci.* **1998**, *864*, 512–516.

123. Venkatesh, R.; Sundaram, P.V. Modulation of stability properties of bovine trypsin after *in vitro* structural changes with a variety of chemical modifiers. *Protein Eng.* **1998**, *11*, 691–698.

124. Arnold, F.H. Combinatorial and computational challenges for biocatalyst design. *Nature* **2001**, *409*, 253–257.

125. Bornscheuer, U.T.; Huisman, G.W.; Kazlauskas, R.J.; Lutz, S.; Moore, J.C.; Robins, K. Engineering the third wave of biocatalysis. *Nature* **2012**, *485*, 185–194.

126. Reetz, M.T. Laboratory evolution of stereoselective enzymes: A prolific source of catalysts for asymmetric reactions. *Angew. Chem. Int. Ed.* **2011**, *50*, 138–174.

127. Kourist, R.; Bornscheuer, U.T. Biocatalytic synthesis of optically active tertiary alcohols. *Appl. Microbiol. Biotechnol.* **2011**, *91*, 505–517.

128. Kourist, R.; Dominguez de Maria, P.; Miyamoto, K. Biocatalytic strategies for the asymmetric synthesis of profens—Recent trends and developments. *Green Chem.* **2011**, *13*, 2607–2618.

129. Eijsink, V.G.; Bjork, A.; Gaseidnes, S.; Sirevag, R.; Synstad, B.; van den Burg, B.; Vriend, G. Rational engineering of enzyme stability. *J. Biotechnol.* **2004**, *113*, 105–120.

130. Siddiqui, K.S.; Cavicchioli, R. Improved thermal stability and activity in the cold-adapted lipase B from *Candida antarctica* following chemical modification with oxidized polysaccharides. *Extremophiles* **2005**, *9*, 471–476.

131. Shu, Z.-Y.; Jiang, H.; Lin, R.-F.; Jiang, Y.-M.; Lin, L.; Huang, J.-Z. Technical methods to improve yield, activity and stability in the development of microbial lipases. *J. Mol. Catal. B: Enzym.* **2010**, *62*, 1–8.

132. Siddiqui, K.S.; Poljak, A.; Guilhaus, M.; De Francisci, D.; Curmi, P.M.; Feller, G.; D'Amico, S.; Gerday, C.; Uversky, V.N.; Cavicchioli, R. Role of lysine *versus* arginine in enzyme

cold-adaptation: Modifying lysine to homo-arginine stabilizes the cold-adapted alpha-amylase from *Pseudoalteramonas haloplanktis*. *Proteins* **2006**, *64*, 486–501.

133. Ferrer, M.; Martinez-Abarca, F.; Golyshin, P.N. Mining genomes and "metagenomes" for novel catalysts. *Curr. Opin. Biotechnol.* **2005**, *16*, 588–593.

134. Handelsman, J. Metagenomics: Application of genomics to uncultured microorganisms. *Microbiol. Mol. Biol. Rev.* **2004**, *68*, 669–685.

135. Lopez-Lopez, O.; Cerdan, M.E.; Gonzalez-Siso, M.I. New extremophilic lipases and esterases from metagenomics. *Curr. Protein Pept. Sci.* **2014**, *15*, 445–455.

136. Kennedy, J.; Flemer, B.; Jackson, S.A.; Lejon, D.P.; Morrissey, J.P.; O'Gara, F.; Dobson, A.D. Marine metagenomics: New tools for the study and exploitation of marine microbial metabolism. *Mar. Drugs* **2010**, *8*, 608–628.

137. Kotera, M.; Moriya, Y.; Tokimatsu, T.; Kanehisa, M.; Goto, S. KEGG and GenomeNet, new developments, metagenomic analysis. In *Encyclopedia of Metagenomics*; Nelson, K.E., Ed.; Springer: New York, NY, USA, 2014; pp. 1–11.

138. Blaser, H.U. Enantioselective catalysis in fine chemicals production. *Chem. Commun.* **2003**, *3*, 293–296.

139. Miyake, R.; Kawamoto, J.; Wei, Y.L.; Kitagawa, M.; Kato, I.; Kurihara, T.; Esaki, N. Construction of a low-temperature protein expression system using a cold-adapted bacterium, *Shewanella* sp. strain Ac10, as the host. *Appl. Environ. Microbiol.* **2007**, *73*, 4849–4856.

140. Tajima, T.; Fuki, K.; Kataoka, N.; Kudou, D.; Nakashimada, Y.; Kato, J. Construction of a simple biocatalyst using psychrophilic bacterial cells and its application for efficient 3-hydroxypropionaldehyde production from glycerol. *AMB Express* **2013**, *3*, 69.

141. Ferrer, M.; Chernikova, T.N.; Yakimov, M.M.; Golyshin, P.N.; Timmis, K.N. Chaperonins govern growth of *Escherichia coli* at low temperatures. *Nat. Biotechnol.* **2003**, *21*, 1266–1267.

142. Adrio, J.L.; Demain, A.L. Microbial enzymes: Tools for biotechnological processes. *Biomolecules* **2014**, *4*, 117–139.

143. Binod, P.; Palkhiwala, P.; Gaikaiwari, R.; Nampoothiri, K.M.; Duggal, A.; Dey, K.; Pandey, A. Industrial enzymes—Present status and future perspectives for India. *J. Sci. Ind. Res. India* **2013**, *72*, 271–286.

144. Kumar, L.; Awasthi, G.; Singh, B. Extremophiles: A novel source of industrially important enzymes. *Biotechnol. Appl. Biochem.* **2011**, *10*, 1–15.

145. NC-IUBMB. *The Enzyme List Class 3—Hydrolases*; ExplorEnz. Avaiable online: http://www.enzyme-database.org/index.php (acessed on 27 November 2014).

146. Zeeman, S.C.; Kossmann, J.; Smith, A.M. Starch: Its metabolism, evolution, and biotechnological modification in plants. *Annu. Rev. Plant Biol.* **2010**, *61*, 209–234.

147. Castro, A.M.; Ribeiro, B.D. Methods for detection of amylolytic activities. In *Methods to Determine Enzymatic Activity*; Vermelho, A.B., Couri, S., Eds.; Bentham Science: Sharjah, UAE, 2013; pp. 100–124.

148. Prakash, O.; Jaiswal, N. alpha-Amylase: An ideal representative of thermostable enzymes. *Appl. Biochem. Biotechnol.* **2010**, *160*, 2401–2414.

149. Feller, G.; Gerday, C. Psychrophilic enzymes: Hot topics in cold adaptation. *Nat. Rev. Microbiol.* **2003**, *1*, 200–208.

150. Kuddus, M.; Roohi; Arif, J.M.; Ramteke, P.W. An overview of cold-active microbial α-amylase: Adaptation strategies and biotechnological potentials. *Biotechnology* **2011**, *10*, 246–258.

151. Niehaus, F.; Bertoldo, C.; Kahler, M.; Antranikian, G. Extremophiles as a source of novel enzymes for industrial application. *Appl. Microbiol. Biotechnol.* **1999**, *51*, 711–729.

152. Sharma, A.; Satyanarayana, T. Microbial acid-stable α-amylases: Characteristics, genetic engineering and applications. *Process Biochem.* **2013**, *48*, 201–211.

153. Brown, I.; Dafforn, T.R.; Fryer, P.J.; Cox, P.W. Kinetic study of the thermal denaturation of a hyperthermostable extracellular alpha-amylase from *Pyrococcus furiosus*. *Biochim. Biophys. Acta* **2013**, *1834*, 2600–2605.

154. Duffner, F.; Bertoldo, C.; Andersen, J.T.; Wagner, K.; Antranikian, G. A new thermoactive pullulanase from *Desulfurococcus mucosus*: Cloning, sequencing, purification, and characterization of the recombinant enzyme after expression in *Bacillus subtilis*. *J. Bacteriol.* **2000**, *182*, 6331–6338.

155. Gomes, I.; Gomes, J.; Steiner, W. Highly thermostable amylase and pullulanase of the extreme thermophilic eubacterium *Rhodothermus marinus*: Production and partial characterization. *Bioresour. Technol.* **2003**, *90*, 207–214.

156. Trincone, A. Marine biocatalysts: Enzymatic features and applications. *Mar. Drugs* **2011**, *9*, 478–499.

157. Gurumurthy, D.M.; Neelagund, S.E. Molecular characterization of industrially viable extreme thermostable novel alpha-amylase of *Geobacillus* sp. Iso5 isolated from geothermal spring. *J. Pure Appl. Microbiol.* **2012**, *6*, 1759–1773.

158. Anto, H.; Trivedi, U.B.; Patel, K.C. Glucoamylase production by solid-state fermentation using rice flake manufacturing waste products as substrate. *Bioresour. Technol.* **2006**, *97*, 1161–1166.

159. Sun, H.; Zhao, P.; Ge, X.; Xia, Y.; Hao, Z.; Liu, J.; Peng, M. Recent advances in microbial raw starch degrading enzymes. *Appl. Biochem. Biotechnol.* **2010**, *160*, 988–1003.

160. Kyaw, N.; de Mesquita, R.F.; Kameda, E.; Neto, J.C.; Langone, M.A.; Coelho, M.A. Characterization of commercial amylases for the removal of filter cake on petroleum wells. *Appl. Biochem. Biotechnol.* **2010**, *161*, 171–180.

161. Sivaramakrishnan, S.; Gangadharan, D.; Nampoothiri, K.M.; Pandey, A. Amylases from microbial sources—An overview on recent developments. *Food Technol. Biotechnol.* **2006**, *44*, 173–184.

162. Callen, W.; Richardson, T.; Frey, G.; Miller, C.; Kazaoka, M.; Mathur, E.; Short, J. Amylases and Methods for Use in Starch Processing. US 8,338,131, 25 December 2012.

163. Nedwin, G.E.; Sharma, V.; Shetty, J.K. Alpha-Amylase Blend for Starch Processing and Method of Use Thereof. US 8,545,907, 1 October 2013.

164. Hii, S.L.; Tan, J.S.; Ling, T.C.; Ariff, A.B. Pullulanase: Role in starch hydrolysis and potential industrial applications. *Enzyme Res.* **2012**, *2012*, 921362.

165. Fernandes, P. Marine enzymes and food industry: Insight on existing and potential interactions. *Front. Mar. Sci.* **2014**, *1*, 46.

166. Lévêque, E.; Janeček, Š.; Haye, B.; Belarbi, A. Thermophilic archaeal amylolytic enzymes. *Enzym. Microb. Tech.* **2000**, *26*, 3–14.

167. Vieille, C.; Zeikus, G.J. Hyperthermophilic enzymes: Sources, uses, and molecular mechanisms for thermostability. *Microbiol. Mol. Biol. Rev.* **2001**, *65*, 1–43.

168. Li, X.; Li, D.; Park, K.H. An extremely thermostable amylopullulanase from *Staphylothermus marinus* displays both pullulan- and cyclodextrin-degrading activities. *Appl. Microbiol. Biotechnol.* **2013**, *97*, 5359–5369.

169. Brown, S.H.; Kelly, R.M. Characterization of amylolytic enzymes, having both alpha-1,4 and alpha-1,6 hydrolytic activity, from the thermophilic Archaea *Pyrococcus furiosus* and *Thermococcus litoralis*. *Appl. Environ. Microbiol.* **1993**, *59*, 2614–2621.

170. Feller, G.; Le Bussy, O.; Gerday, C. Expression of psychrophilic genes in mesophilic hosts: Assessment of the folding state of a recombinant alpha-amylase. *Appl. Environ. Microbiol.* **1998**, *64*, 1163–1165.

171. Srimathi, S.; Jayaraman, G.; Feller, G.; Danielsson, B.; Narayanan, P.R. Intrinsic halotolerance of the psychrophilic alpha-amylase from *Pseudoalteromonas haloplanktis*. *Extremophiles* **2007**, *11*, 505–515.

172. Qin, Y.; Huang, Z.; Liu, Z. A novel cold-active and salt-tolerant alpha-amylase from marine bacterium *Zunongwangia profunda:* Molecular cloning, heterologous expression and biochemical characterization. *Extremophiles* **2014**, *18*, 271–281.

173. Pancha, I.; Jain, D.; Shrivastav, A.; Mishra, S.K.; Shethia, B.; Mishra, S.; V, P.M.; Jha, B. A thermoactive alpha-amylase from a *Bacillus* sp. isolated from CSMCRI salt farm. *Int. J. Biol. Macromol.* **2010**, *47*, 288–291.

174. Uzyol, K.S.; Sarıyar-Akbulut, B.; Denizci, A.A.; Kazan, D. Thermostable α-amylase from moderately halophilic *Halomonas* sp. AAD21 *Turk. J. Biol.* **2012**, *36*, 327–338.

175. Fukushima, T.; Mizuki, T.; Echigo, A.; Inoue, A.; Usami, R. Organic solvent tolerance of halophilic alpha-amylase from a Haloarchaeon, *Haloarcula* sp. strain S-1. *Extremophiles* **2005**, *9*, 85–89.

176. Laderman, K.A.; Asada, K.; Uemori, T.; Mukai, H.; Taguchi, Y.; Kato, I.; Anfinsen, C.B. Alpha-amylase from the hyperthermophilic archaebacterium *Pyrococcus furiosus*. Cloning and sequencing of the gene and expression in *Escherichia coli*. *J. Biol. Chem.* **1993**, *268*, 24402–24407.

177. Bertoldo, C.; Duffner, F.; Jorgensen, P.L.; Antranikian, G. Pullulanase type I from *Fervidobacterium pennavorans* Ven5: Cloning, sequencing, and expression of the gene and biochemical characterization of the recombinant enzyme. *Appl. Environ. Microbiol.* **1999**, *65*, 2084–2091.

178. Bhattacharya, A.; Pletschke, B.I. Review of the enzymatic machinery of *Halothermothrix orenii* with special reference to industrial applications. *Enzyme Microb. Technol.* **2014**, *55*, 159–169.

179. Perez-Pomares, F.; Bautista, V.; Ferrer, J.; Pire, C.; Marhuenda-Egea, F.C.; Bonete, M.J. Alpha-amylase activity from the halophilic archaeon *Haloferax mediterranei*. *Extremophiles* **2003**, *7*, 299–306.

180. Tomme, P.; Warren, R.A.; Gilkes, N.R. Cellulose hydrolysis by bacteria and fungi. In *Advances in Microbial Physiology*; Stafford, R., Ed.; Academic Press: London, UK, 1995; Volume 37, pp. 1–81.

181. Wilson, D.B. Aerobic microbial cellulase systems. In *Biomass Recalcitrance: Deconstructing the Plant Cell Wall for Bioenergy*; Himmel, M.E., Ed.; Blackwell Publishing: Oxford, UK, 2008; pp. 374–392.

182. Collins, T.; Gerday, C.; Feller, G. Xylanases, xylanase families and extremophilic xylanases. *FEMS Microbiol. Rev.* **2005**, *29*, 3–23.

183. Ratanakhanokchai, K.; Waeonukul, R.; Pason, P.; Pason, C.; Kyu, K.L.; SakkaK.; Kosugi, A.; Mori, Y. *Paenibacillus curdlanolyticus* strain B-6 multienzyme complex: A novel system for biomass utilization. In *Biomass Now—Cultivation and Utilization*; Matovic, M.D., Ed.; Intech: Rijeka, Croatia, 2013; pp. 369–394.

184. Maki, M.; Leung, K.T.; Qin, W. The prospects of cellulase-producing bacteria for the bioconversion of lignocellulosic biomass. *Int. J. Biol. Sci.* **2009**, *5*, 500–516.

185. Naik, S.N.; Goud, V.; Rout, P.K.; Dalai, A.K. Production of first and second generation biofuels: A comprehensive review. *Renew. Sust. Energ. Rev.* **2010**, *14*, 578–597.

186. Kasana, R.C. Proteases from psychrotrophs: An overview. *Crit. Rev. Microbiol.* **2010**, *36*, 134–145.

187. Kumar, V.; Satyanarayana, T. Thermoalkaliphilic microbes. In *Polyextremophiles: Life under Multiple Forms of Stress*; Seckbach, J., Oren, A., Stan-Latter, H., Eds.; Springer Science +Business Media: Dordrecht, The Netherlands, 2013; Volume 27, pp. 271–298.

188. Duffaud, G.D.; McCutchen, C.M.; Leduc, P.; Parker, K.N.; Kelly, R.M. Purification and characterization of extremely thermostable beta-mannanase, beta-mannosidase, and alpha-galactosidase from the hyperthermophilic eubacterium *Thermotoga neapolitana* 5068. *Appl. Environ. Microbiol.* **1997**, *63*, 169–177.

189. Matsui, I.; Sakai, Y.; Matsui, E.; Kikuchi, H.; Kawarabayasi, Y.; Honda, K. Novel substrate specificity of a membrane-bound beta-glycosidase from the hyperthermophilic archaeon *Pyrococcus horikoshii*. *FEBS Lett.* **2000**, *467*, 195–200.

190. Goyal, K.; Selvakumar, P.; Hayashi, K. Characterization of a thermostable beta-glucosidase (Bg1B) from *Thermotoga maritima* showing transglycosylation activity. *J. Mol. Catal. B-Enzym.* **2001**, *15*, 45–53.

191. Subramaniyan, S.; Prema, P. Biotechnology of microbial xylanases: Enzymology, molecular biology, and application. *Crit. Rev. Biotechnol.* **2002**, *22*, 33–64.

192. Ooteghem, V.S. Process for Generation of Hydrogen Gas from Various Feedstocks Using Thermophilic Bacteria. US 6,942,998, 13 September 2005.

193. Jiang, Z.Q.; Deng, W.; Zhu, Y.P.; Li, L.T.; Sheng, Y.J.; Hayashi, K. The recombinant xylanase B of T*hermotoga maritima* is highly xylan specific and produces exclusively xylobiose from xylans, a unique character for industrial applications. *J. Mol. Catal. B-Enzym.* **2004**, *27*, 207–213.

194. Kang, H.J.; Uegaki, K.; Fukada, H.; Ishikawa, K. Improvement of the enzymatic activity of the hyperthermophilic cellulase from *Pyrococcus horikoshii*. *Extremophiles* **2007**, *11*, 251–256.

195. Mao, X.; Hong, Y.; Shao, Z.; Zhao, Y.; Liu, Z. A novel cold-active and alkali-stable beta-glucosidase gene isolated from the marine bacterium *Martelella mediterranea*. *Appl. Biochem. Biotechnol.* **2010**, *162*, 2136–2148.

196. Chen, S.; Hong, Y.; Shao, Z.; Liu, Z. A cold-active β-glucosidase (Bg1C) from a sea bacteria *Exiguobacterium oxidotolerans* A011. *World J. Microb. Biot.* **2010**, *26*, 1427–1435.

197. Garsoux, G.; Lamotte, J.; Gerday, C.; Feller, G. Kinetic and structural optimization to catalysis at low temperatures in a psychrophilic cellulase from the Antarctic bacterium *Pseudoalteromonas haloplanktis*. *Biochem. J.* **2004**, *384*, 247–253.

198. Collins, T.; Hoyoux, A.; Dutron, A.; Georis, J.; Genot, B.; Dauvrin, T.; Arnaut, F.; Gerday, C.; Feller, G. Use of glycoside hydrolase family 8 xylanases in baking. *J. Cereal Sci.* **2006**, *43*, 79–84.

199. Dornez, E.; Verjans, P.; Arnaut, F.; Delcour, J.A.; Courtin, C.M. Use of psychrophilic xylanases provides insight into the xylanase functionality in bread making. *J. Agric. Food Chem.* **2011**, *59*, 9553–9562.

200. Hildebrandt, P.; Wanarska, M.; Kur, J. A new cold-adapted beta-D-galactosidase from the Antarctic *Arthrobacter* sp. 32c—Gene cloning, overexpression, purification and properties. *BMC Microbiol.* **2009**, *9*, 151.

201. Varbanets, L.D.; Avdeeva, L.V.; Borzova, N.V.; Matseliukh, E.V.; Gudzenko, A.V.; Kiprianova, E.A.; Iaroshenko, L.V. The Black Sea bacteria—Producers of hydrolytic enzymes. *Mikrobiol. Z* **2011**, *73*, 9–15.

202. Rawlings, N.D.; Waller, M.; Barrett, A.J.; Bateman, A. MEROPS: The database of proteolytic enzymes, their substrates and inhibitors. *Nucleic Acids. Res.* **2014**, *42*, 503–509.

203. Lopez-Otin, C.; Matrisian, L.M. Emerging roles of proteases in tumour suppression. *Nat. Rev. Cancer* **2007**, *7*, 800–808.

204. Rawlings, N.D.; Bateman, A. Pepsin homologues in bacteria. *BMC Genomics* **2009**, *10*, 437.

205. Gunbin, K.V.; Afonnikov, D.A.; Kolchanov, N.A. Molecular evolution of the hyperthermophilic archaea of the *Pyrococcus* genus: Analysis of adaptation to different environmental conditions. *BMC Genomics* **2009**, *10*, 639.

206. Stetter, K.O. History of discovery of the first hyperthermophiles. *Extremophiles* **2006**, *10*, 357–362.

207. Halio, S.B.; Bauer, M.W.; Mukund, S.; Adams, M.; Kelly, R.M. Purification and characterization of two functional forms of intracellular protease PfpI from the hyperthermophilic archaeon *Pyrococcus furiosus*. *Appl. Environ. Microbiol.* **1997**, *63*, 289–295.

208. Ward, D.E.; Shockley, K.R.; Chang, L.S.; Levy, R.D.; Michel, J.K.; Conners, S.B.; Kelly, R.M. Proteolysis in hyperthermophilic microorganisms. *Archaea* **2002**, *1*, 63–74.

209. Atomi, H. Recent progress towards the application of hyperthermophiles and their enzymes. *Curr. Opin. Chem. Biol.* **2005**, *9*, 166–173.

210. Cavicchioli, R.; Siddiqui, K.S.; Andrews, D.; Sowers, K.R. Low-temperature extremophiles and their applications. *Curr. Opin. Biotechnol.* **2002**, *13*, 253–261.

211. Gerday, C.; Aittaleb, M.; Bentahir, M.; Chessa, J.P.; Claverie, P.; Collins, T.; D'Amico, S.; Dumont, J.; Garsoux, G.; Georlette, D.; *et al.* Cold-adapted enzymes: From fundamentals to biotechnology. *Trends Biotechnol.* **2000**, *18*, 103–107.

212. Cristobal, H.A.; Lopez, M.A.; Kothe, E.; Abate, C.M. Diversity of protease-producing marine bacteria from sub-antarctic environments. *J. Basic Microbiol.* **2011**, *51*, 590–600.

213. Kulakova, L.; Galkin, A.; Kurihara, T.; Yoshimura, T.; Esaki, N. Cold-active serine alkaline protease from the psychrotrophic bacterium *Shewanella* strain ac10: Gene cloning and enzyme purification and characterization. *Appl. Environ. Microbiol.* **1999**, *65*, 611–617.

214. Huston, A.L.; Methe, B.; Deming, J.W. Purification, characterization, and sequencing of an extracellular cold-active aminopeptidase produced by marine psychrophile *Colwellia psychrerythraea* strain 34H. *Appl. Environ. Microbiol.* **2004**, *70*, 3321–3328.

215. Sastre, D.E.; Paggi, R.A.; de Castro, R.E. The Lon protease from the haloalkaliphilic archaeon *Natrialba magadii* is transcriptionally linked to a cluster of putative membrane proteases and displays DNA-binding activity. *Microbiol. Res.* **2011**, *166*, 304–313.

216. Cerletti, M.; Martinez, M.J.; Gimenez, M.I.; Sastre, D.E.; Paggi, R.A.; de Castro, R.E. The LonB protease controls membrane lipids composition and is essential for viability in the extremophilic haloarchaeon *Haloferax volcanii*. *Environ. Microbiol.* **2014**, *16*, 1779–1792.

217. Gimenez, M.I.; Studdert, C.A.; Sanchez, J.J.; De Castro, R.E. Extracellular protease of *Natrialba magadii*: Purification and biochemical characterization. *Extremophiles* **2000**, *4*, 181–188.

218. Kamekura, M.; Seno, Y.; Dyall-Smith, M. Halolysin R4, a serine proteinase from the halophilic archaeon *Haloferax mediterranei*; gene cloning, expression and structural studies. *Biochim. Biophys. Acta* **1996**, *1294*, 159–167.

219. De Castro, R.E.; Maupin-Furlow, J.A.; Gimenez, M.I.; Herrera Seitz, M.K.; Sanchez, J.J. Haloarchaeal proteases and proteolytic systems. *FEMS Microbiol. Rev.* **2006**, *30*, 17–35.

220. Vidyasagar, M.; Prakash, S.; Mahajan, V.; Shouche, Y.S.; Sreeramulu, K. Purification and characterization of an extreme halothermophilic protease from a halophilic bacterium *Chromohalobacter* sp. TVSP101. *Braz. J. Microbiol.* **2009**, *40*, 12–19.

221. Hough, D.W.; Danson, M.J. Extremozymes. *Curr. Opin. Chem. Biol.* **1999**, *3*, 39–46.

222. Asker, D.; Awad, T.S.; Beppu, T.; Ueda, K. *Deinococcus misasensis* and *Deinococcus roseus*, novel members of the genus *Deinococcus*, isolated from a radioactive site in Japan. *Syst. Appl. Microbiol.* **2008**, *31*, 43–49.

223. Asker, D.; Awad, T.S.; McLandsborough, L.; Beppu, T.; Ueda, K. *Deinococcus depolymerans* sp. nov., a gamma- and UV-radiation-resistant bacterium, isolated from a naturally radioactive site. *Int. J. Syst. Evol. Microbiol.* **2011**, *61*, 1448–1453.

224. Kampfer, P.; Lodders, N.; Huber, B.; Falsen, E.; Busse, H.J. *Deinococcus aquatilis* sp. nov., isolated from water. *Int. J. Syst. Evol. Microbiol.* **2008**, *58*, 2803–2806.

225. Asker, D.; Awad, T.S.; Beppu, T.; Ueda, K. *Deinococcus aquiradiocola* sp. nov., isolated from a radioactive site in Japan. *Int. J. Syst. Evol. Microbiol.* **2009**, *59*, 144–149.

226. Suresh, K.; Reddy, G.S.; Sengupta, S.; Shivaji, S. *Deinococcus indicus* sp. nov., an arsenic-resistant bacterium from an aquifer in West Bengal, India. *Int. J. Syst. Evol. Microbiol.* **2004**, *54*, 457–461.

227. Kimura, H.; Asada, R.; Masta, A.; Naganuma, T. Distribution of microorganisms in the subsurface of the manus basin hydrothermal vent field in Papua New Guinea. *Appl. Environ. Microbiol.* **2003**, *69*, 644–648.

228. Liedert, C.; Peltola, M.; Bernhardt, J.; Neubauer, P.; Salkinoja-Salonen, M. Physiology of resistant *Deinococcus geothermalis* bacterium aerobically cultivated in low-manganese medium. *J. Bacteriol.* **2012**, *194*, 1552–1561.

229. Pietrow, O.; Panek, A.; Synowiecki, J. Extracellular proteolytic activity of *Deinococcus geothermalis*. *Afr. J. Biotechnol.* **2013**, *12*, 4020–4027.

230. Friedrich, A.B.; Antranikian, G. Keratin degradation by *Fervidobacterium pennavorans*, a novel thermophilic anaerobic species of the order thermotogales. *Appl. Environ. Microbiol.* **1996**, *62*, 2875–2882.

231. Burlini, N.; Magnani, P.; Villa, A.; Macchi, F.; Tortora, P.; Guerritore, A. A heat-stable serine proteinase from the extreme thermophilic archaebacterium *Sulfolobus solfataricus*. *Biochim. Biophys. Acta* **1992**, *1122*, 283–292.

232. Klingeberg, M.; Hashwa, F.; Antranikian, G. Properties of extremely thermostable proteases from anaerobic hyperthermophilic bacteria. *Appl. Microbiol. Biotechnol.* **1991**, *34*, 715–719.

233. Jang, H.J.; Kim, B.C.; Pyun, Y.R.; Kim, Y.S. A novel subtilisin-like serine protease from *Thermoanaerobacter yonseiensis* KB-1: Its cloning, expression, and biochemical properties. *Extremophiles* **2002**, *6*, 233–243.

234. Lloyd, K.G.; Schreiber, L.; Petersen, D.G.; Kjeldsen, K.U.; Lever, M.A.; Steen, A.D.; Stepanauskas, R.; Richter, M.; Kleindienst, S.; Lenk, S.; *et al.* Predominant archaea in marine sediments degrade detrital proteins. *Nature* **2013**, *496*, 215–218.

235. Seth-Pasricha, M.; Bidle, K.A.; Bidle, K.D. Specificity of archaeal caspase activity in the extreme halophile *Haloferax volcanii*. *Environ. Microbiol. Rep.* **2013**, *5*, 263–271.

236. Arpigny, J.L.; Jaeger, K.E. Bacterial lipolytic enzymes: Classification and properties. *Biochem. J.* **1999**, *343*, 177–183.

237. de Pascale, D.; Cusano, A.M.; Autore, F.; Parrilli, E.; di Prisco, G.; Marino, G.; Tutino, M.L. The cold-active Lip1 lipase from the Antarctic bacterium *Pseudoalteromonas haloplanktis* TAC125 is a member of a new bacterial lipolytic enzyme family. *Extremophiles* **2008**, *12*, 311–323.

238. Gupta, R.; Gupta, N.; Rathi, P. Bacterial lipases: An overview of production, purification and biochemical properties. *Appl. Microbiol. Biotechnol.* **2004**, *64*, 763–781.

239. Hasan, F.; Shah, A.A.; Hameed, A. Industrial applications of microbial lipases. *Enzyme Microb. Technol.* **2006**, *39*, 235–251.

240. Jaeger, K.E.; Eggert, T. Lipases for biotechnology. *Curr. Opin. Biotechnol.* **2002**, *13*, 390–397.

241. Sangeetha, R.; Arulpandi, I.; Geetha, A. Bacterial lipases as potential industrial biocatalysts: An overview. *Res. J. Microbiol.* **2011**, *6*, 1–24.

242. Charpe, T.W.; Rathod, V.K. Biodiesel production using waste frying oil. *Waste Manag.* **2011**, *31*, 85–90.

243. Gupta, P.; Upadhyay, L.S.B.; Shrivastava, R. Lipase catalyzed-transesterification of vegetable oils by lipolytic bacteria. *Res. J. Microbiol.* **2011**, *6*, 281–288.

244. Bajaj, A.; Lohan, P.; Jha, P.N.; Mehrotra, R. Biodiesel production through lipase catalyzed transesterification: An overview. *J. Mol. Catal. B: Enzym.* **2010**, *62*, 9–14.

245. Reetz, M.T.; Jaeger, K.E. Overexpression, immobilization and biotechnological application of *Pseudomonas* lipases. *Chem. Phys. Lipids* **1998**, *93*, 3–14.

246. Fucinos, P.; Gonzalez, R.; Atanes, E.; Sestelo, A.B.; Perez-Guerra, N.; Pastrana, L.; Rua, M.L. Lipases and esterases from extremophiles: Overview and case example of the production and purification of an esterase from *Thermus thermophilus* HB27. *Methods Mol. Biol.* **2012**, *861*, 239–266.

247. Andualema, B.; Gessesse, A. Microbial lipases and their industrial applications: Review. *Biotechnol. Appl. Biochem.* **2012**, *11*, 100–118.

248. Schafer, T.; Antranikian, G.; Royter, M.; Hoff, T. Lipases from Thermophilic Anaerobes. US 7,972,831, 5 July 2011.

249. Vind, J.; Knötzel, J.C.F.; Borch, K.; Svendsen, A.; Callisen, T.H.; Yaver, D.; Bjornvad, M.E.; Hansen, P.K.; Lamsa, M. Lipase Variants. US 8,187,854, 29 May 2012.

250. Mardanov, A.V.; Ravin, N.V.; Svetlitchnyi, V.A.; Beletsky, A.V.; Miroshnichenko, M.L.; Bonch-Osmolovskaya, E.A.; Skryabin, K.G. Metabolic versatility and indigenous origin of the archaeon *Thermococcus sibiricus*, isolated from a siberian oil reservoir, as revealed by genome analysis. *Appl. Environ. Microbiol.* **2009**, *75*, 4580–4588.

251. Kim, S.B.; Lee, W.; Ryu, Y.W. Cloning and characterization of thermostable esterase from *Archaeoglobus fulgidus*. *J. Microbiol.* **2008**, *46*, 100–107.

252. Hotta, Y.; Ezaki, S.; Atomi, H.; Imanaka, T. Extremely stable and versatile carboxylesterase from a hyperthermophilic archaeon. *Appl. Environ. Microbiol.* **2002**, *68*, 3925–3931.

253. Park, Y.J.; Yoon, S.J.; Lee, H.B. A novel thermostable arylesterase from the archaeon *Sulfolobus solfataricus* P1: Purification, characterization, and expression. *J. Bacteriol.* **2008**, *190*, 8086–8095.

254. Arpigny, J.L.; Jendrossek, D.; Jaeger, K.E. A novel heat-stable lipolytic enzyme from *Sulfolobus acidocaldarius* DSM 639 displaying similarity to polyhydroxyalkanoate depolymerases. *FEMS Microbiol. Lett.* **1998**, *167*, 69–73.

255. Huddleston, S.; Yallop, C.A.; Charalambous, B.M. The identification and partial characterisation of a novel inducible extracellular thermostable esterase from the archaeon *Sulfolobus shibatae*. *Biochem. Biophys. Res. Commun.* **1995**, *216*, 495–500.

256. Ikeda, M.; Clark, D.S. Molecular cloning of extremely thermostable esterase gene from hyperthermophilic archaeon *Pyrococcus furiosus* in *Escherichia coli*. *Biotechnol. Bioeng.* **1998**, *57*, 624–629.

257. Ryu, H.S.; Kim, H.K.; Choi, W.C.; Kim, M.H.; Park, S.Y.; Han, N.S.; Oh, T.K.; Lee, J.K. New cold-adapted lipase from *Photobacterium lipolyticum* sp. nov. that is closely related to filamentous fungal lipases. *Appl. Microbiol. Biotechnol.* **2006**, *70*, 321–326.

258. Bosley, J.A.; Peilow, A.D. Preparation of Immobilized Lipase by Adsorption of Lipase and a Non-Lipase Protein on a Support. US 5232843, 3 August 1993.

259. Yang, K.S.; Sohn, J.H.; Kim, H.K. Catalytic properties of a lipase from *Photobacterium lipolyticum* for biodiesel production containing a high methanol concentration. *J. Biosci. Bioeng.* **2009**, *107*, 599–604.

260. Gotor-Fernández, V.; Busto, E.; Gotor, V. *Candida antarctica* Lipase B: An ideal biocatalyst for the preparation of nitrogenated organic compounds. *Adv. Synth. Catal.* **2006**, *348*, 797–812.

261. Qin, B.; Liang, P.; Jia, X.; Zhang, X.; Mu, M.; Wang, X.-Y.; Ma, G.-Z.; Jin, D.-N.; You, S. Directed evolution of *Candida antarctica* lipase B for kinetic resolution of profen esters. *Catal. Commun.* **2013**, *38*, 1–5.

262. Singh, N.; Jha, M.K.; Sarma, A.K. A critical review of enzymatic transesterification: A sustainable technology for biodiesel production. In *Recent Advances in Bioenergy Research*; Kumar, S., Sarma, A.K., Tyagi, S.K., Yadav, Y.K., Eds.; SSS-NIRE: Kapurthala, India, 2014; Volume 3, pp. 298–312.

263. Nielsen, T.B.; Ishii, M.; Kirk, O. Lipases A and B from the yeast *Candida antarctica*. In *Biotechnological Applications of Cold-Adapted Organisms*; Margesin, R., Schinner, F., Eds.; Springer-Verlag: Heidelberg, Germany, 1999; pp. 49–61.

264. Widersten, M. Protein engineering for development of new hydrolytic biocatalysts. *Curr. Opin. Chem. Biol.* **2014**, *21*, 42–47.

265. Lo Giudice, A.; Michaud, L.; de Pascale, D.; de Domenico, M.; di Prisco, G.; Fani, R.; Bruni, V. Lipolytic activity of Antarctic cold-adapted marine bacteria (Terra Nova Bay, Ross Sea). *J. Appl. Microbiol.* **2006**, *101*, 1039–1048.

266. Al Khudary, R.; Venkatachalam, R.; Katzer, M.; Elleuche, S.; Antranikian, G. A cold-adapted esterase of a novel marine isolate, *Pseudoalteromonas arctica*: Gene cloning, enzyme purification and characterization. *Extremophiles* **2010**, *14*, 273–285.

267. Rabus, R.; Ruepp, A.; Frickey, T.; Rattei, T.; Fartmann, B.; Stark, M.; Bauer, M.; Zibat, A.; Lombardot, T.; Becker, I.; *et al.* The genome of *Desulfotalea psychrophila*, a sulfate-reducing bacterium from permanently cold Arctic sediments. *Environ. Microbiol.* **2004**, *6*, 887–902.

268. Zeng, X.; Xiao, X.; Wang, P.; Wang, F. Screening and characterization of psychrotrophic lipolytic bacteria from deep-sea sediments. *J. Microbiol. Biotechnol.* **2004**, *14*, 952–958.

269. Do, H.; Lee, J.H.; Kwon, M.H.; Song, H.E.; An, J.Y.; Eom, S.H.; Lee, S.G.; Kim, H.J. Purification, characterization and preliminary X-ray diffraction analysis of a cold-active lipase (CpsLip) from the psychrophilic bacterium *Colwellia psychrerythraea* 34H. *Acta Crystallogr. Sect. F. Struct. Biol. Cryst. Commun.* **2013**, *69*, 920–924.

270. Amoozegar, M.A.; Salehghamari, E.; Khajeh, K.; Kabiri, M.; Naddaf, S. Production of an extracellular thermohalophilic lipase from a moderately halophilic bacterium, *Salinivibrio* sp. strain SA-2. *J. Basic Microbiol.* **2008**, *48*, 160–167.

271. Córdova-López, J.; Camacho-Córdova, D.I.; Camacho-Ruiz, R.M.; Mateos, J.C.; Rodríguez, J. *Haloarcula marismortui*, eighty-four years after its discovery in the Dead Sea, Review. *IJERT* **2014**, *3*, 1257–1267.

272. Muller-Santos, M.; de Souza, E.M.; Pedrosa Fde, O.; Mitchell, D.A.; Longhi, S.; Carriere, F.; Canaan, S.; Krieger, N. First evidence for the salt-dependent folding and activity of an esterase from the halophilic archaea *Haloarcula marismortui*. *Biochim. Biophys. Acta* **2009**, *1791*, 719–729.

273. Corral, P.; Gutierrez, M.C.; Castillo, A.M.; Dominguez, M.; Lopalco, P.; Corcelli, A.; Ventosa, A. *Natronococcus roseus* sp. nov., a haloalkaliphilic archaeon from a hypersaline lake. *Int. J. Syst. Evol. Microbiol.* **2013**, *63*, 104–108.

2

Genomic Sequence and Experimental Tractability of a New Decapod Shrimp Model, *Neocaridina denticulata*

Nathan J. Kenny [1,2,†], **Yung Wa Sin** [1,†], **Xin Shen** [1], **Qu Zhe** [1], **Wei Wang** [1], **Ting Fung Chan** [1], **Stephen S. Tobe** [3], **Sebastian M. Shimeld** [2], **Ka Hou Chu** [1] and **Jerome H. L. Hui** [1,*]

[1] School of Life Sciences, The Chinese University of Hong Kong, Shatin, Hong Kong, China;
     E-Mails: nathanjameskenny@gmail.com (N.J.K.); yungwa.sin@cuhk.edu.hk (Y.W.S.);
     shenthin@163.com (X.S.); quzheouc@gmail.com (Q.Z.); wangweinbu@126.com (W.W.);
     tf.chan@cuhk.edu.hk (T.F.C.); kahouchu@cuhk.edu.hk (K.H.C.)

[2] Department of Zoology, University of Oxford, Oxford OX1 3PS, UK;
     E-Mail: sebastian.shimeld@zoo.ox.ac.uk

[3] Department of Cell and Systems Biology, University of Toronto, Toronto M5S 3G5, Canada;
     E-Mail: stephen.tobe@utoronto.ca

[†] These authors contributed equally to this work.

[*] Author to whom correspondence should be addressed; E-Mail: jeromehui@cuhk.edu.hk

Abstract: The speciose Crustacea is the largest subphylum of arthropods on the planet after the Insecta. To date, however, the only publically available sequenced crustacean genome is that of the water flea, *Daphnia pulex*, a member of the Branchiopoda. While *Daphnia* is a well-established ecotoxicological model, previous study showed that one-third of genes contained in its genome are lineage-specific and could not be identified in any other metazoan genomes. To better understand the genomic evolution of crustaceans and arthropods, we have sequenced the genome of a novel shrimp model, *Neocaridina denticulata*, and tested its experimental malleability. A library of 170-bp nominal fragment size was constructed from DNA of a starved single adult and sequenced using the Illumina HiSeq2000 platform. Core eukaryotic genes, the mitochondrial genome, developmental patterning genes (such as Hox) and microRNA processing pathway genes are all present in this animal, suggesting it has not undergone massive genomic loss. Comparison with the published genome of *Daphnia pulex* has allowed us to reveal 3750 genes that are indeed specific to the lineage containing malacostracans and branchiopods, rather than

Daphnia-specific (*E*-value: 10^{-6}). We also show the experimental tractability of *N. denticulata*, which, together with the genomic resources presented here, make it an ideal model for a wide range of further aquacultural, developmental, ecotoxicological, food safety, genetic, hormonal, physiological and reproductive research, allowing better understanding of the evolution of crustaceans and other arthropods.

Keywords: genomics; evolution; biotechnology; arthropods; crustaceans; decapod; shrimp

1. Introduction

Crustaceans are found worldwide in marine and terrestrial environments and are of great scientific and commercial importance. However, they are relatively underrepresented at the genomic level [1,2]. The Crustacea is conventionally divided into six classes [3], the Branchiopoda, Cephalocarida, Maxillopoda, Ostracoda, Remipedia and Malacostraca (which includes decapods, isopods, amphipods and stomatopods) (Figure 1), with an approximate number of 67,000 described living species [4]. Recent phylogenetic investigation has revealed that the Hexapoda, a group that includes the Insecta, is in fact nested within the crustaceans [5,6]. This renders the subphylum "Crustacea" paraphyletic, and the number of extant insect species is excluded from the number of crustacean species given above.

Figure 1. Simplified pancrustacean phylogeny, after [5]. Note some smaller, cryptic clades are not shown, including some members of the Maxillopoda, which are paraphyletically grouped with the Ostracoda (e.g., Branchiura and Pentastomida). Phylogenetic analysis is presently conflicted on the closest sister group to the Hexapoda: [5] places the Xenocarida as the outgroup to the Hexapoda, whereas [6] places the Branchiopods in this position.

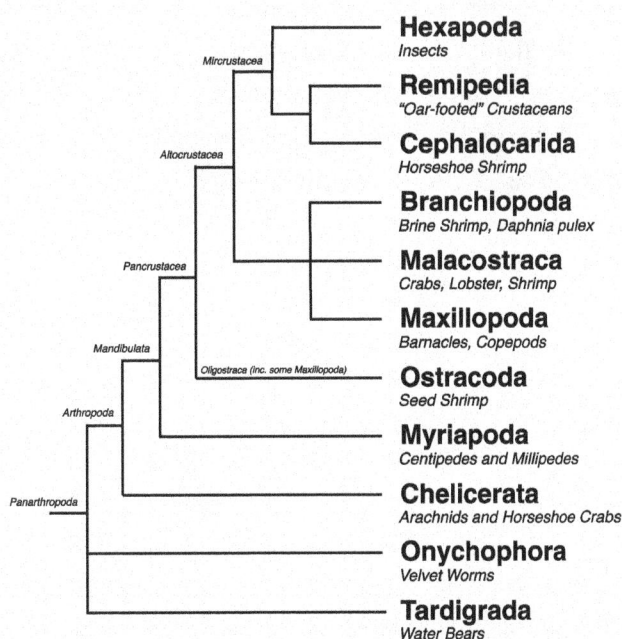

Extant published sequence resources in the Crustacea outside the Insecta are limited to that of a branchiopod, *Daphnia pulex* [7]. Although this animal is an important ecotoxicological model, its genome exhibits apparently high levels of gene duplication and loss and thus is not always suitable for use as an outgroup for comparison to the Insecta. Further, a single genome in the Crustacea also severely limits the conclusions that can be drawn as to the gain and loss of characters across the Pancrustacea, as current transcriptomic and expressed sequence tag (EST) resources will always only reflect the transcriptomic content of the specific tissues and time points sampled [1]. The provision of additional models to compare within the Crustacea is therefore a priority, given the importance of arthropods to the economy and a range of scientific spheres of investigation.

Malacostraca contains a number of orders, including the Amphipoda, Isopoda and Decapoda. A species within the Amphipoda, *Parhyale hawaiensis*, has already been used in developmental investigations [8], and isopods are familiar, due to the ubiquity of the common woodlouse, which is often used as a behavioural and environmental model [9]. The most economically important malacostracan crustaceans, however, belong to the order Decapoda. Decapods are both wild-caught and farmed and provide an important global food source. They are also ecologically vital as detritovores for environmental stability [10,11]. With an estimated 15,000 living crab, crayfish, shrimp, lobster and related species [12], the diversity of body plans and novelties seen in the Decapoda, including appendages, feeding mouthparts and segments, make them interesting models to study in evolutionary and developmental biology.

The cherry shrimp, *Neocaridina denticulata* (De Haan, 1844), is suggested as an excellent laboratory model within Decapoda—an experimentally tractable [13–15], cosmopolitan [16] and phylogenetically well-placed species [17]. It also has a limited history of use as a crustacean ecotoxicological model [13,14] and as a model of recent invasion [18]. The draft genome sequence of *N. denticulata denticulata* (De Haan, 1844) is presented here as a resource of benefit to a wide range of scientific investigations, including genomic, developmental, ecotoxicological, evolutionary, physiological and reproductive research. As another outgroup to the Insecta, it provides another lineage to that of the water flea, *D. pulex*, for comparison, and will allow further understanding of crustacean and arthropod biology. In addition, this species can be easily cultured and maintained in the laboratory [15], where insights gleaned from this species will be applicable to marine decapods and of much commercial utility for crustacean aquaculture and food safety. As such, this genomic resource will be of service to a range of scientific investigations worldwide.

2. Results and Discussion

2.1. Animal Culture and Lifecycle

Neocaridina shrimp are native to many freshwater areas of East and Southeast Asia, tolerant of a wide range of conditions, commercially used for human consumption as a food flavouring agent and include a variety of colour forms (Figure 2c), and have spread around the world courtesy of the aquarium trade [19]. We have chosen *N. denticulata* for further study as it has much potential to become a model for aquacultural, developmental, ecotoxicological, food safety, genetic, hormonal, physiological and reproductive studies in the laboratory.

The start-up of culture and maintenance of *N. denticulata* is straightforward. Animals can be acquired commercially from the aquarium trade in a variety of colour morphs (Figure 2c). They can be kept in small freshwater tanks at room temperature, with simple filtration and aeration facilities sufficient for survival and reproduction. The native environments of *N. denticulata* generally are medium-soft and slightly alkaline (pH 7–7.5). However, *N. denticulata* can survive in a pH of 6.5–8.0 [14,19,20]. Optimum water temperature is 22–24 °C, and while shrimp are tolerant of a range of five degrees above and below this [15], temperatures should not be allowed to change rapidly or markedly. Low-powered filtration systems are recommended to aid the survival of juvenile shrimp, and a sponge filter is adequate, provided uneaten food is removed from tanks at regular intervals. Similar to other arthropods, *N. denticulata* is, however, sensitive to heavy metals and the use of insecticides, and care should be taken not to expose the culture tanks to airborne pollutants [15]. *N. denticulata* will thrive on a variety of foodstuffs [19], and we have had good results with several commercially available shrimp feed formulations. The provision of hiding spaces or plant material, such as Java Moss (Hypnaceae) is recommended, as decapods are known to be cannibalistic of tank mates during ecdysis [21], and plants can provide an alternative food source for the shrimp, both directly and as a substrate for algal growth.

Sexual maturity of the female can be observed through the transparent carapace and body instead of sacrificing the animal to measure the gonadosomatic index (ovary weight divided by total body weight) (Figure 2b). Breeding occurs in summer in subspecies in colder climates, whereas warm-weather subpopulations breed year-round [19]. Mating occurs shortly after ecdysis, following which the female lays her eggs, fertilizing them as they are laid; they are then attached to her pleopods (swimmerets). Approximately 20–30 eggs with sizes ranging from 0.57 to 1.08 mm [22] are laid simultaneously and are carried externally [15], where they are amenable to injection or manipulation (see Section 2.8). The eggs hatch at around 30 days post-mating, depending on water temperature [23]. *N. denticulata* grow to 7.3–28.5 mm in length, with both male and female *N. denticulata* attaining sexually maturity at 4–6 months of age [15,22]. Generation time is therefore relatively short compared to other shrimps or decapods and is similar to that of other model organisms, such as zebrafish and medaka, but faster than that of the frog, *Xenopus laevis*.

Figure 2. (a) The lifecycle of *N. denticulata*; and (b) the appearance of gravid female (**left**, dorsal view; **right**, side view) compared to a female without eggs (**centre**, dorsal view). The scale bar represents 5 mm on three adult shrimp in (b). (c) Some of the colour forms available commercially. (i) Red patched; (ii) punctate red patterning; and (iii) blue.

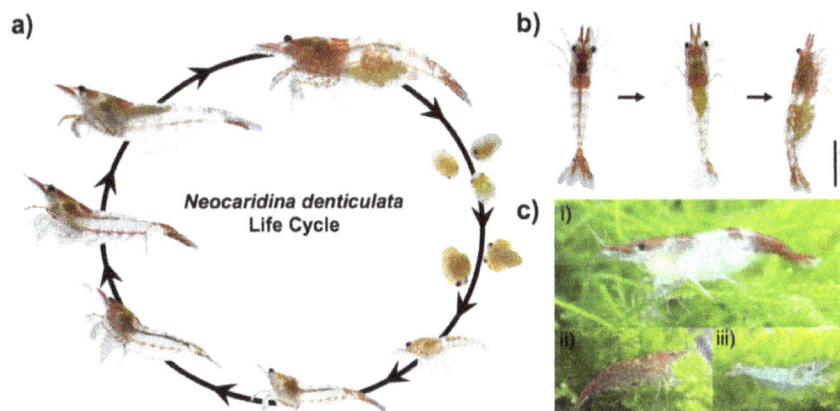

2.2. Genomic Sequencing

Genomic DNA from a single adult *N. denticulata denticulata* (De Haan, 1844) was extracted and sequenced on a single lane of the Illumina HiSeq2000 platform, as described in the experimental section. Basic read metrics relating to this sequencing are shown in Table 1. FastQC was run to ascertain read quality, with excellent results, and median Phred quality scores greater than 34 through to the last base in both reads (Supplementary Figure S1). No over-represented sequences were detected in our analysis. Raw sequence data have been uploaded to NCBI's SRA (Bioproject PRJNA224755, Biosample SAMN02384679, experiment SRX375172, reads SRR1027643). After an initial assembly trial, Bowtie [24] was used to determine actual fragment size and the standard deviation for future use.

Table 1. Basic metrics relating to raw reads.

Platform	Illumina HiSeq2000
Number of Reads	364,013,140
Read Length (bp)	100
Average GC %	36
Fragment Size	167.22
Fragment Size SD (bp)	12.01

2.3. Genomic Assembly

Genomic assembly procedures are summarized in Figure 3a. After initial trials using a range of assembly software, including Velvet [25] and SOAPdenovo [26], raw reads were assembled using the abyss-pe script in ABySS [27] with a *k*-mer size of 51. Read cleaning using Sickle [28] and Musket [29] was assayed, but found to impair assembly by conventional metrics. Results of the empirically-determined "best" assembly are shown in Table 2. This assembly can be downloaded from [30] or can be supplied by the authors upon request.

While the genome size of *N. denticulata* has not been measured, a closely related species in the same family, Atyidae, *Antecaridina* sp., has been determined to have a *C*-value of 3.30 pg, or approximately 3.2 Gbp [31]. Such large genomes are known to be difficult to assemble and traditionally exhibit a large amount of repetitive sequences. Our efforts have recovered sequences totalling 1.2 Gbp. If the genome size of *N. denticulata* is close to 3 Gb, one possibility could be that the short fragment length used for library construction constrains the contiguity of our sequences across repetitive regions and, thus, also accounts for the relatively short N50 (Table 2). Assuming a 3 Gb genome, our sequence data provide approximately 12x coverage. A small amount of contamination with bacterial DNA (three large contigs greater than 30 kb in length, similar to the *Novosphingobium* sp. bacterial DNA sequence) without high similarity to the known *Wolbachia* sequences was removed manually after BLASTN detection. The availability of funding for additional long mate pair data for scaffolding in the future would greatly enhance contiguity and allow the exploration of the content of non-coding regions, which we suspect are poorly recovered in this assembly.

Figure 3. (**a**) Schematic diagram of the genomic assembly of shrimp *N. denticulata.* (**b**) The summary statistics relating to the comparison of *D. pulex* and *N. denticulata* genomes, compared to the non-redundant (nr) database. *D. pulex* image is courtesy of Paul Hebert [32].

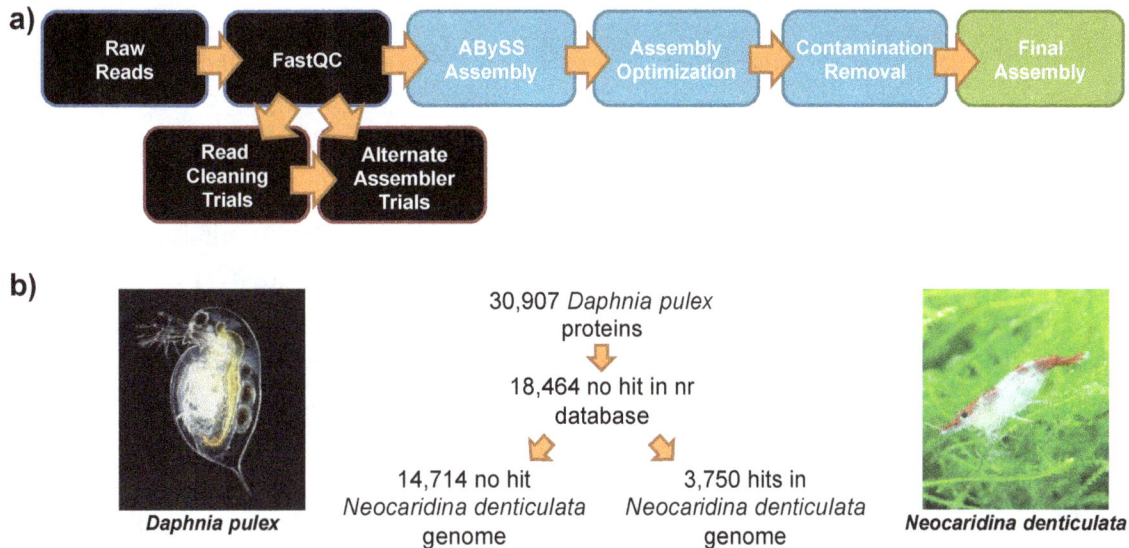

Table 2. Metrics relating to final assembly.

Criteria	Value (base pairs)
Min. contig length	200
Max. contig length	124,746
Mean contig length	383.84
Standard deviation of contig length	285.33
Median contig length	302
N50 contig length	400
Number of contigs	3,346,358
Number of contigs ≥1 kb	97,432
Number of contigs in N50	987,201
Number of bases in all contigs	1,284,468,468
Number of bases in contigs ≥1 kb	132,397,543
GC Content of contigs (%)	35.21

2.4. Comparison of Core Eukaryotic Genes

Despite the scaffold size being relatively short, our data contain a great deal of useful information concerning the coding regions of this genome. We used the Core Eukaryotic Gene Mapping Approach (CEGMA) dataset [33], which consists of 458 single-copy genes found in almost every eukaryote genome, as an assay of the completeness of the coding sequence coverage contained in our sequence data. Using TBLASTN [34] with a cut-off of 10^{-3}, of the 458 genes, only three (ribosome biogenesis protein RLP24, ribosomal 60S subunit protein L24A and the HSP binding protein, YER156C) did not possess a recognizable hit in our sequence. This *E*-value cut-off was selected empirically after several trials and, at this stringency, represents 455 genes or 99.3% recovery of the expected coding sequences, which suggests that our dataset is excellent as a starting point for assaying the presence of genes in decapod crustaceans. Of these contigs annotated with CEGMA, the mean size of the contigs identified is 2500.01 bp and the median is 534 bp, which are longer than our N50 and mean/median contig sizes.

2.5. Mitochondrial Genomic Characteristics

Retrieval of the *N. denticulata denticulata* (Crustacea: Caridea) mitochondrial genome from our dataset in a single, well-assembled contig revealed a circular molecule of 15,565 bp that encodes the typical set of 37 metazoan genes (13 protein-coding genes, 22 transfer RNA genes and two ribosomal RNA genes). The majority-strand (α) and minority-strand (β) encode 23 and 14 genes, respectively (Figure 4 and Supplementary Information: Table S1 in mtDNA). The result is comparable to the mitogenome of a related subspecies, *N. denticulata sinensis*, which differs by 4 bp in length and has slight differences in amino acid and codon usage [35]. Due to the compactness of the mitochondrial genome, ten instances of gene overlaps were found. A total of 841 bp non-coding nucleotides were found, with 153 bp in 13 intergenic regions and a 688 bp-long non-coding region between the *srRNA* and *trnIle* (Supplementary Information: Table S1 in mtDNA).

Figure 4. The *N. denticulata denticulata* mitochondrial genome. The orientation of genes is represented by the position on outside circle (transcription clockwise or anticlockwise is represented outside or inside the form, respectively). Local nucleotide identity (GC, dark blue) is represented on the inner ring as implemented by OrganellarGenomeDRAW (OGDRAW) [36].

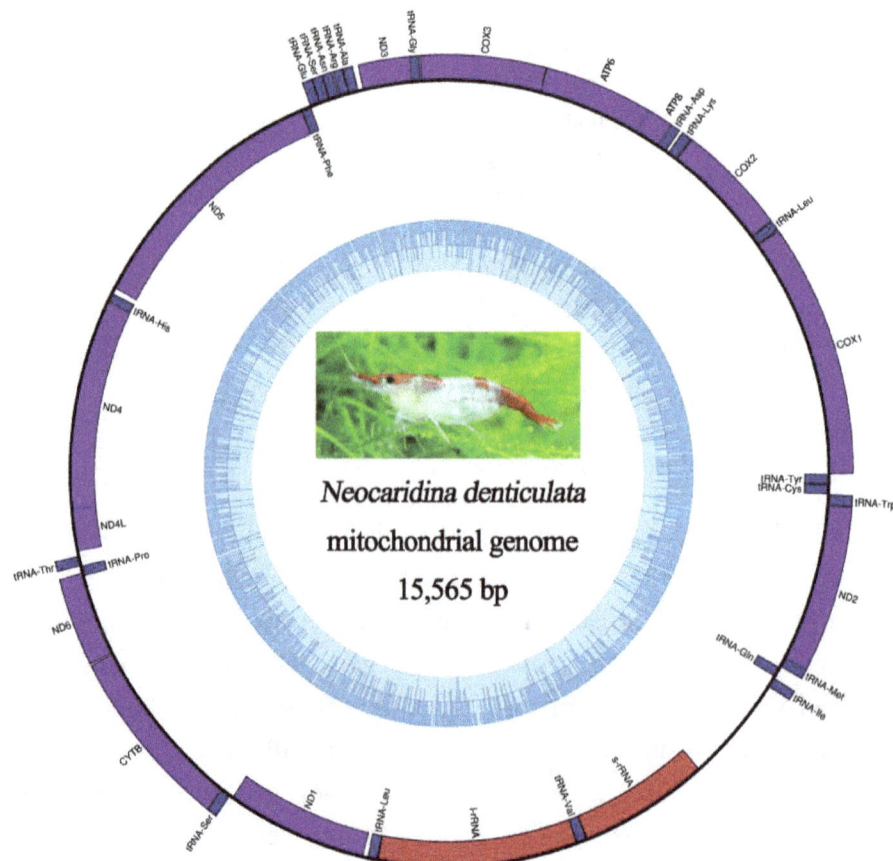

The typical metazoan initiation codon for transcription "ATN" is used by 12 out of 13 protein-coding genes, whereas *cox1* employs "AAG" as the start codon, which is similar to other caridean mitochondrial genomes (Supplementary Information: Table S2 in mtDNA). The open-reading frames of 11 protein-coding genes are terminated by the typical stop codon (TAA or TAG), while the remaining two genes (*cox2* and *nad4*) have an incomplete stop codon "T-". All protein-coding genes and both rRNAs have skewed T *vs.* A (AT skew ranging from −0.041 to −0.293). The majority of protein-coding genes have a skew of C *vs.* G, but both rRNAs have a skew of G *vs.* C (the GC skews are 0.316 and 0.273 for *srRNA* and *lrRNA*, respectively) (Supplementary Information: Table S3 in mtDNA). In total, there are 3696 codons in all 13 mitochondrial protein-coding genes, excluding incomplete termination codons, and the most frequently used amino acids are Leu (15.58%), followed by Ser (9.60%), Ile (8.39%), Phe (8.12%) and Val (7.06%) (Supplementary Information: Table S4 in mtDNA).

Using the nucleotide sequences of the mitogenome of *N. denticulata denticulata* presented here, a Bayesian phylogenetic reconstruction of malacostracan inter-relationships was performed and summarized in Figure 5, which reinforces our knowledge of the inter-relationships of the Caridea.

Figure 5. The Bayesian phylogenetic tree showing inter-relationships of a variety of malacostracan crustacean species, including the position of *N. denticulata* (underlined in red), based on concatenated nucleotide sequences from mitochondrial genomes. Numbers at nodes represent the posterior probability expressed out of 100.

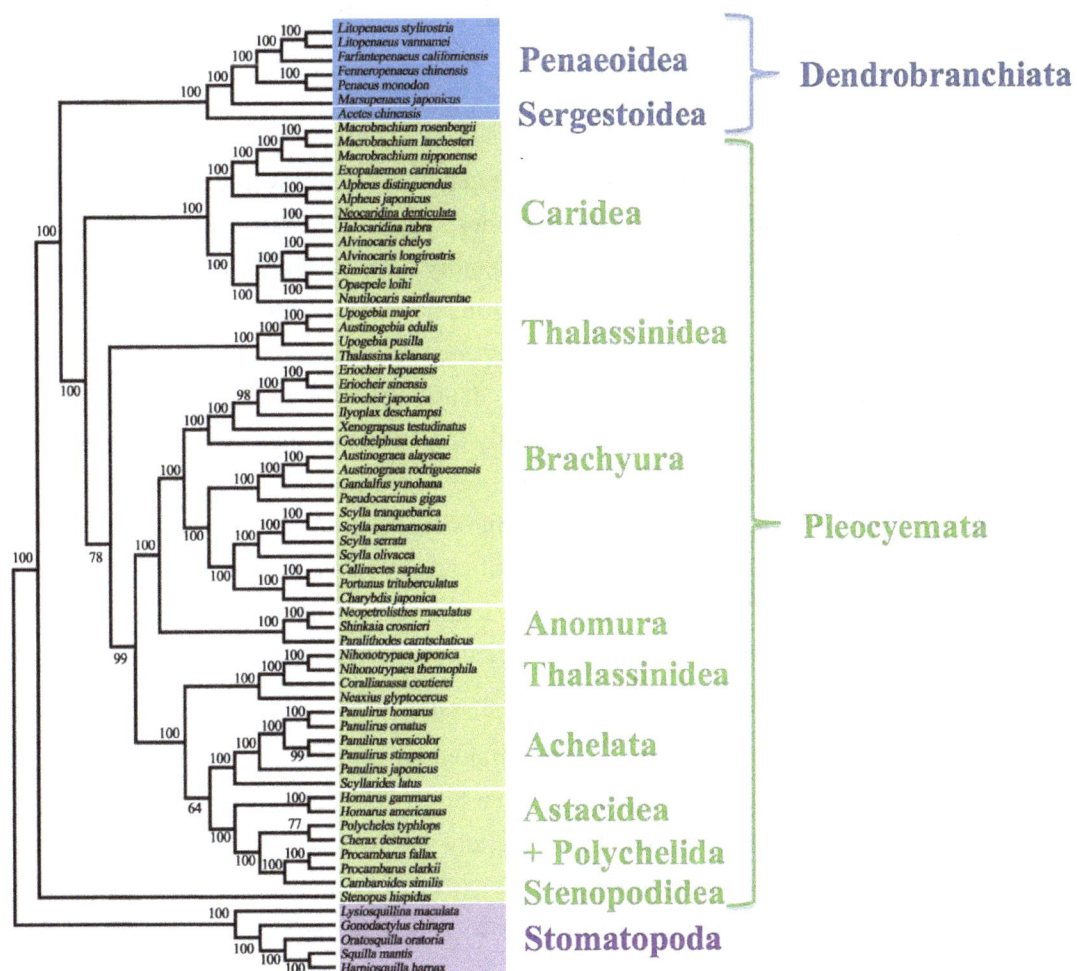

Further in our analysis, the arrangement of the *N. denticulata* mitochondrial genes is found to be identical to the hypothetical pancrustacean ground pattern (Figure 6), whereas some other members of infraorder Caridea, such as *Exopalaemon carinicauda* (Palaemonidae), *Alpheus japonicus* and *A. distinguendus* (Alpheidae), all have their mitochondrial gene orders rearranged ([37]; Figure 6).

Figure 6. *N. denticulata* mitochondrial genome organisation compared to that of other crustaceans. *N. denticulata* possesses the stereotypical pancrustacean mitochondrial gene order as described first in *Limulus polyphemus* [38]; the orders of closely related species are provided for ease of comparison.

Neocaridina denticulata and *Halocaridina rubra* (Atyidae)
Macrobrachium lanchesteri, M. nipponense and *M. rosenbergii* (Palaemonidae)
Rimicaris kairei, Nautilocaris saintlaurentae, Alvinocaris longirostris, A. chelys and *Opaepele loihi* (Alvinocarididae)
Pancrustacean ground pattern

Alpheus japonicus and *A. distinguendus* (Alpheidae)

Exopalaemon carinicauda (Palaemonidae)

2.6. Recovery of Hox Genes and Other Families

To confirm the coverage of the coding regions of this genome, several well-annotated and catalogued developmental gene families were assayed. Our searches suggest that most, if not all, of the coding regions of the genome were recovered in our assembly. For example, Hox genes pattern the developing anteroposterior axis of animals, and 12 families of Hox genes are commonly described in pancrustaceans [39]. In our analysis, nine of the 12 Hox gene members could be identified (Figure 7). Of the three missing families, *zen2* and *bcd* have not been identified in any crustaceans outside the Insecta to date [39], so their absence in our dataset very probably indicates that these genes are insect novelties. Only one Hox gene absent from our dataset, *pb*, could be a consequence of the poor recovery of this genomic locus or the first loss of this gene reported in the Crustacea *sensu stricto*. The identification of Hox gene *zen1* in *N. denticulata* provides the first identification of this gene in the decapods (Figure 7). Unfortunately, these sequences are predominantly found on short contigs (see Supplementary Information: HoxL tab), and therefore, no syntenic relationship information can be gleaned from the data as it stands.

Similarly, our recovery of other families of well-catalogued genes is equally impressive (Table 3). The Fox genes, which are helix-turn-helix genes with an 80 to 100 amino acid "Forkhead Box" motif, are separated into 23 classes, which perform a variety of roles in metabolism and embryonic development [40,41]. We find 21 homologues of these genes in our dataset, from 16 families, with almost all those missing probably restricted to clades to which *N. denticulata* does not belong and, hence, was not expected to be found in the genome sequence data. The one exception to this is *Fox L1*, which appears to be absent from our dataset, while being described in other protostomes.

T-box genes are also well-catalogued and perform a similarly wide range of vital evolutionarily conserved roles [42]. Of these genes, we recovered 10 sequences corresponding to five families. The families missing, *T-Brain*, *Tbx4/5* and *Tbx 15/18/22*, are genes limited to other Superphyla [42–44] within the Metazoa and, hence, are not expected to be found in *N. denticulata*.

Another gene repertoire examined to show the conservation of key pathways was the core microRNA/miRNA processing cassette [45]. Sequences (such as *Ago*, *Dicer*, *Exportin-5*) from the totality of the expected pathway were recovered. These findings further suggest that the coding sequences of the *N. denticulata* genome are well-recovered in our genomic build.

Figure 7. Hox cluster gene recovery in *N. denticulata*, compared to a number of other pancrustacean species (diagram after [39]). Sequences provided in Supplementary Information, HoxL tab.

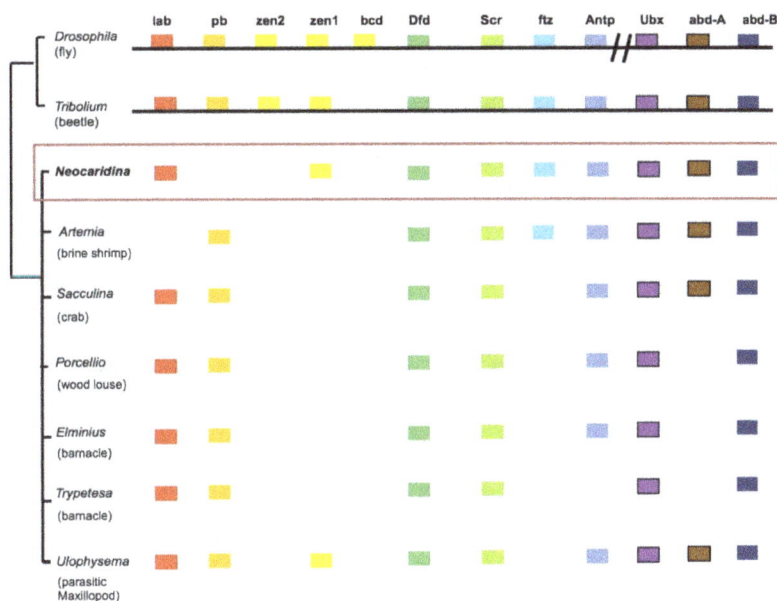

Table 3. Recovery of developmentally important gene families. Details and sequences provided in Supplementary Information. Missing genes are generally absent ancestrally, rather than in our assembly, as discussed in the main text.

Gene Classes	Homologues Recovered	Missing
Homeobox Genes (HoxL)	20	Pb
Fox Genes	21	Fox E, H, I, J2/3, L1, M, Q1
T-box Genes	10	T-Brain, Tbx 4/5, Tbx 15/18/22
miRNA processing genes	8	-

2.7. Comparison to the Lineage-Specifically Gained Genes of Daphnia Pulex

To gain an understanding of crustacean genome evolution, we compared the genome composition of *N. denticulata* with that of the branchiopod, *D. pulex*, the only publically available crustacean genome sequence to date. The genome of *D. pulex* is noted for its large number of gene duplications and a rapid rate of genomic evolution, and a previous study has suggested 17,424 new and 1079 lost genes in the branch leading to *D. pulex* [7].

To determine whether the genes gained in *D. pulex* represent a Crustacea-wide gain or are truly limited to *D. pulex* alone, we used BLASTX to compare the *D. pulex* proteome to the *N. denticulata* genome (Figure 3b). As significant numbers of sequences have been added to the non redundant (nr) database since the publication of the *D. pulex* genome in 2011, the 30,907-sequence proteome of *D. pulex* was blasted against the nr database using Blast2GO (database as of October, 2013, BLASTP, cut-off e^{-6}, crustacean genes (excluding the Insecta) excluded from blast hits). At this threshold, 18,464 *D. pulex* genes were found to have no hits in nr.

The *D. pulex* proteome was then compared to our genomic build using BLASTX. As *E*-values cannot be directly compared between datasets, as they depend on the size of the database and the search used, and a number of *E*-values were trialled for the initial comparison of the *D. pulex* proteome and *N. denticulata* genome. By way of example, 16,640 *D. pulex* genes were found to have no hits in *N. denticulata* at an *E*-value of 10^{-3}, whereas 18,739 *D. pulex* genes have no hits at an *E*-value of 10^{-6}. As this latter figure represented a similar number to the unannotatable gene complement of *D. pulex*, this was taken as our cut-off for further comparison.

Of the 18,464 *D. pulex* genes with no hit in the nr database, 14,714 had no identifiable homologue in shrimp either. However, 3750 *D. pulex* genes, unidentifiable previously, were found to have hits in *N. denticulata* above the cut-off threshold of an *E*-value of 10^{-6} (Figure 3b). These 3750 genes may represent novelties gained in the maxillopod, branchiopod and malacostracan lineages or pancrustacean genes lost in sequenced insects. These genes will be key targets for future work on crustacean novelties and are likely of interest to a range of researchers. The list of 3750 *D. pulex* genes, along with the lists of all other genes and appropriate sequences, are summarized in Supplementary Information, under the Details of Hits and Sequences tabs.

Additionally, 1927 of the 12,443 *D. pulex* genes identifiable in the nr database have no hit in the shrimp, which either represents genes not assembled in our dataset or losses on the malacostracan lineage leading to *N. denticulata*.

2.8. MS-222 Treatment

In addition to describing the culture and sequencing of the draft genome of *N. denticulata*, we have also established an anaesthesia technique using tricaine methanesulfonate (MS-222) for future genetic manipulation requiring microinjection of eggs or adults.

The average induction time for the adult shrimp in 1500, 2000, 2500 and 3000 mg L^{-1} baths of MS-222 were 27 min 5 s (Standard error (SE) = 3 min 41 s), 12 min 42 s (SE = 1 min 15 s), 10 min 58 s (SE = 53 s) and 6 min 27 s (SE = 47 s), respectively (Figure 8). As only two out of ten individuals entered anaesthesia in 1000 mg L^{-1} bath of MS-222 after 40 min, we concluded this concentration is too low to induce anaesthesia in *N. denticulata*, and no further analysis was performed at this or a lower concentration. Both the anaesthetic concentration and bath duration clearly affect the recovery time of *N. denticulata* (Figure 8a,b). At higher MS-222 concentrations, shrimp took a longer time for their first movement (Figure 8a; General Linear Model (GLM): $F_{3,30} = 26.1$, $p < 0.001$) and complete recovery

(Figure 8b; GLM: $F_{3,24} = 21.4$, $p < 0.001$) was observed. Similarly, a longer bath duration also led to a longer time to their first movement (Figure 8a; GLM: $F_{3,30} = 11.0$, $p < 0.001$) and complete recovery (Figure 8b; GLM: $F_{3,24} = 16.8$, $p < 0.001$). The effect of bath duration was more obvious in higher rather than in lower MS-222 concentrations (the interactive effect of concentration and duration on time until first movement: $F_{3,30} = 3.2$, $p < 0.05$). At a concentration of 3000 mg L^{-1} MS-222, one/four individuals of five sampled were dead after 10 and 20 min of anaesthetic bath, respectively. Therefore, a 3000 mg L^{-1} concentration is too high for safe anaesthesia in *N. denticulata*. Following the treatment of animals with 2000 mg L^{-1} for 30 min, one individual of the three sampled died, suggesting that a long bath duration could be lethal to more susceptible individuals, even at a lower anaesthetic concentration. In our experiments, all other individuals treated with MS-222 at less than 3000 mg L^{-1} could recover completely, and none died in the following three days.

Figure 8. Boxplot of recovery time when (**a**) the first movement and (**b**) complete recovery was observed in *N. denticulata* individuals after anaesthesia in different MS-222 concentrations and bath durations. The horizontal line shows the median, boxes the first and third quartiles, whiskers the highest and lowest values that are within the 1.5 inter-quartile range, black dots the outliers and numbers above boxes the sample size. [a] A ten-minute bath duration is insufficient to induce anaesthesia at a concentration of 1500 mg L^{-1}. [b] A thirty-minute bath duration is lethal to some individuals at a concentration of 2000 mg L^{-1} and was not performed for higher concentrations. [c] The forty minute bath duration was not performed for concentrations higher than 1500 mg L^{-1}. [d] Some individuals that showed signs of movement were not fully recovered at the end of the assayed period.

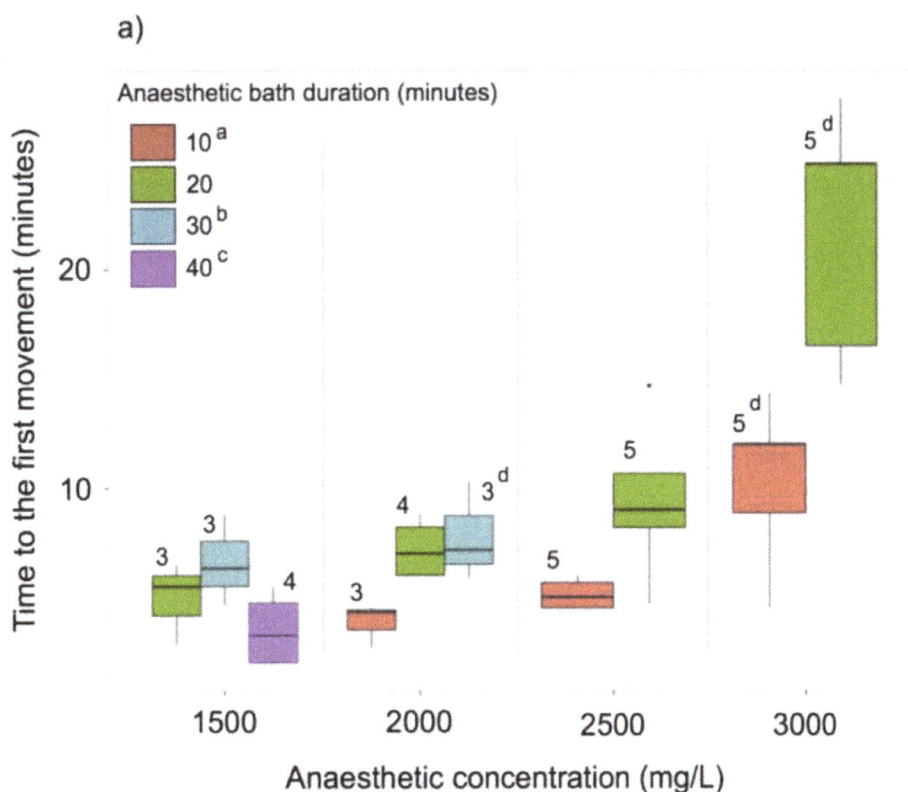

a)

Figure 8. *Cont.*

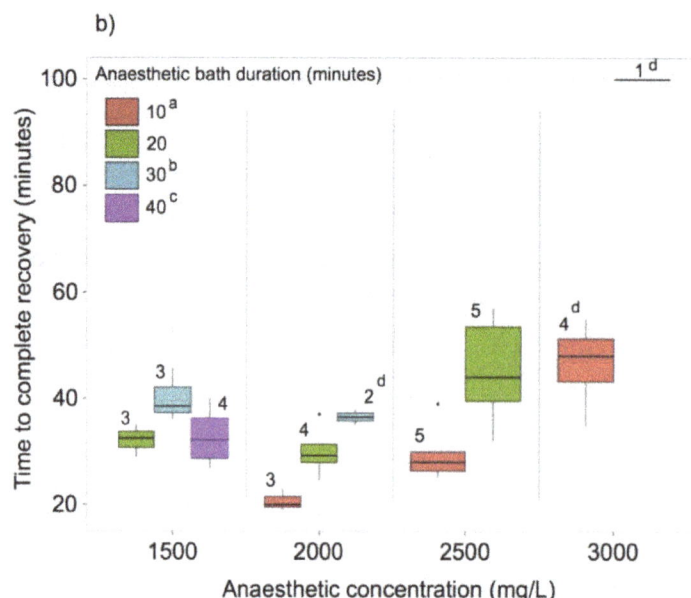

To allow sufficient time for experimental manipulation, we have determined that bathing *N. denticulata* for 20 min in 2500 mg L^{-1} MS-222 is optimal for longest non-lethal dose and duration. Under these conditions, the recovery time until first movement is 9 min 26 s (SE = 1 min 38 s). Compared to other aquatic animals, the MS-222 concentration we suggest is higher than those for fish anaesthesia in general, which vary from 50–400 mg L^{-1} [46] However, the effective MS-222 concentration for crustacean anaesthesia is generally found to be high, for example, 500 mg/L on the microcrustacean *Eucypris virens* [47], and could vary among different crustaceans with 100 mg L^{-1} and 1000 mg L^{-1} MS-222 being found ineffective on the shrimp, *Macrobrachium rosenbergii* [48], Chinese mitten crab, *Eriocheir sinensis* [49], and crayfish, *Orconectes virilis* [50].

3. Experimental Section

3.1. Animal Husbandry and Genomic DNA Extraction

N. denticulata denticulata (red patched strain) were sourced from local suppliers and kept in a recirculating freshwater aquarium at room temperature, at approximately 25 °C. A single adult was starved for 2 days prior to gut dissection, and whole animal genomic DNA was extracted using DNeasy Blood & Tissue Kit (Qiagen, Hilden, Germany), according to the manufacturer's protocol.

3.2. Illumina Hi-Seq and Assembly

The *N. denticulata* genomic DNA sample was sequenced on a single lane on the Illumina HiSeq2000 platform. DNA was prepared for sequencing using a TruSeq DNA Sample Preparation Kit by BGI Hong Kong. Reads were first filtered by BGI according to their internal protocol, including the removal of the adaptor sequence, contamination and low-quality reads. Raw sequence data were uploaded to NCBI's SRA (Bioproject PRJNA224755, Biosample SAMN02384679, experiment SRX375172, reads

SRR1027643). ABySS 1.3.3 [27] was used to assemble the genome at a *k*-mer length of 51 and all default settings. A minimum scaffold length of 200 bp was imposed after the assembly was complete.

3.3. mtDNA/ Nuclear Gene Retrieval

Gene sequences were identified using TBLASTN [34] searches using known gene sequences of confirmed homology downloaded from the NCBI nr database as queries. Genes thus putatively identified were then reciprocally blasted against the NCBI nr database using BLASTX to further confirm their identity. Characteristic domains and trees constructed in MrBayes 3.1 [51] using genes of known homology downloaded from the nr database and aligned using MAFFT [52] were used to further confirm the homology for developmental gene families (data not shown).

3.4. Gene Comparison

For the comparison of *D. pulex* and *N. denticulata* sequences, the *D. pulex* proteome (FilteredModelsv1.1.aa) was downloaded from the Joint Genome Institute (JGI) website [53]. Standalone ncbi-blast-2.2.23+ [34] and Blast2GO [54] were used to perform blasts locally, as described in the text, with the latter used only for the comparison of the *D. pulex* dataset with the nr database.

3.5. mtDNA Annotation and Display

The locations of the 13 protein-coding genes (PCGs) and 2 rRNAs were determined with Dual Organellar GenoMe Annotator (DOGMA) [55] and subsequent alignments with caridean mitochondrial genes. All tRNA genes were identified by tRNAscan-SE 1.21 [56]. The gene map of the mitochondrial genome was drawn by OrganellarGenomeDRAW (OGDRAW) [36]. Codon usage in 13 PCGs of the mitochondrial genome was estimated with DnaSP 5.10.1 [57]. The nucleic acid sequences of 13 PCGs were aligned using Clustal W [58]. To determine the best fitting model of sequence evolution for the nucleic acid dataset, a nested likelihood ratio test was performed using jModelTest 2 [59]. After the evolutionary model (GTR + G + I) was determined, the phylogenetic relationship was inferred by MrBayes 3.1 [51]. The Markov Chain Monte Carlo analyses were run for 1,000,000 generations (sampling every 1000 generations) to allow adequate time for convergence. After omitting the first 250 sampled trees as "burn-in", the remaining 750 sampled trees were used to estimate the Bayesian posterior probabilities.

3.6. MS222 Anaesthesia

Anaesthesia was performed by dissolving MS-222 salt (Sigma, St. Louis, MO, USA) in culture tank water. Sodium bicarbonate was added to neutralize the pH to the original culture tank water pH (*i.e.*, 7.2). Ten individual adult shrimp were tested for each of the five MS-222 concentrations (1000, 1500, 2000, 2500 and 3000 mg/L). Shrimp were kept in an aerated anaesthetic bath for durations as stated (10, 20, 30 or 40 min). The induction time (the time from immersion in the anaesthetic bath until the shrimp showed no movement and no reaction to touch stimuli) was recorded. After the anaesthetic bath, the shrimps were immediately put in aerated water from the original tank for recovery. The recovery time for each individual was also recorded, including: (a) the time until the first movement of pleopods and/or pereiopods; and (b) the time until complete recovery at which the individual resumed balanced movements and normal feeding behaviour. A general linear model (GLM) was used to test the recovery time difference between different anaesthetic conditions, with the recovery time being used as the dependent variable and anaesthetic concentration and duration as independent factors.

4. Conclusions

To date, genomic sampling in Crustacea remains depauperate, but resources such as the one presented here will aid in the further study of this neglected taxon. As suggested by the recovery of the majority of core eukaryotic genes, the complete and un-rearranged mitogenome and considerable fractions of several developmental gene cassettes, the *N. denticulata* sequences presented here represent an excellent crustacean model for comparison to other arthropods. Our analysis also allows the identification of 3750 putatively crustacean-specific genes that could shed light on the understanding of the crustacean genome evolution. A non-lethal anaesthesia protocol has also been established for use in future genetic manipulation of this species. Given the lack of an easily-cultivable decapod laboratory model, we propose the shrimp, *N. denticulata*, as an experimentally tractable, easily grown species for a wide variety of future investigations in aquacultural, developmental, ecotoxicological, evolution, food safety, genetic, hormonal, physiological and reproductive research of decapod crustaceans.

Author Contributions

Jerome H.L. Hui conceived and supervised the project. Nathan J. Kenny assembled the next-generation sequencing reads, and identified genes. Nathan J. Kenny, Xin Shen, Sebastian M. Shimeld and Jerome H.L. Hui analysed and compared the genes. Yung Wa Sin, Qu Zhe, Jerome H.L. Hui maintained animal culture. Qu Zhe and Wei Wang prepared the genomic DNA for sequencing. Yung Wa Sin carried out the MS-222 anaesthesia. Nathan J. Kenny, Yung Wa Sin, Xin Shen, Qu Zhe, Wei Wang, Ting Fung Chan, Stephen S. Tobe , Sebastian M. Shimeld , Ka Hou Chu and Jerome H.L. Hui wrote the manuscript. All authors read and approved the final manuscript.

Acknowledgements

The authors would like to thank members of the CUHK Laboratory of Evolution and Development, whose discussions contributed to this manuscript, and BGI Hong Kong for the help with the sequencing. We thank two anonymous reviewers for their time in commenting on and improving this work. NJ Kenny's work on this manuscript was carried out under the Global Scholarship Programme for Research Excellence, CUHK. This study was supported by a direct grant (4053034) of the Chinese University of Hong Kong (JHL Hui).

References

1.   Stillman, J.H.; Colbourne, J.K.; Lee, C.E.; Patel, N.H.; Phillips, M.R.; Towle, D.W.; Eads, B.D.; Gelembuik, G.W.; Henry, R.P.; Johnson, E.A.; *et al.* Recent advances in crustacean genomics. *Integr. Comp. Biol.* **2008**, *48*, 852–868.

2. Kenny, N.J.; Quah, S.; Holland, P.W.H.; Tobe, S.S.; Hui, J.H.L. How are comparative genomics and the study of microRNAs changing our views on arthropod endocrinology and adaptations to the environment? *Gen. Comp. Endocrinol.* **2013**, *188*, 16–22.

3. Martin, J.W.; Davis, G.E. An updated classification of the recent Crustacea. *Sci. Ser.* **2001**, *39*, 1–124.

4. Zhang, Z.Q. Phylum Arthropoda von Siebold, 1848. *Zootaxa* **2011**, *3148*, 99–103.

5. Regier, J.C.; Shultz, J.W.; Zwick, A.; Hussey, A.; Ball, B.; Wetzer, R.; Martin, J.W.; Cunningham, C.W. Arthropod relationships revealed by phylogenomic analysis of nuclear protein-coding sequences. *Nature* **2010**, *463*, 1079–1083.

6. Rota-Stabelli, O.; Campbell, L.; Brinkmann, H.; Edgecombe, G.D.; Longhorn, S.J.; Peterson, K.J.; Pisani, D.; Philippe, H.; Telford, M.J. A congruent solution to arthropod phylogeny: Phylogenomics, microRNAs and morphology support monophyletic Mandibulata. *Proc. Biol. Sci.* **2011**, *278*, 298–306.

7. Colbourne, J.K.; Pfrender, M.E.; Gilbert, D.; Thomas, W.K.; Tucker, A.; Oakley, T.H.; Tokishita, S.; Aerts, A.; Arnold, G.J.; Basu, M.K.; *et al.* The ecoresponsive genome of *Daphnia pulex*. *Science* **2011**, *331*, 555–561.

8. Rehm, E.J.; Hannibal, R.L.; Chaw, R.C.; Vargas-Vila, M.A.; Patel, N.H. The crustacean *Parhyale hawaiensis*: A new model for arthropod development. *Cold Spring Harb. Protoc.* **2009**, *1*, 373–404.

9. Sloof, W.; de Kruijf, H.; Hopkin, S.P.; Jones, D.T.; Dietrich, D. The isopod *Porcellio scaber* as a monitor of the bioavailability of metals in terrestrial ecosystems: Towards a global "woodlouse watch" scheme. *Sci. Total Environ.* **1993**, *134*, 357–365.

10. Hart, R.C. Population dynamics and production of the tropical freshwater shrimp *Caridina nilotica* (Decapoda: Atyidae) in the littoral of Lake Sibaya. *Freshw. Biol.* **1981**, *11*, 531–547.

11. De Silva, P.K.; de Silva, K.H.G.M. Aspects of the population ecology of a tropical freshwater atyid shrimp *Caridina fernandoi* Arud. & Costa, 1962 (Crustacea: Decapoda: Caridea). *Arch. Hydrobiol.* **1962**, *117*, 237–253.

12. De Grave, S.; Pentcheff, N.D.; Ahyong, S.T.; Chan, T.Y.; Crandall, K.A.; Dworschak, P.C.; Felder, D.L.; Feldmann, R.M.; Fransen, C.H.J.M.; Goulding, L.Y.D.; *et al.* A classification of living and fossil genera of decapod crustaceans. *Raffles Bull. Zool.* **2009**, *21*, 1–109.

13. Huang, D.; Chen, H. Effects of chlordane and lindane on testosterone and vitellogenin levels in green neon shrimp (*Neocaridina denticulata*). *Int. J. Toxicol.* **2004**, *23*, 91–95.

14. Huang, D.-J.; Chen, H.-C.; Wu, J.-P.; Wang, S.-Y. Reproduction obstacles for the female green neon shrimp (*Neocaridina denticulata*) after exposure to chlordane and lindane. *Chemosphere* **2006**, *64*, 11–16.

15. Mizue, K.; Iwamoto, Y. On the development and growth of *Neocaridina denticulata* de Haan. *Bull. Fac. Fish. Nagasaki Univ.* **1961**, *10*, 15–24.

16. Liang, X. Crustacea: Decapoda: Atyidae. In *Fauna Sinica: Invertebrata*; Science Press: Beijing, China, 2004; Volume 36.

17. Bracken, H.; de Grave, S.; Felder, D. Phylogeny of the infraorder Caridea based on mitochondrial and nuclear genes (Crustacea: Decapoda). *Decapod Crustac.* **2009**, *18*, 281–305.

18. Englund, R.; Cai, Y. Occurrence and description of *Neocaridina denticulata sinensis* (Kemp, 1918) (Crustacea: Decapoda: Atyidae), a new introduction to the Hawaiian Islands. *Bish. Museum Occas.*

Pap. **1999**, *58*, 58–65.

19. Oh, C.-W.; Ma, C.-W.; Hartnoll, R.G.; Suh, H.-L. Reproduction and population dynamics of the temperate freshwater shrimp, *Neocaridina denticulata denticulata* (De Haan, 1844), in a Korean stream. *Crustaceana* **2003**, *76*, 993–1015.

20. Dudgeon, D. The population dynamics of some freshwater carideans (Crustacea: Decapoda) in Hong Kong, with special reference to *Neocaridina serrata* (Atyidae). *Hydrobiologia* **1985**, *120*, 141–149.

21. Marshall, S.; Warburton, K.; Paterson, B.; Mann, D. Cannibalism in juvenile blue-swimmer crabs *Portunus pelagicus* (Linnaeus, 1766): Effects of body size, moult stage and refuge availability. *Appl. Anim. Behav.* **2005**, *90*, 65–82.

22. Hung, M.; Chan, T.; Yu, H. Atyid shrimps (Decapoda: Caridea) of Taiwan, with descriptions of three new species. *J. Crustac. Biol.* **1993**, *13*, 481–503.

23. Shy, J.; Ho, P.; Yu, H. Complete larval development of *Neocaridina denticulata* (De Haan, 1884) (Crustacean, Decapoda, Caridea) reared in the laboratory. *Ann. Taiwan Mus.* **1992**, *35*, 75–89.

24. Langmead, B.; Trapnell, C.; Pop, M.; Salzberg, S.L. Ultrafast and memory-efficient alignment of short DNA sequences to the human genome. *Genome Biol.* **2009**, *10*, R25.

25. Zerbino, D.R.; Birney, E. Velvet: Algorithms for *de novo* short read assembly using de Bruijn graphs. *Genome Res.* **2008**, *18*, 821–829.

26. Luo, R.; Liu, B.; Xie, Y.; Li, Z.; Huang, W.; Yuan, J.; He, G.; Chen, Y.; Pan, Q.; Liu, Y.; *et al.* SOAPdenovo2: An empirically improved memory-efficient short-read *de novo* assembler. *Gigascience* **2012**, *1*, 18.

27. Simpson, J.T.; Wong, K.; Jackman, S.D.; Schein, J.E.; Jones, S.J.; Birol, I. ABySS: A parallel assembler for short read sequence data. *Genome Res.* **2009**, *19*, 1117–1123.

28. Sickle Github Repository. Available online: https://github.com/najoshi/sickle (accessed on 21 April 2013).

29. Liu, Y.; Schröder, J.; Schmidt, B. Musket: A multistage k-mer spectrum-based error corrector for Illumina sequence data. *Bioinformatics* **2013**, *29*, 308–315.

30. *Neocaridina denticulata* Genome Website. Available online: http://tiny.cc/shrimpgenome/ (accessed on 18 July 2013).

31. Bachmann, K.; Rheinsmith, E.L. Nuclear DNA amounts in Pacific Crustacea. *Chromosoma* **1973**, *43*, 225–236.

32. Gewin, V. Functional genomics thickens the biological plot. *PLoS Biol.* **2005**, *3*, e219.

33. Parra, G.; Bradnam, K.; Korf, I. CEGMA: A pipeline to accurately annotate core genes in eukaryotic genomes. *Bioinformatics* **2007**, *23*, 1061–1067.

34. Altschul, S.F.; Gish, W.; Miller, W.; Myers, E.W.; Lipman, D.J. Basic local alignment search tool. *J. Mol. Biol.* **1990**, *215*, 403–410.

35. Yu, Y. Q.; Yang, W. J.; Yang, J. S. The complete mitogenome of the Chinese swamp shrimp *Neocaridina denticulata sinensis* Kemp 1918 (Crustacea: Decapoda: Atyidae). *Mitochondrial DNA* **2013**, doi:10.3109/19401736.2013.796465.

36. Lohse, M.; Drechsel, O.; Bock, R. OrganellarGenomeDRAW (OGDRAW): A tool for the easy generation of high-quality custom graphical maps of plastid and mitochondrial genomes. *Curr. Genet.* **2007**, *52*, 267–274.

37. Shen, X.; Li, X.; Sha, Z.; Yan, B.; Xu, Q. Complete mitochondrial genome of the Japanese snapping shrimp *Alpheus japonicus* (Crustacea: Decapoda: Caridea): Gene rearrangement and phylogeny within Caridea. *Sci. China Life Sci.* **2012**, *55*, 591–598.

38. Lavrov, D.V.; Boore, J.L.; Brown, W.M. The complete mitochondrial DNA sequence of the horseshoe crab *Limulus polyphemus*. *Mol. Biol. Evol.* **2000**, *17*, 813–824.

39. Cook, C.E.; Smith, M.L.; Telford, M.J.; Bastianello, A.; Akam, M. Hox genes and the phylogeny of the arthropods. *Curr. Biol.* **2001**, *11*, 759–763.

40. Kaestner, K.H.; Knoechel, W.; Martinez, D.E. Unified nomenclature for the winged helix/forkhead transcription factors. *Genes Dev.* **2000**, *14*, 142–146.

41. Shimeld, S.M.; Degnan, B.; Luke, G.N. Evolutionary genomics of the Fox genes: Origin of gene families and the ancestry of gene clusters. *Genomics* **2010**, *95*, 256–260.

42. Papaioannou, V.E.; Silver, L.M. The T-box gene family. *Bioessays* **1998**, *20*, 9–19.

43. Tagawa, K.; Humphreys, T.; Satoh, N. *T-brain* expression in the apical organ of hemichordate tornaria larvae suggests its evolutionary link to the vertebrate forebrain. *J. Exp. Zool.* **2000**, *288*, 23–31.

44. Kenny, N.J.; Shimeld, S.M. Additive multiple k-mer transcriptome of the keelworm *Pomatoceros lamarckii* (Annelida; Serpulidae) reveals annelid trochophore transcription factor cassette. *Dev. Genes Evol.* **2012**, *222*, 325–339.

45. Winter, J.; Jung, S.; Keller, S.; Gregory, R.I.; Diederichs, S. Many roads to maturity: microRNA biogenesis pathways and their regulation. *Nat. Cell Biol.* **2009**, *11*, 228–34.

46. Sneddon, L.U. Clinical anesthesia and analgesia in fish. *J. Exot. Pet. Med.* **2012**, *21*, 32–43.

47. Schmit, O.; Mezquita, F. Experimental test on the use of MS-222 for ostracod anaesthesia: concentration, immersion period and recovery time. *J. Limnol.* **2010**, *69*, 350–352.

48. Coyle, S.D.; Dasgupta, S.; Tidwell, J.H.; Beavers, T.; Bright, L.A.; Yasharian, D.K. Comparative efficacy of anesthetics for the freshwater prawn *Macrobrachiurn rosenbergii*. *J. World Aquac. Soc.* **2007**, *36*, 282–290.

49. Hajek, G.; Choczewski, M.; Dziaman, R.; Klyszejko, B. Evaluation of immobilizing methods for the Chinese mitten crab, *Eriocheir sinensis*. *Electron. J. Pol. Agric. Univ.* **2009**, *12*, 1.

50. Brown, P.; White, M. Evaluation of three anesthetic agents for crayfish (*Orconectes virilis*). *J. Shellfish Res.* **1996**, *15*, 433–436.

51. Ronquist, F.; Huelsenbeck, J.P. MrBayes 3: Bayesian phylogenetic inference under mixed models. *Bioinformatics* **2003**, *19*, 1572–1574.

52. Katoh, K.; Standley, D.M. MAFFT multiple sequence alignment software version 7: Improvements in performance and usability. *Mol. Biol. Evol.* **2013**, *30*, 772–780.

53. JGI *Daphnia pulex* Genome Resources. Available online:. http://genome.jgi-psf.org/Dappu1/Dappu1.home.html (accessed on 25 September 2013).

54. Conesa, A.; Gotz, S.; Garcia-Gomez, J.M.; Terol, J.; Talon, M.; Robles, M. Blast2GO: A universal tool for annotation, visualization and analysis in functional genomics research. *Bioinformatics* **2005**, *21*, 3674–3676.

55. Wyman, S.K.; Jansen, R.K.; Boore, J.L. Automatic annotation of organellar genomes with DOGMA. *Bioinformatics* **2004**, *20*, 3252–3255.

56. Lowe, T.M.; Eddy, S.R. tRNAscan-SE: A program for improved detection of transfer RNA genes in genomic sequence. *Nucleic Acids Res.* **1997**, *25*, 955–964.

57. Librado, P.; Rozas, J. DnaSP v5: A software for comprehensive analysis of DNA polymorphism data. *Bioinformatics* **2009**, *25*, 1451–1452.

58. Thompson, J.D.; Higgins, D.G.; Gibson, T.J. Clustal-W: Improving the sensitivity of progressive multiple sequence alignment through sequence weighting, position-specific gap penalties and weight matrix choice. *Nucleic Acids Res.* **1994**, *22*, 4673–4680.

59. Darriba, D.; Taboada, G.L.; Doallo, R.; Posada, D. jModelTest 2: More models, new heuristics and parallel computing. *Nat. Methods* **2012**, *9*, 772.

3

Characterization of a Novel *Conus bandanus* Conopeptide Belonging to the M-Superfamily Containing Bromotryptophan

Bao Nguyen [1,2], **Jean-Pierre Le Caer** [3], **Gilles Mourier** [4], **Robert Thai** [4], **Hung Lamthanh** [1,2], **Denis Servent** [4], **Evelyne Benoit** [1] **and Jordi Molgó** [1,*]

[1]   Neurobiology and Development Laboratory, Research Unit # 3294, Institute of Neurobiology Alfred Fessard # 2118, National Center for Scientific Research, Gif sur Yvette Cedex 91198, France; E-Mails: bao.nguyen@inaf.cnrs-gif.fr or nguyenbaocbp@yahoo.com (B.N.); lamthanh5hung@yahoo.fr (H.L.); Evelyne.Benoit@inaf.cnrs-gif.fr (E.B.)

[2]   Institute of Biotechnology and Environment, University of Nha Trang, Nha Trang, Khanh Hoa 57000, Vietnam

[3]   Research Unit # 2301, Natural Product Chemistry Institute, National Center for Scientific Research, Gif sur Yvette Cedex 91198, France; E-Mail: jean-pierre.lecaer@icsn.cnrs-gif.fr

[4]   Molecular Engineering of Proteins, Institute of Biology and Technology Saclay, Atomic Energy Commission, Gif sur Yvette Cedex 91191, France; E-Mails: gilles.mourier@cea.fr (G.M.); robert.thai@cea.fr (R.T.); denis.servent@cea.fr (D.S.)

*   Author to whom correspondence should be addressed; E-Mail: jordi.molgo@inaf.cnrs-gif.fr

---

Abstract: A novel conotoxin (conopeptide) was biochemically characterized from the crude venom of the molluscivorous marine snail, *Conus bandanus* (Hwass in Bruguière, 1792), collected in the south-central coast of Vietnam. The peptide was identified by screening bromotryptophan from chromatographic fractions of the crude venom. Tandem mass spectrometry techniques were used to detect and localize different post-translational modifications (PTMs) present in the BnIIID conopeptide. The sequence was confirmed by Edman's degradation and mass spectrometry revealing that the purified BnIIID conopeptide had 15 amino acid residues, with six cysteines at positions 1, 2, 7, 11, 13, and 14, and three PTMs: bromotryptophan, γ-carboxy glutamate, and amidated aspartic acid, at positions "4", "5", and "15", respectively. The BnIIID peptide was synthesized for comparison with the native peptide. Homology comparison with conopeptides having the III-cysteine framework ($-CCx_1x_2x_3x_4Cx_1x_2x_3Cx_1CC-$) revealed that BnIIID belongs to the M-1 family of conotoxins. This is the first report of a member of the M-superfamily containing bromotryptophan as PTM.

Keywords: *Conus bandanus*; cone snail venom; mass spectrometry; BnIIID; bromotryptophan; γ-carboxy glutamate; post-translational modifications

1. Introduction

The peptides from the venom of cone snails (conotoxins or conopeptides) constitute a rich source of useful pharmacological tools and peptide probes for ion channels, transporters, and neurotransmitter receptors with a high degree of diversity, specificity, and potency [1–3]. Each *Conus* species is estimated to have a large number of species-specific conotoxins comprising cysteine poor-/rich-peptides [4–6]. Most mature peptides range between 8 and 40 amino acid residues with various cysteine-framework, and inter-cyteine variations in the number and kind of amino acids. Moreover, they have a high degree of post-translational modifications (PTMs) [7], whereby the modifications serve to create efficiently new conotoxin structures and pharmacological properties [4]. In addition to extensive disulphide linkages, which are a common feature of conotoxins, common PTMs include *C*-terminal amidation, proline hydroxylation, γ-carboxylation of glutamic acid, pyroglutamic acid, tyrosine sulfation, tryptophan bromination, and glycosylations [8–11]. Thus, there is an important role to entirely characterize the complement of mature peptides from cone snails.

In the search of new conopeptides from the venom of *Conus bandanus*, collected from the coast of Vietnam, we have found an unusual peptide containing bromine. We used tandem mass spectrometry to characterize the primary peptide sequence and localize the positions of tryptophan bromination, and γ-carboxylation of glutamic acid. The amidation of aspartic acid of the *C*-terminus was confirmed by the combination of theoretical mass calculation, Edman degradation, and homology comparison. We compared also the peptide sequence similarity of BnIIID to that of other conopeptides from *Conus marmoreus*, considered a close-relative interspecies with the molluscivorous *Conus bandanus* [12,13]. In addition, the BnIIID peptide was synthesized for comparison with the native peptide. This is the first report of an M-superfamily conopeptide containing a bromotryptophan.

2. Results and Discussion

2.1. Venom Fractionation and Purification

C. bandanus dissected crude venom was fractionated, sub-fractionated, and purified by reversed-phase chromatography on Vydac semi-preparative and analytical columns (Figure 1). Each HPLC fraction containing conopeptides (Figure 1A) was analyzed by Matrix-Assisted Laser Desorption/Ionization Time-Of-Flight (MALDI-TOF) mass spectrometry. The fraction 3.2 highlighted in black (Figure 1B) shows several different molecules with major intensity (Figure S1 in the Supplementary Information) containing an unusual monoisotopic ion. Figure 1C,D show further purifications from the previous fraction 3.2. An asterisk tags the subfraction that exhibited an unusual isotopic distribution for a peptide. This fraction was purified to near homogeneity on the same reversed-phase analytical column, by optimizing the elution gradient slope, hence, avoiding the use of a two dimensional chromatography (Figure 1D) and was named BnIIID.

Figure 1. HPLC purification profile of dissected *C. bandanus* venom, extract, and BnIIID purification. (**A**) Fractionation was carried out with a Vydac semi-preparative C_{18} column (300 Å, 5 μm, 10 mm i.d. × 250 mm) and eluted at 3 mL min^{-1} with gradient 0%–56% of B buffer/50 min; (**B**) Fractionation from zone 3 with a Vydac analytical C_{18} column (300 Å, 5 μm, 4.6 mm i.d. × 250 mm) and eluted at 1 mL min^{-1} with gradient (0%–0.5% B/5 min, 0.5%–15% B/1 min and 15%–40% of B/40 min); (**C**) Further purification of fraction 3.2, from the previous step, was further purified with gradient (10%–15% B/7.5 min, 15%–20% B/42.5 min and 20%–35% B/45 min); (**D**) The asterisk indicates BnIIID. The purity of BnIIID peak was checked by MALDI-TOF-MS.

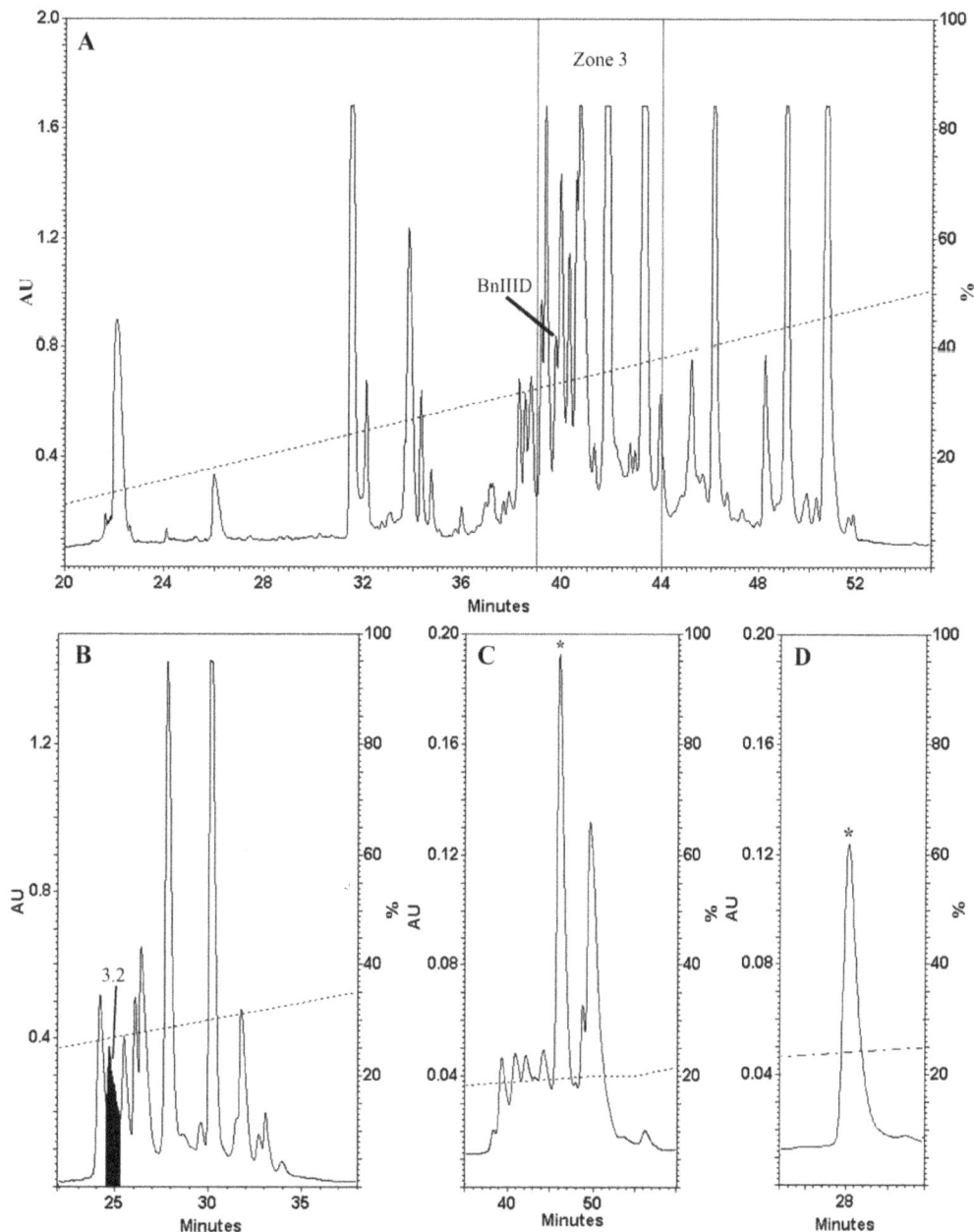

2.2. Determination of the Number of Disulphide Bonds

Analysis of the native BnIIID by MALDI-TOF/TOF MS showed two single-charged ion distributions (Figure 2A), which yielded two monoisotopic ion signals at *m/z* 1819.129 and *m/z* 1863.159. The BnIIID isotopic distributions indicated the possible presence of one bromine atom [14,15]. The abundance of these peaks "doublet" at *m/z* (1819.129; 1821.21) and *m/z* (1863.159; 1865.159) is typical of a molecule containing the equally abundant bromine isotopes (50.69% [79]Br and 49.31% [81]Br, respectively). Following reduction with TCEP, BnIIID signals were shifted by 6 u indicating the presence of three disulfide bonds (Figure 2B).

Figure 2. Identification of bromination using isotopic distributions of native BnIIID (**A**); and reduced BnIIID form (**B**); from off-line LC/MALDI-TOF MS. A shift of 6 u is observed characterizing the reduction of three S-S bonds.

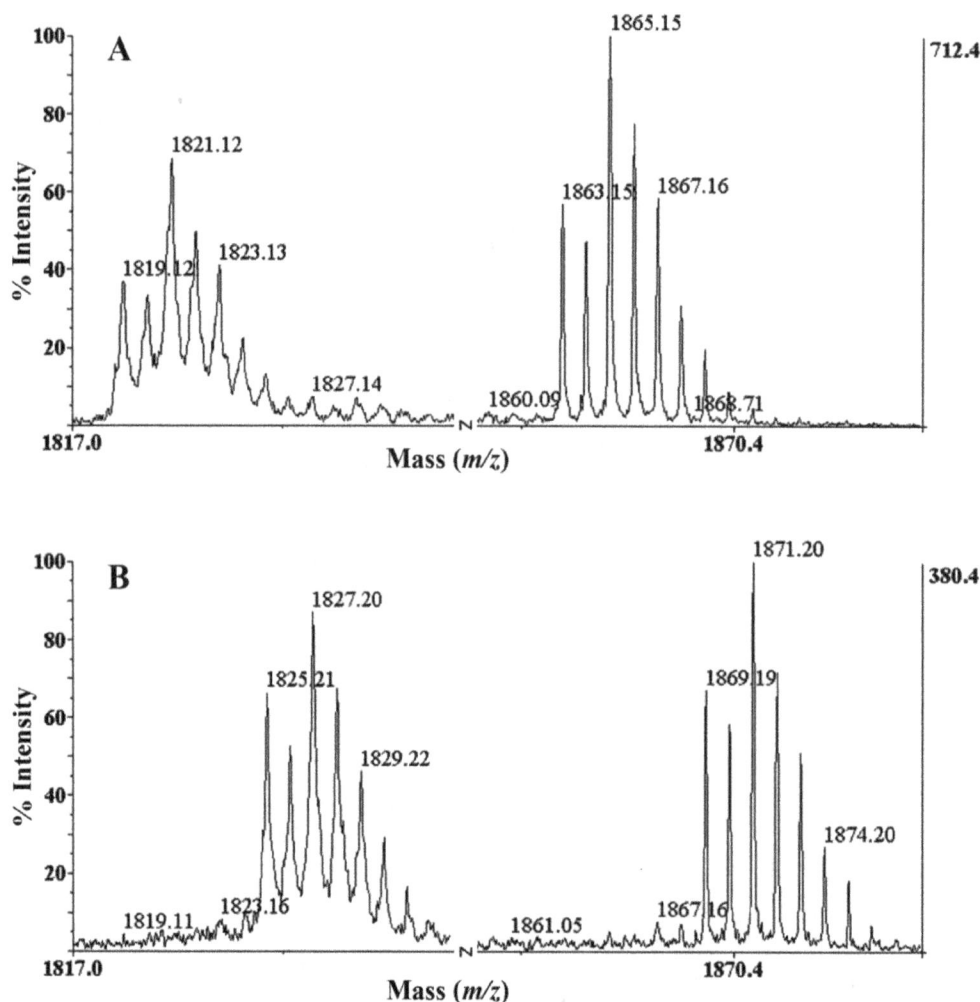

2.3. Presence of Gamma-Carboxylate Glutamate Residue

To elucidate the two different major masses of the pure fraction, we reduced and alkylated the cysteines to avoid disulfide reformation and to improve fragmentation yield during ESI-MSMS measurements (Figure 3). The CID fragmentation of the major compound in the fraction, 2214.55 Da, produce the decarboxylated fragment. In order to elucidate the sequence, the decarboxylated fragment was submitted to MS³ CID fragmentation. Figure 3C illustrate the *b* and *y* series obtained, confirming again the presence of a bromotryptophan. Thus, we could predict BnIIID having one γ-carboxylated glutamate residue and a bromotryptophan.

Figure 3. Analysis of reduced and alkylated BnIIID by high resolution MS and MSMS mass spectrometry: (**A**) Deconvoluted spectrum of the fraction shown in Figure 1D after reduction and alkylation with IAA; (**B**) Deconvoluted spectrum of the CID fragmentation of the 2214.55 Da ion; (**C**) MS³ product (2168.55 Da) corresponding to the decarboxylation of the precursor ion. Note: C_1: carbamidomethyl-cysteine; alkylated cystein by IAA; <u>W</u>: bromotryptophan.

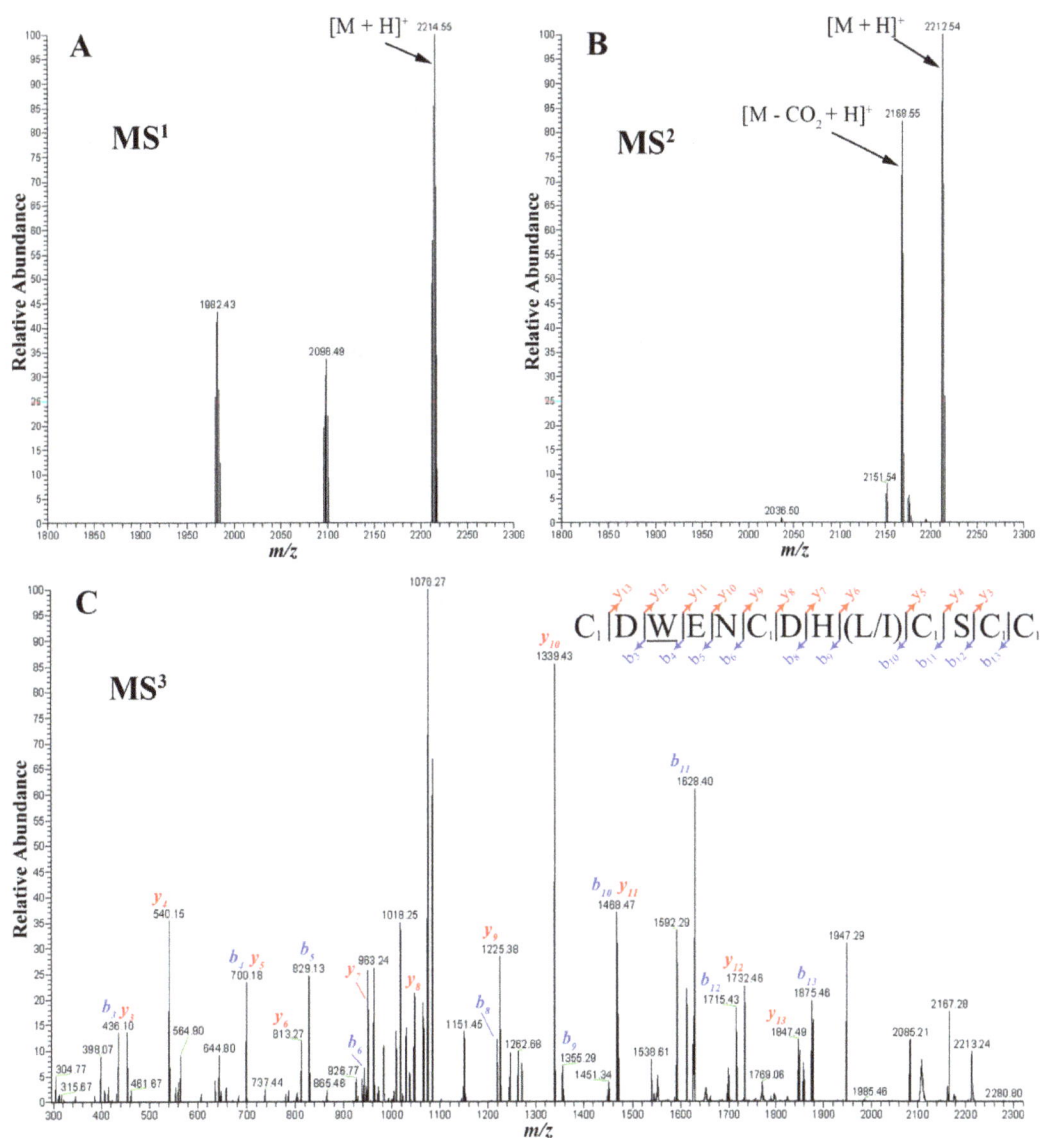

2.4. Peptide Sequencing

To determine the amino acid sequence of BnIIID, we also employed the CID fragmentation technique using a MALDI TOF/TOF mass spectrometer that generated predominantly *b* and *y*-type product ions. Figure 4 shows two CID mass spectra of IAA-labeled BnIIID, with and without-carboxylation. Moreover, the product ions that contained a bromotryptophan residue showed unusual isotope patterns. The MS/MS spectrum of the parent ion at m/z 2167.245, without carboxylation (Figure 4A), showed complete series of b and y-type ions from position 1 to 15. We clearly observed a shift corresponding to the bromotryptophan residue (+260/262 u) between b_3 and b_4 ions or y_{11} and y_{12} ions. The glutamic acid residue is also identified after the bromotryptophan residue using the b_4/b_5 and y_{10}/y_{11} ions. Furthermore, the m/z increment of 160 u between (b_1/b_2; b_6/b_7; b_{10}/b_{11}; b_{12}/b_{13}; b_{13}/b_{14}) ions confirmed the presence of an IAA-modified cysteine. Therefore, the mass of an IAA-modified cysteine at first position is inferred from m/z 161 b_1 ion and t corresponding m/z 2007 y_{14} ion. Thus, BnIIID linear sequence exhibits a cysteine framework III, with the following pattern CC–C–C–CC. There are limits in distinguishing leucine/isoleucine residue (mass 113 Da) at position "10", and amidated aspartic acid/asparagine residue (mass of 114 Da) at the C-terminus (C-terminal amidation is a common PTM due to 1 Da reduction in mass). Therefore, the initial sequence assignment of m/z 2167.245 parent ion was CCD$\underline{W}$ENCDH(L/I)CSCC(D*/N), as shown in Figure 4A.

Figure 4. CID MS/MS spectrum of TCEP-reduced and IAA-labeled BnIIID, recorded with the MALDI-TOF/TOF 4800 mass spectrometer. MS/MS fragmentation of BnIIID without (**A**), and with carboxylation of E (**B**). Insets show the sequences derived from these MS/MS spectra. Note: C_1: carbamidomethyl-cysteine; alkylated cystein by IAA; $\underline{W}$: bromotryptophan; $\underline{E}$: γ-carboxylic glutamic acid; *: C-terminal amidation.

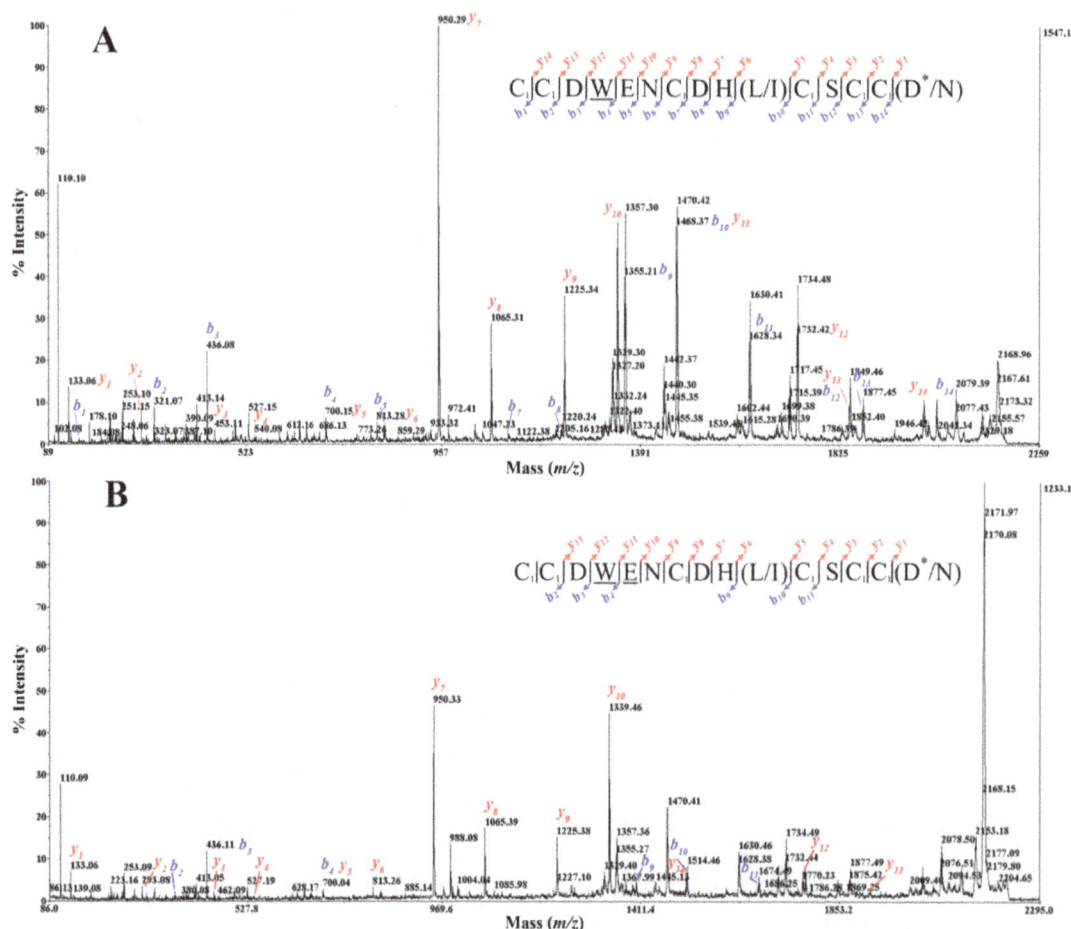

Figure 4B shows the spectrum of the parent ion at m/z 2211.198 with predicted carboxylation. Here we focused on the identification of the γ-carboxylic glutamic acid residue. The m/z 2211.198- CID MS/MS spectrum is not so informative than the m/z 2167.245-spectrum due to the preferential loss of the carboxylic group of the γ-carboxylic glutamic acid residue during the fragmenting process. Though, we determined a γ-carboxy glutamate (E) according to the difference of 173 u at the same position "5", *i.e.*, between y_{10} and y_{11} ions (Figure 5). We also found a perfect agreement with ESI-MS/MS data using the series of *y*-type ions. Thus, the initial BnIIID sequence is CCDWENCDH(L/I)CSCC(D*/N) with two confirmed PTMs: bromotryptophan (W) and γ-carboxy glutamate (E). Nair *et al.* have used the Fourier transform ion cyclotron resonance MS/MS technique of electron capture dissociation, infrared multiphoton dissociation and CID to detect, and localize the bromotryptophan of a conopeptide, named Mo 1274, from *Conus monile* venom [16]. There was difficulty in the initial sequence assignment of CID mass spectrum of the Mo 1274 peptide. In our case, the MALDI-TOF CID MS/MS was used to characterize simply the initial sequence, and also positions of bromotryptophan and γ-carboxy glutamate.

Figure 5. A zoom of CID MS/MS mass spectrum profile of IAA-labeled BnIIID with carboxylation of E (Figure 4B). Insets show the sequences derived from these MS/MS spectra. Note: C_1: carbamidomethyl-cysteine; alkylated cystein by IAA; W: bromotryptophan; E: γ-carboxylic glutamic acid; *: *C*-terminal amidation.

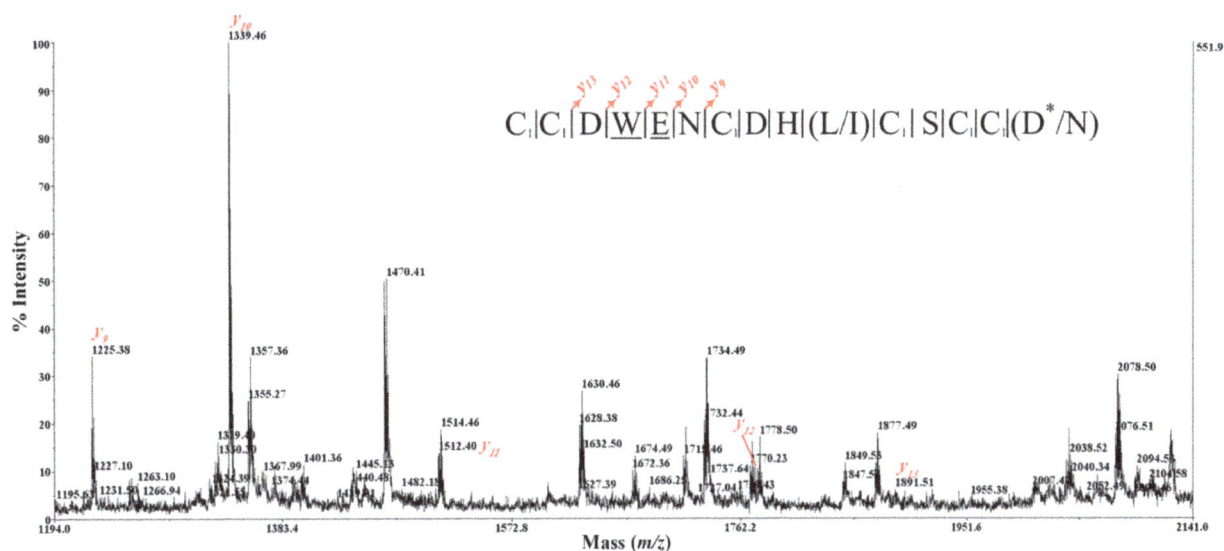

To confirm the complete amino acid sequence of BnIIID, this native conopeptide was submitted to Edman's degradation. Figure 6 presents the release of BnIIID PTH-amino acid residues for each cycle. BnIIID sequence has totally 15 amino acid residues with six cysteines at positions "1; 2; 7; 11; 13; 14" which supports well the CID MS/MS obtained data. Additionally, there are no amounts of PTH-amino acid residues at cycle "4" and "5", and it explains clearly that PTH-amino acid residue of bromotryptophan and γ-carboxy glutamate are not noted in Edman's degradation profile of the standard program. At position "10", we can confirm the leucine residue having an amount of 4.3 pmol in place of isoleucine. No data are observed at the last cycle, which may propose that there is a PTM in the amino acid residue. The MS/MS spectrum data and theoretical calculation offered previously two possibilities: an amidated aspartic acid (D*) or an asparagine (N) residue. Therefore, it may be inferred an aspartic acid amidation at the *C*-terminus. Thus, the complete linear BnIIID sequence is CCDWENCDHLCSCCD* with three PTMs: bromotryptophan, γ-carboxy glutamate, and amidated aspartic acid, at positions "4", "5", and "15", respectively. This sequence is also in agreement with the experimental mass obtained in Figure 3B for BnIIID; theoretical mass of BnIIIID (C78 H111 N26 O33 S6 Br) = 2210.5309 Da, experimental mass: 2210.5293 Da (Δ = 1.6 ppm).

Figure 6. Solid-phase Edman degradation of native BnIIID. Note: W̲: bromotryptophan; E̲: γ-carboxylic glutamic acid; *: *C*-terminal amidation.

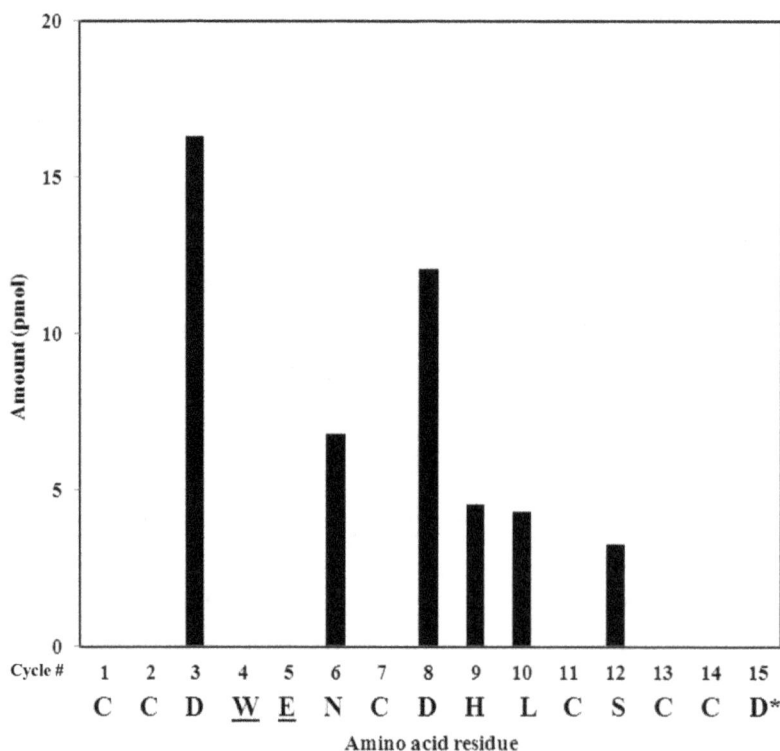

2.5. Peptide Synthesis

To confirm the refolding of the peptide like the native product, a synthetic BnIIID peptide was prepared (containing 6-D/L-6 bromotryptophan, γ-carboxy glutamate and amidated aspartic acid at positions "4", "5", and "15", respectively). Figure 7 demonstrates the near identity of RP-HPLC chromatogram profiles from the same retention time, co-elution of refolded synthetic and native peptides. It is worth noting that the synthetic BnIIID was purified from the racemic mixture containing both D and L forms of 6-bromotryptophan. We deduced that the isomer position occurring in BnIIID is 6-bromotryptophan, following the earlier study by Craig *et al.* [14]. Remarkably, 6-bromotryptophan has also been found, mostly in other marine animals. The peptide component that co-eluted with the native peptide was less hydrophobic (data do not shown). This phenomenon is similar in the synthetic bromoheptapeptide purification of *Conus imperialis*, which demonstrated L-6-bromotryptophan from digestion with α-chymotrypsin [14]. Thus, it could be deduced that the native BnIIID contains L-6-bromotryptophan.

Figure 7. Verification of the synthetic peptide folding. The HPLC chromatograms of the synthetic BnIIID (**A**); the native peptide (**B**); and the co-elution of the mixture of synthetic and native BnIIID (**C**). The experiments were carried out with a X-Bridge analytical C_{18} column and eluted at 1 mL·min^{-1} with a gradient 0%–45% of acetonitrile in 40 min.

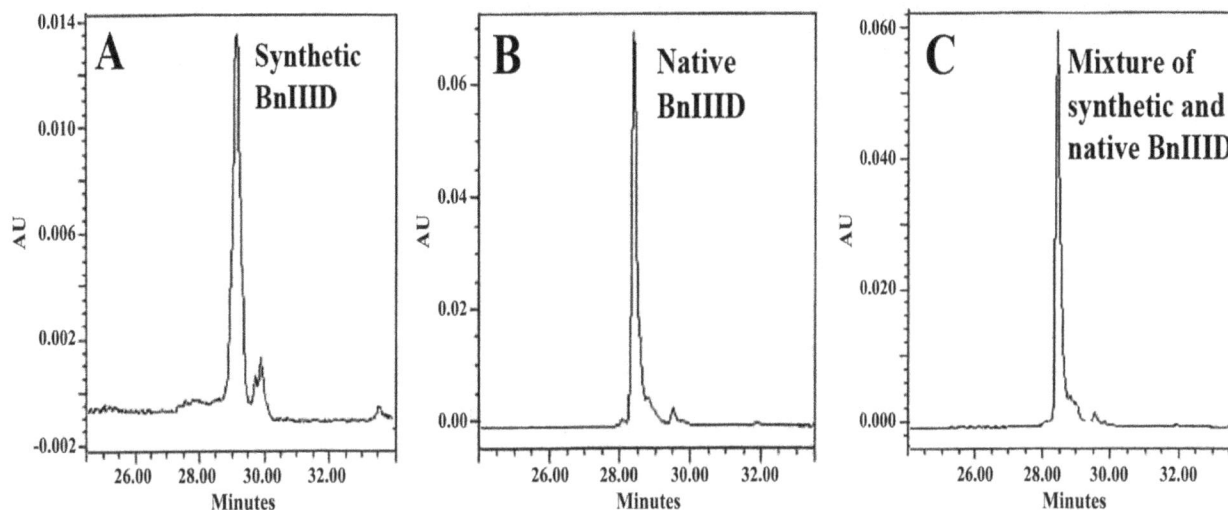

2.6. Sequence Similarity Analysis

As shown in Table 1, the BnIIID sequence was compared to other conopeptide sequences having the same "4/3/1", III-cysteine framework (like $-CCx_1x_2x_3x_4Cx_1x_2x_3Cx_1CC-$) in three different *Conus* species (piscivorous, vermivorous, and molluscivorous) on the ConoServer [17]. The BnIIID peptide shares some homology with two conopeptides (S3-S01, S3-S02) of the piscivorous *Conus striatus*, and two other peptides (Ts3.6, LtIIID) of the vermivorous *Conus tessulatus* and *Conus litteratus*, respectively. Interestingly, BnIIID exhibits high homology with five *C. marmoreus* conopeptide sequences (e.g., Mr3.16, MrIIIE, MrIIIF, Mr3.18, and Mr3.8). The MrIIIE peptide of *C. marmoreus* has been shown to have (C1–C5, C2–C4, C3–C6) disulfide linkage arrangement using Edman analysis of the partially reduced peptide [18]. We suggest that the BnIIID conopeptide could share the same disulfide connectivity as the MrIIIE peptide. However, further work will be necessary to definitively attribute the disulfide linkage of BnIIID. All compared conopeptides presented in Table 1 belong to M-1 family of the M-superfamily. Thus, it can be inferred that BnIIID also belongs to the same family. Moreover, the bomotryptophan residue in BnIIID is found as a PTM for the first time, not only in a conopeptide having the cysteine framework III, but also in the M superfamily sequences.

Craig *et al.* (1997) [14] were the first to report the bromination of tryptophan residues in *Conus* venom components. The initial work was focused on two conopeptides: the bromosleeper conopeptide isolated from the piscivorous *Conus radiatus*, and a heptapeptide isolated from the vermivorous *Conus imperialis*. A mechanism for tryptophan bromination in cone snails has been proposed, such as the bromination of tryptophan to L-6-bromotryptophan by a bromo peroxidase [10,19,20]. The relative importance and the role of bromination in conopeptides are currently unknown.

Table 1. BnIIID sequence similarity with other conopeptides, from piscivorous (p), vermivorous (v), and molluscivorous (m) cone snails, belonging to the M superfamily. Note: <u>W</u>: bromotryptophan; <u>E</u>: γ-carboxylic glutamic acid; *: C-terminal amidation.

Name	Cone snail	Diet	Sequence	Reference
S3-S01	*C. striatus*	p	- C C P K E W C N R D C S C C T -	[17,21]
S3-S02	*C. striatus*	p	- C C P A R M C M A A C S C C D -	[17,21]
Ts3.6	*C. tessulatus*	v	Q C C D W Q W C D G A C D C C A -	[22]
LtIIID	*C. litteratus*	v	- C C D W E W C D E L C S C C W -	[23]
Mr3.16	*C. marmoreus*	m	V C C S F G S C D S L C Q C C D * -	[5]
MrIIIE	*C. marmoreus*	m	V C C P F G G C H E L C Y C C D * -	[18]
MrIIIF	*C. marmoreus*	m	V C C P F G G C H E L C L C C D * -	[18]
Mr3.18	*C. marmoreus*	m	- C C H R N W C D H L C S C C G S	[5]
Mr3.8	*C. marmoreus*	m	- C C H W N W C D H L C S C C G S	[18]
BnIIID	*C. bandanus*	m	- C C D <u>W</u> <u>E</u> N C D H L C S C C D * -	This work

Conus are well-known for using many different strategies to capture their own preys through the use of a variety of conotoxins. The capacity to generate conopeptides, in which the number and nature of amino-acid residues can vary within putative inter-cysteines, is one of their most important and efficient strategies. To these characteristics it should be added that conopeptides may become still more diverse and complex considering the reported PTMs in *Conus* venoms [7,17]. Here, BnIIID was analyzed as belonging to the M-superfamily of conotoxins [18,21,24,25], which shares the same cysteine framework (–CC–C–C–CC–) pattern (Table 2). The conopeptide framework-III, according to available data, can be represented by μ-, ι-, ψ-, and κ-conotoxins, with pharmacological properties, such as blockage of voltage-gated sodium channels (μ-conotoxins), activation of voltage-gated sodium channels, without delaying channel inactivation (ι-conotoxins), modulation or blockage of nicotinic acetylcholine receptors (ψ-conotoxins), and blockage of voltage-gated potassium channels (κ-conotoxins). Studies are in progress to determine the molecular target(s) of BnIIID.

The –CC$x_1x_2x_3x_4$C$x_1x_2x_3$Cx_1CC– conopeptides have been poorly explored for their specific pharmacological targets. The conopeptide LtIIIA from *C. litteratus*, among ~63 other peptides, was evaluated as belonging to the ι-conotoxin class [26]. Remarkably, this pattern exists mostly in mollusk- and worm-hunting cone snails (e.g., eight conopeptides in *C. marmoreus*). The synthesis of BnIIID conopeptide will enable us to investigate in detail its pharmacological properties, and, what is more important, the eventual role of the PTMs, since three other peptides (BnIIIA, BnIIIB, and BnIIIC) have been just isolated by us, and such peptides lack some of the PTMs (bromotrytophan and γ-carboxyglutamate) here reported.

Table 2. Inter-cysteine variation of "framework III" conopeptides *vs.* their pharmacological properties. Note: <u>W</u>: bromotryptophan; <u>E</u>: γ-carboxylic glutamic acid; *: C-terminal amidation; p: piscivorous; m: molluscivorous; v: vermivorous.

Name	Organism	Diet	Sequence	Pharmacology	Reference
BuIIIA	*C. bullatus*	p		μ-conotoxin	[27]
BuIIIB	*C. bullatus*	p		μ-conotoxin	[27]
CIIIA	*C. catus*	p		μ-conotoxin	[28]
CnIIIA	*C. consors*	p		μ-conotoxin	[28]
CnIIIB	*C. consors*	p		μ-conotoxin	[28]
GIIIA	*C. geographus*	p		μ-conotoxin	[29]
GIIIB	*C. geographus*	p		μ-conotoxin	[30]
GIIIC	*C. geographus*	p		μ-conotoxin	[31]
KIIIA	*C. kinoshitai*	p		μ-conotoxin	[32]
LtIIIA	*C. litteratus*	v		τ-conotoxin	[26]
MIIIA	*C. magus*	p		μ-conotoxin	[33]
PIIIA	*C. purpurascens*	p		μ-conotoxin	[34]
PIIIE	*C. purpurascens*	p		ψ-conotoxin	[35]
PIIIF	*C. purpurascens*	p		ψ-conotoxin	[36]
PrIIIE	*C. parius*	p		ψ-conotoxin	[37]
RIIIJ	*C. radiatus*	p		κ-conotoxin	[38]
RIIIK	*C. radiatus*	p		κ-conotoxin	[39]
SIIIA	*C. striatus*	p		μ-conotoxin	[32]
SIIIB	*C. striatus*	p		μ-conotoxin	[40]
SmIIIA	*C. stercusmuscarum*	p		μ-conotoxin	[41]
SxIIIA	*C. striolatus*	p		μ-conotoxin	[42]
SxIIIB	*C. striolatus*	p		μ-conotoxin	[42]
TIIIA	*C. tulipa*	p		μ-conotoxin	[43]
BnIIID	*C. bandanus*	m		unknown	This work

3. Experimental Section

3.1. Isolation and Purification of Native Conopeptides

Twelve adult specimens (≥72 mm shell length) of *C. bandanus* were collected in the Nha Trang bay in the South central coast of Vietnam. We have used the same methods for venom extraction as previously described [13]. The extract was lyophilized and stored at −80 °C for later fractionation. All subfractions, purifications, and analyses were performed on a HPLC (System Gold with programmable solvent module 126 pumps using 32 Karat software; Beckman Coulter, Fullerton, CA, USA) at room temperature. Fractions were collected based on absorbance at 220 nm (System Gold 166 detector; Beckman Coulter, Fullerton, CA, USA). The elution buffers used for all HPLC were the following: A buffer (1000 mL H_2O/1 mL trifluoroacetic acid (TFA; Merck, Darmstadt, Germany); B buffer (900 mL CH_3CN/100 mL H_2O/1 mL TFA). The lyophilized *C. bandanus* venom (~300 mg) was dissolved in A buffer, and the extract was loaded in batches of ~10 mg on a Vydac semi-preparative C_{18} column (300 Å, 5 µm, 10 mm i.d. × 250 mm). The gradient program for the semi-preparative column was 0% of B buffer/10 min, then 0%–56% of B buffer/50 min with the flow rate 3 mL/min. Further purification steps were carried out using a Vydac analytical C_{18} column (300 Å, 5 µm, 4.6 mm i.d. × 250 mm) at flow rate 1 mL/min.

3.2. Reduction-Alkylation Procedures

The purified fraction was reduced by incubation for 10 min at 65 °C in a solution of 20 mM tris-2-carboxyethyl-phosphine (TCEP) in 0.5 M HEPES. Alkylation was then performed by addition of iodoacetamide (IAA) 0.05 M and incubation for 20 min at room temperature in darkness. The mixture was finally desalted by solid phase extraction on a Zip Tip C18 column (Millipore, Billerica, MA, USA).

3.3. Mass Spectrometry Analysis

MALDI analysis—all bioactive fractions were analyzed with a 5800 MALDI-TOF/TOF mass spectrometer (AB Sciex, Les Ulis, France). The instrument was equipped with an Nd:YAG laser operating at 355 nm wavelength. Aliquots of 0.5 µL of a purified fraction were mixed with 0.5 µL of a solution of 4 mg/mL of cyano-4-hydroxycinnamic acid. Acquisitions were performed on positive reflection mode. Instrument calibration was done using a peptide mixt (peptide calibration 1 and 2 from ABSciex between 700 and 3700 Da). For MS/MS experiments, precursor ions were accelerated at 8 keV and the MS/MS spectra were acquired using 2 keV collision energy, with CID gas (air) at a pressure of 3.5×10^{-6} Torr. MS and MS/MS data were processed using DataExplorer 4.9 (AB Sciex).

Nano ESI orbitrap analysis—some important fractions were verified on their accurate mass on a LTQ Orbitrap mass spectrometer equipped with a nano-electrospray source (Thermo Scientific, Bremen, Germany). Few microliters of chromatographic fraction were loaded onto metal-coated borosilicate emitters (Thermo Scientific). The 1.2 KV were applied to the emitter and the acquisition was monitored on the orbitrap set with a theoretical resolution of 30,000 at m/z 400. For MS2 and MS3, normalized collision energy of 36 eV was applied. Spectra were treated with Xcalibur 2.1, the multiplied-charged species were recalculated into its singly-charged form using the Xtract software (Thermo Scientific, San Jose, CA, USA).

3.4. Automatic Amino Acid Sequencing

Pure peptide fractions were dissolved in buffer A and their concentration were measured by scan mode "Absorbance" (range 240–350 nm wavelength) in a DU 800 Spectrophotometer (Beckman-Coulter, Brea, CA, USA). The 2 μL native peptide (~190 μM) was sequenced using an automated Edman degradation using a Procise protein sequencer (Applied Biosystem model 492, Applied Biosystem, Foster City, CA, USA).

3.5. Chemical Synthesis

Materials—Fmoc-amino acids and 2-(6-Chloro-1-H-benzotriazole-1-yl)-1,1,3,3,-tetramethylaminium hexafluorophosphate (HCTU) were obtained from Activotec (Cambridge, UK), Fmoc-Gla(OtBu)2-OH from Iris Biotech (Marktredwitz, Germany) and Fmoc-DL-6-Bromotryptophan from AnaSpec (Fremont, CA, USA). The Chemmatrix resin and all the peptide synthesis grade reagents (N-methylpyrrolidone (NMP), N-methylmorpholine (NMM), piperidine, trifluoroacetic acid (TFA), anisole, thioanisole and triisopropylsilane (TPS) were purchased from Sigma Aldrich (Saint-Quentin Fallavier, France).

Synthesis—peptide synthesis was performed on a Protein Technologies, Inc Prelude synthesizer at a 50 μmole scale using fivefold excess of Fmoc-amino acid relative to the rink amide chemmatrix resin (0.4–0.6 mmol/g). All the amino acids were coupled twice 5 min using 1:1:2 amino acid/HCTU/ NMM in N-methylpyrrolidone (NMP) with the exception of Fmoc-DL-6-bromotryptophane and Fmoc-Gla(OtBu)2-OH where a single 30 min coupling was used. Fmoc deprotection was performed using 20% piperidine in NMP, and NMP top washes were performed between deprotection and coupling steps. All cysteines were protected with S-trityl groups. Following chain assembly, the peptidyl-resin was treated with a mixture of TFA/thioanisole/anisole/TPS/water (82:5:5:2.5:5) for 2 h. The crude peptide was obtained after precipitation and washes in cold ethyl ether followed by dissolution in 10% acetic acid and lyophilisation.

Folding and characterization—refolding of the peptide was performed by stirring the crude reduced toxin in aqueous 0.1 M Tris buffer, pH 8.2 containing reduced and oxidized glutathione at 2 mM and 1 mM respectively for 24 and 3 h, respectively. The solution was then acidified using TFA 20% followed by purification by semi-preparative RP-HPLC using a X-Bridge (Waters, Milford, MA, USA) C_{18} column with a linear gradient from 0% to 60% buffer B in A at 4mL/min during 40 min, (buffer A, 0.1% TFA in water; Buffer B, acetonitrile, 0.1% TFA). The main fraction was collected and lyophilized. The homogeneity of the peptide was confirmed by MALDI-TOF mass spectrometry (AB SCIEX 4800, Les Ulis, France). The co-elution of the synthetic and native peptides was checked by analytical RP-HPLC.

4. Conclusions

In this work, we describe the characterization of a novel conopeptide named BnIIID, isolated from the venom of the molluscivorous snail species *C. bandanus* collected from the south central coast of Vietnam. The determination of the primary conopeptide structure on the basis of CID MS/MS mass-spectra analysis and Edman degradation revealed that BnIIID contained PTMs, such as bromotryptophan, γ-carboxy glutamate, and amidated aspartic acid at positions "4", "5", and "15", respectively. The complete amino acid sequence of the conopeptide (CCD<u>WE</u>NCDHLCSCCD*) and its III-cysteine framework allowed to categorize BnIIID in the M-1 family of conotoxins, belonging to the M-superfamily.

Acknowledgments

We are grateful to the Institute of Biotechnology and Environment (Nha Trang University, Vietnam) for the specimen preparation. Bao NGUYEN was supported in part by a pre-doctoral fellowship from the French government (Embassy of France in Vietnam), and in part by the National Center for Scientific Research. We are grateful to P. Villeneuve for technical assistance. We thank David Touboul from the Natural Product Chemistry Institute of CNRS for helpful comments and suggestions.

Author Contributions

J.M., J.P.L.C., G.M., and D.S. designed research; B.N., J.P.L.C., G.M., and R.T. performed research; B.N., J.P.L.C., G.M., H.L., and E.B. analyzed data; J.M., B.N., J.P.L.C., and G.M. wrote the manuscript. All authors discussed the results and commented on the manuscript.

References

1. Olivera, B.M. *Conus* peptides: Biodiversity-based discovery and exogenomics. *J. Biol. Chem.* **2006**, *281*, 31173–31177.

2. Lewis, R.J.; Dutertre, S.; Vetter, I.; Christie, M.J. *Conus* venom peptide pharmacology. *Pharmacol. Rev.* **2012**, *64*, 259–298.

3. Favreau, P.; Stocklin, R. Marine snail venoms: Use and trends in receptor and channel neuropharmacology. *Curr. Opin. Pharmacol.* **2009**, *9*, 594–601.

4. Olivera, B.M.E.E. Just Lecture, 1996. *Conus* venom peptides, receptor and ion channel targets, and drug design: 50 million years of neuropharmacology. *Mol. Biol. Cell* **1997**, *8*, 2101–2109.

5. Dutertre, S.; Jin, A.H.; Kaas, Q.; Jones, A.; Alewood, P.F.; Lewis, R.J. Deep venomics reveals the mechanism for expanded peptide diversity in cone snail venom. *Mol. Cell. Proteomics* **2013**, *12*, 312–329.

6. Terlau, H.; Olivera, B.M. Conus venoms: A rich source of novel ion channel-targeted peptides. *Physiol. Rev.* **2004**, *84*, 41–68.

7. Craig, A.G.; Bandyopadhyay, P.; Olivera, B.M. Post-translationally modified neuropeptides from *Conus* venoms. *Eur. J. Biochem.* **1999**, *264*, 271–275.

8. Jakubowski, J.A.; Kelley, W.P.; Sweedler, J.V. Screening for post-translational modifications in conotoxins using liquid chromatography/mass spectrometry: An important component of conotoxin discovery. *Toxicon* **2006**, *47*, 688–699.

9. Gerwig, G.J.; Hocking, H.G.; Stocklin, R.; Kamerling, J.P.; Boelens, R. Glycosylation of conotoxins. *Mar. Drugs* **2013**, *11*, 623–642.

10. Buczek, O.; Bulaj, G.; Olivera, B.M. Conotoxins and the posttranslational modification of secreted gene products. *Cell. Mol. Life Sci.* **2005**, *62*, 3067–3079.

11. Yates, J.R., III. Mass spectrometry. From genomics to proteomics. *Trends Genet.* **2000**, *16*, 5–8.

12. Nam, H.H.; Corneli, P.S.; Watkins, M.; Olivera, B.; Bandyopadhyay, P. Multiple genes elucidate the evolution of venomous snail-hunting *Conus* species. *Mol. Phylogenet. Evol.* **2009**, *53*, 645–652.

13. Nguyen, B.; Molgó, J.; Lamthanh, H.; Benoit, E.; Khuc, T.A.; Ngo, D.N.; Nguyen, N.T.; Millares, P.; le Caer, J.P. High accuracy mass spectrometry comparison of *Conus bandanus* and *Conus marmoreus* venoms from the South Central Coast of Vietnam. *Toxicon* **2013**, *75*, 148–159.

14. Craig, A.G.; Jimenez, E.C.; Dykert, J.; Nielsen, D.B.; Gulyas, J.; Abogadie, F.C.; Porter, J.; Rivier, J.E.; Cruz, L.J.; Olivera, B.M.; *et al*. A novel post-translational modification involving bromination of tryptophan. Identification of the residue, L-6-bromotryptophan, in peptides from *Conus imperialis* and *Conus radiatus* venom. *J. Biol. Chem.* **1997**, *272*, 4689–4698.

15. England, L.J.; Imperial, J.; Jacobsen, R.; Craig, A.G.; Gulyas, J.; Akhtar, M.; Rivier, J.; Julius, D.; Olivera, B.M. Inactivation of a serotonin-gated ion channel by a polypeptide toxin from marine snails. *Science* **1998**, *281*, 575–578.

16. Nair, S.S.; Nilsson, C.L.; Emmett, M.R.; Schaub, T.M.; Gowd, K.H.; Thakur, S.S.; Krishnan, K.S.; Balaram, P.; Marshall, A.G. *De novo* sequencing and disulfide mapping of a bromotryptophan-containing conotoxin by Fourier transform ion cyclotron resonance mass spectrometry. *Anal. Chem.* **2006**, *78*, 8082–8088.

17. Kaas, Q.; Yu, R.; Jin, A.H.; Dutertre, S.; Craik, D.J. ConoServer: updated content, knowledge, and discovery tools in the conopeptide database. *Nucleic Acids Res.* **2012**, *40*, D325–D330.

18. Han, Y.H.; Wang, Q.; Jiang, H.; Liu, L.; Xiao, C.; Yuan, D.D.; Shao, X.X.; Dai, Q.Y.; Cheng, J.S.; Chi, C.W. Characterization of novel M-superfamily conotoxins with new disulfide linkage. *FEBS J.* **2006**, *273*, 4972–4982.

19. Jimenez, E.C.; Craig, A.G.; Watkins, M.; Hillyard, D.R.; Gray, W.R.; Gulyas, J.; Rivier, J.E.; Cruz, L.J.; Olivera, B.M. Bromocontryphan: Post-translational bromination of tryptophan. *Biochemistry* **1997**, *36*, 989–994.

20. Jimenez, E.C.; Watkins, M.; Olivera, B.M. Multiple 6-bromotryptophan residues in a sleep-inducing peptide. *Biochemistry* **2004**, *43*, 12343–12348.

21. Zhou, M.; Wang, L.; Wu, Y.; Zhu, X.; Feng, Y.; Chen, Z.; Li, Y.; Sun, D.; Ren, Z.; Xu, A. Characterizing the evolution and functions of the M-superfamily conotoxins. *Toxicon* **2013**, *76*, 150–159.

22. Conticello, S.G.; Gilad, Y.; Avidan, N.; Ben-Asher, E.; Levy, Z.; Fainzilber, M. Mechanisms for evolving hypervariability: The case of conopeptides. *Mol. Biol. Evol.* **2001**, *18*, 120–131.

23. Pi, C.; Liu, J.; Peng, C.; Liu, Y.; Jiang, X.; Zhao, Y.; Tang, S.; Wang, L.; Dong, M.; Chen, S.; Xu, A. Diversity and evolution of conotoxins based on gene expression profiling of *Conus litteratus*. *Genomics* **2006**, *88*, 809–819.

24. Corpuz, G.P.; Jacobsen, R.B.; Jimenez, E.C.; Watkins, M.; Walker, C.; Colledge, C.; Garrett, J.E.; McDougal, O.; Li, W.; Gray, W.R.; *et al.* Definition of the M-conotoxin superfamily: Characterization of novel peptides from molluscivorous *Conus* venoms. *Biochemistry* **2005**, *44*, 8176–8186.

25. Jacob, R.B.; McDougal, O.M. The M-superfamily of conotoxins: A review. *Cell. Mol. Life Sci.* **2010**, *67*, 17–27.

26. Wang, L.; Liu, J.; Pi, C.; Zeng, X.; Zhou, M.; Jiang, X.; Chen, S.; Ren, Z.; Xu, A. Identification of a novel M-superfamily conotoxin with the ability to enhance tetrodotoxin sensitive sodium currents. *Arch. Toxicol.* **2009**, *83*, 925–932.

27. Holford, M.; Zhang, M.M.; Gowd, K.H.; Azam, L.; Green, B.R.; Watkins, M.; Ownby, J.P.; Yoshikami, D.; Bulaj, G.; Olivera, B.M. Pruning nature: Biodiversity-derived discovery of novel sodium channel blocking conotoxins from *Conus bullatus*. *Toxicon* **2009**, *53*, 90–98.

28. Zhang, M.M.; Fiedler, B.; Green, B.R.; Catlin, P.; Watkins, M.; Garrett, J.E.; Smith, B.J.; Yoshikami, D.; Olivera, B.M.; Bulaj, G. Structural and functional diversities among mu-conotoxins targeting TTX-resistant sodium channels. *Biochemistry* **2006**, *45*, 3723–3732.

29. Wakamatsu, K.; Kohda, D.; Hatanaka, H.; Lancelin, J.M.; Ishida, Y.; Oya, M.; Nakamura, H.; Inagaki, F.; Sato, K. Structure-activity relationships of mu-conotoxin GIIIA: Structure determination of active and inactive sodium channel blocker peptides by NMR and simulated annealing calculations. *Biochemistry* **1992**, *31*, 12577–12584.

30. Hill, J.M.; Alewood, P.F.; Craik, D.J. Three-dimensional solution structure of mu-conotoxin GIIIB, a specific blocker of skeletal muscle sodium channels. *Biochemistry* **1996**, *35*, 8824–8835.

31. Cruz, L.J.; Gray, W.R.; Olivera, B.M.; Zeikus, R.D.; Kerr, L.; Yoshikami, D.; Moczydlowski, E. *Conus geographus* toxins that discriminate between neuronal and muscle sodium channels. *J. Biol. Chem.* **1985**, *260*, 9280–9288.

32. Bulaj, G.; West, P.J.; Garrett, J.E.; Watkins, M.; Zhang, M.M.; Norton, R.S.; Smith, B.J.; Yoshikami, D.; Olivera, B.M. Novel conotoxins from *Conus striatus* and *Conus kinoshitai* selectively block TTX-resistant sodium channels. *Biochemistry* **2005**, *44*, 7259–7265.

33. Wilson, M.J.; Yoshikami, D.; Azam, L.; Gajewiak, J.; Olivera, B.M.; Bulaj, G.; Zhang, M.M. mu-Conotoxins that differentially block sodium channels NaV1.1 through 1.8 identify those responsible for action potentials in sciatic nerve. *Proc. Natl. Acad. Sci. USA* **2011**, *108*, 10302–10307.

34. Shon, K.J.; Olivera, B.M.; Watkins, M.; Jacobsen, R.B.; Gray, W.R.; Floresca, C.Z.; Cruz, L.J.; Hillyard, D.R.; Brink, A.; Terlau, H.; *et al.* mu-Conotoxin PIIIA, a new peptide for discriminating among tetrodotoxin-sensitive Na channel subtypes. *J. Neurosci.* **1998**, *18*, 4473–4481.

35. Shon, K.J.; Grilley, M.; Jacobsen, R.; Cartier, G.E.; Hopkins, C.; Gray, W.R.; Watkins, M.; Hillyard, D.R.; Rivier, J.; Torres, J.; *et al.* A noncompetitive peptide inhibitor of the nicotinic acetylcholine receptor from *Conus purpurascens* venom. *Biochemistry* **1997**, *36*, 9581–9587.

36. Van Wagoner, R.M.; Jacobsen, R.B.; Olivera, B.M.; Ireland, C.M. Characterization and three-dimensional structure determination of psi-conotoxin Piiif, a novel noncompetitive

antagonist of nicotinic acetylcholine receptors. *Biochemistry* **2003**, *42*, 6353–6362.

37. Lluisma, A.O.; Lopez-Vera, E.; Bulaj, G.; Watkins, M.; Olivera, B.M. Characterization of a novel psi-conotoxin from *Conus parius* Reeve. *Toxicon* **2008**, *51*, 174–180.

38. Chen, P.; Dendorfer, A.; Finol-Urdaneta, R.K.; Terlau, H.; Olivera, B.M. Biochemical characterization of kappaM-RIIIJ, a Kv1.2 channel blocker: evaluation of cardioprotective effects of kappaM-conotoxins. *J. Biol. Chem.* **2010**, *285*, 14882–14889.

39. Ferber, M.; Sporning, A.; Jeserich, G.; DeLaCruz, R.; Watkins, M.; Olivera, B.M.; Terlau, H. A novel *Conus* peptide ligand for K$^+$ channels. *J. Biol. Chem.* **2003**, *278*, 2177–2183.

40. Schroeder, C.I.; Ekberg, J.; Nielsen, K.J.; Adams, D.; Loughnan, M.L.; Thomas, L.; Adams, D.J.; Alewood, P.F.; Lewis, R.J. Neuronally micro-conotoxins from *Conus striatus* utilize an alpha-helical motif to target mammalian sodium channels. *J. Biol. Chem.* **2008**, *283*, 21621–21628.

41. West, P.J.; Bulaj, G.; Garrett, J.E.; Olivera, B.M.; Yoshikami, D. Mu-conotoxin SmIIIA, a potent inhibitor of tetrodotoxin-resistant sodium channels in amphibian sympathetic and sensory neurons. *Biochemistry* **2002**, *41*, 15388–15393.

42. Walewska, A.; Skalicky, J.J.; Davis, D.R.; Zhang, M.M.; Lopez-Vera, E.; Watkins, M.; Han, T.S.; Yoshikami, D.; Olivera, B.M.; Bulaj, G. NMR-based mapping of disulfide bridges in cysteine-rich peptides: Application to the mu-conotoxin SxIIIA. *J. Am. Chem. Soc.* **2008**, *130*, 14280–14286.

43. Lewis, R.J.; Schroeder, C.I.; Ekberg, J.; Nielsen, K.J.; Loughnan, M.; Thomas, L.; Adams, D.A.; Drinkwater, R.; Adams, D.J.; Alewood, P.F. Isolation and structure-activity of mu-conotoxin TIIIA, a potent inhibitor of tetrodotoxin-sensitive voltage-gated sodium channels. *Mol. Pharmacol.* **2007**, *71*, 676–685.

4

Omega-3 Polyunsaturated Fatty Acids Protect Neural Progenitor Cells against Oxidative Injury

Qiang Liu [1,†], Di Wu [1,†], Na Ni [1], Huixia Ren [1], Chuanming Luo [1], Chengwei He [1], Jing-Xuan Kang [2], Jian-Bo Wan [1] and Huanxing Su [1,*]

[1] State Key Laboratory of Quality Research in Chinese Medicine, Institute of Chinese Medical Sciences, University of Macau, Macao 999078, China; E-Mails: yb37525@umac.mo (Q.L.); mb35801@umac.mo (D.W.); mb15844@umac.mo (N.N.); yb27526@umac.mo (H.R.); 15989166366@163.com (C.L.); chengweihe@umac.mo (C.H.); jbwan@umac.mo (J.-B.W.)

[2] Laboratory for Lipid Medicine and Technology, Massachusetts General Hospital and Harvard Medical School, Boston, MA 02114, USA; E-Mail: jxkang@mgh.harvard.edu (J.-X.K.)

† These two authors contributed equally to this work.

* Author to whom correspondence should be addressed; E-Mail: huanxingsu@umac.mo

Abstract: The omega-3 polyunsaturated fatty acids (ω-3 PUFAs), eicosapentaenoic acid (EPA) and docosahexaenoic acid (DHA), derived mainly from fish oil, play important roles in brain development and neuroplasticity. Here, we reported that application of ω-3 PUFAs significantly protected mouse neural progenitor cells (NPCs) against H_2O_2-induced oxidative injury. We also isolated NPCs from transgenic mice expressing the *Caenorhabditis elegans fat-1* gene. The *fat-1* gene, which is absent in mammals, can add a double bond into an unsaturated fatty acid hydrocarbon chain and convert ω-6 to ω-3 fatty acids. Terminal deoxynucleotidyl transferase dUTP nick end labeling (TUNEL) staining showed that a marked decrease in apoptotic cells was found in *fat-1* NPCs after oxidative injury with H_2O_2 as compared with wild-type NPCs. Quantitative RT-PCR and Western blot analysis demonstrated a much higher expression of nuclear factor erythroid 2-related factor 2 (Nrf2), a master transcriptional factor for antioxidant genes, in *fat-1* NPCs. The results of the study provide evidence that ω-3 PUFAs resist oxidative injury to NPCs.

Keywords: oxidative stress; DHA; fat-1; neural progenitor cells

1. Introduction

Oxidative stress-induced neuronal apoptosis plays a critical role in the pathogenesis of stroke and neurodegenerative diseases [1,2]. Oxidants, such as hydrogen peroxide and free radicals, produce cell damage by inducing production of reactive oxygen species (ROS) and activate an inflammatory response [3]. Several studies confirmed the transcription factor nuclear factor erythroid 2-related factor 2 (Nrf2) represents an important cellular protective mechanism against oxidative stress over the Nrf2-ARE pathway [4,5]. Activation of Nrf2 signaling induces the transcriptional regulation of ARE-dependent expression of various antioxidant and phase II detoxification enzymes, which include hemeoxygenase-1 (HO-1), NAD(P)H quinine oxidoreductase 1 (NQO-1), glutamate-cysteine ligase modifier subunit (GCLM), and glutamate-cysteine ligase catalytic subunit (GCLC) [6].

The long-chain omega-3 polyunsaturated fatty acids (ω-3 PUFAs) from fish oil, for example, Docosahexaenoic acid (DHA), are highly enriched in the brain and play a key role in brain development and repair under many conditions [7,8]. Dietary DHA has been suggested to improve neuronal development [9], restore and enhance cognitive functions [10–12], and protect against beta-amyloid production, accumulation, and potential downstream toxicity in an aged Alzheimer mouse model [13]. DHA have also been shown to exert a beneficial effect on ROS related cellular damage [14]. These studies indicate that omega-3, such as DHA, can increase neural resistance to various types of insults.

Mammals are unable to synthesize ω-3 PUFAs de novo and must rely on a dietary source of these essential fatty acids. The *C. elegans fat-1* gene encodes an *n*-3 fatty acid desaturase that converts ω-6 to ω-3 PUFA, which could significantly reduce the omega-6/omega-3 fatty acid ratio [15]. The *fat-1* transgenic mice are rich in endogenous ω-3 PUFAs, specifically in the brain, with a reduction in ω-6 fatty acids, which provides an optimal model to evaluate the actions of ω-3 PUFAs [16,17].

In the present study, we investigated whether ω-3 PUFAs could protect neural progenitor cells (NPCs) against oxidative injury. NPCs are multipotent with a broad self-renewing potential and with the capacity to generate neurons, astrocytes and oligodendrocytes. Their inherent biological properties of NPCs provide multiple potentials to treat various neurological dysfunctions. Our results provide evidence that both exogenous and endogenous ω-3 PUFAs can resist oxidative injury to NPCs.

2. Results

2.1. In Vitro *Characterization of NPCs^{WT} and NPCs^{fat-1}*

Neural progenitor cells (NPCs) are self-renewing, multipotent cells that could be effectively differentiated into neurons, astrocytes, and oligodendrocytes [18]. With bFGF-supplemented culture medium, both NPCs^{WT} and NPCs^{fat-1} cells showed bipolar or multipolar morphology with small cell bodies. Nestin is an intermediate filament protein and widely used as a specific marker for NPCs. Immunostaining showed that more than 95% cells in both NPCs^{WT} and NPCs^{fat-1} culture were nestin-positive, confirming that the majority of NPCs^{WT} and NPCs^{fat-1} were immature (Figure 1A). PCR analysis demonstrated the high expression of *fat-1* in NPCs^{fat-1} (lanes 3 and 4) while no expression in NPCs^{WT} (lanes 1 and 2) (Figure 1B). To study differentiation potential of NPCs^{WT} and NPCs^{fat-1} *in vitro*, bFGF was replaced with 1% FBS in the cell culture medium and NPCs began to differentiate. At the 5th day, cultures with this differentiating medium, both NPCs^{WT} and NPCs^{fat-1} were successfully differentiated into Tuj1-positive neurons, GFAP-positive astrocytes and Rip-positive oligodendrocytes with similar differentiation capacities (NPCs^{WT}: 14.7% neurons, 61.3% astrocytes, and 11.5% oligodendrocytes; NPCs^{fat-1}: 15.4% neurons, 65.6% astrocytes, and 12.1% oligodendrocytes) (Figure 1C,D).

Figure 1. Characterization on NPCs^{WT} and NPCs^{fat-1}. (**A**) The purity of neural progenitor cells (NPCs) was identified by Nestin staining and nuclei were counter-stained with DAPI. More than 95% of NPCs^{WT} or NPCs^{fat-1} were nestin-positive cells; (**B**) Gel electrophoresis of PCR products using primers for *fat-1* gene. Wild-type controls (lanes 1 and 2) and positive *fat-1* specimens (lanes 3 and 4); (**C**) Both NPCs^{WT} and NPCs^{fat-1} were shown to successfully differentiate into Tuj1-positive neurons, GFAP-positive astrocytes, and Rip-positive oligodendrocytes; (**D**) The bar graph showing the percentage of neural cells differentiated from NPCs at the 5th day in the differentiating medium. Scale bar: 250 μm in (**A**) and 100 μm in (**C**).

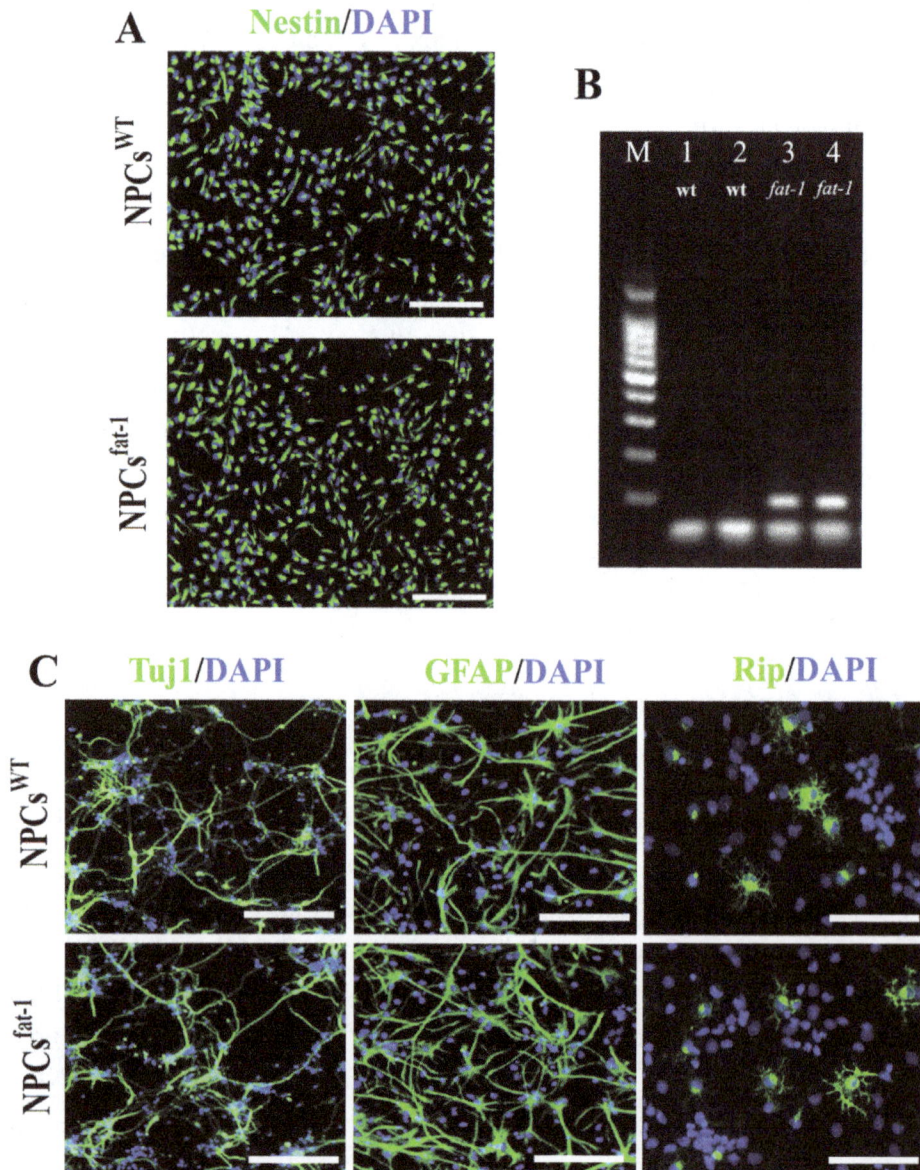

Figure 1. *Cont.*

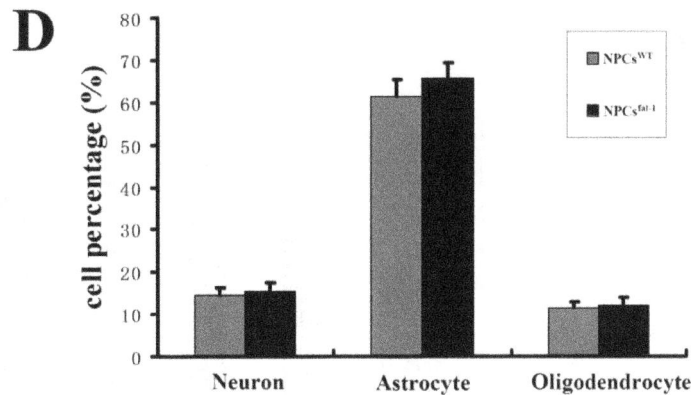

2.2. DHA Protected NPCs against H_2O_2-Mediated Apoptosis

To evaluate the effect of H_2O_2 on cell viability, we first incubated NPCs with 200 μM H_2O_2 and investigated cell viability at different time points. Cell viability was measured using the WST-8 assay. A 200 μM concentration of H_2O_2 in culture was used to establish the oxidative injury model according to a previous study reporting that cultured NPCs exposed to H_2O_2 at this concentration was sufficiently induced to undergo apoptosis [19]. As shown in Figure 2A, cell viability was significantly reduced in a time-dependent manner. Incubation of 200 μM H_2O_2 for 6 h caused an approximate 50% cell loss, which was considered to be an optimal oxidative injury model for investigating drug effects.

Figure 2. DHA pretreatment reduced oxidative stress on cultured NPCs. (**A**) WST-8 assays revealed that incubation of 200 μM H_2O_2 has caused a significant cytotoxicity in a time-dependent manner. (**B**) DHA prevented H_2O_2-induced cell death in a concentration-dependent manner. Cell viability was presented as a percentage of control, and each value represents the mean ± SD of three independent experiments; * $p < 0.05$ and ** $p < 0.001$ *versus* control.

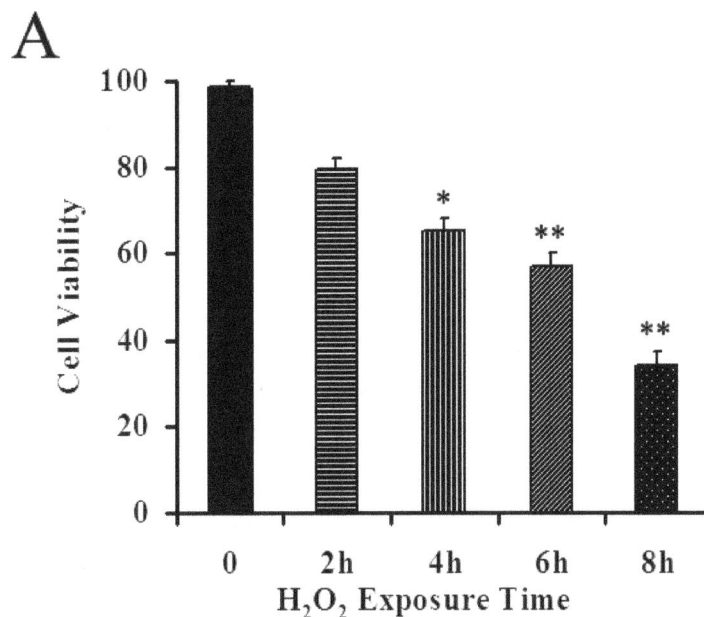

Figure 2. *Cont.*

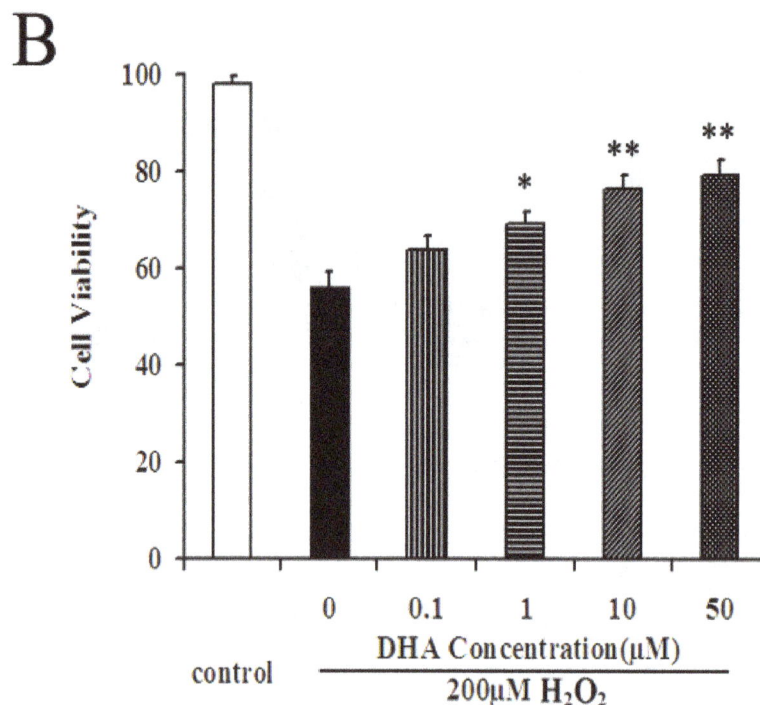

We then investigated the neuroprotective effects of exogenous DHA on H_2O_2-mediated apoptosis. $NPCs^{WT}$ at the confluence of 80% was pretreated with DHA (0, 0.1, 1, 10, and 50 μM) for 2 h and then suffered an oxidative injury induced by incubation of 200 μM H_2O_2 for 6 h. WST-8 assay revealed that the cell viability increased in a concentration-dependent manner: The pretreatment of 1 μM DHA increased the cell viability by 22.1% as compared to vehicle control ($p < 0.05$, Figure 2B), while the pretreatment of 10 μM and 50 μM increased the cell viability by 35.6% and 36.2%, respectively, as compared to the vehicle control ($p < 0.001$, Figure 2B).

2.3. NPCs^{fat-1} Prevented H_2O_2-Mediated Apoptosis

We further investigated anti-oxidative effects of endogenous ω-3 PUFAs in NPCs. $NPCs^{fat-1}$ were isolated from *fat-1* mice, which are rich in endogenous ω-3 PUFAs, specifically in the brain [16,17]. WST-8 assay showed that $NPCs^{fat-1}$ exhibited a potent anti-oxidative effect similar to that found in the DHA-treated $NPCs^{WT}$ group when exposed to H_2O_2 for 6 h. The cell viability of these two groups was significantly increased as compared to the vehicle control (Figure 3A). Terminal deoxynucleotidyl transferase-mediated UTP end-labeling (TUNEL) staining was also performed to detect H_2O_2-mediated apoptosis. Only a very small proportion of intrinsic apoptosis were detected in cultured $NPCs^{WT}$ (Figure 3B,F). Incubation with 200 μM H_2O_2 for 6 h induced approximately 40% $NPCs^{WT}$ to undergo apoptosis (Figure 3C,F), while pretreatment of $NPCs^{WT}$ with 10 μM DHA significantly attenuated H_2O_2-mediated apoptosis to less than 30% (Figure 3D,F). $NPCs^{fat-1}$ exhibited potent anti-oxidative properties, as shown by a significant decrease in apoptosis compared to $NPCs^{WT}$ when exposed to H_2O_2 for 6 h (Figure 3E,F). These findings indicated that both exogenous and endogenous ω-3 PUFAs could protect NPCs against H_2O_2-mediated oxidative injury.

Figure 3. NPCs^{fat-1} attenuated H$_2$O$_2$-mediated apoptosis. (**A**) The cell viability of NPCs was assessed after exposure to H$_2$O$_2$ for 6 h by WST-8 analysis. Each value represents the mean ± SD of three independent experiments (n = 3, ** p < 0.01 *versus* other groups); (**B–E**) Representative photomicrographs of TUNEL assay; (**F**) Quantitative analysis was carried out by measuring TUNEL-positive cells in each group. Figures were selected as representative data from three independent experiments. Cell apoptosis was significantly reduced in DHA-pretreated NPCs^{WT} and NPCs^{fat-1}. Each value represents the mean ± SD of three independent experiments (n = 3, ** p < 0.01 *versus* other groups). Scale bar: 75 μm.

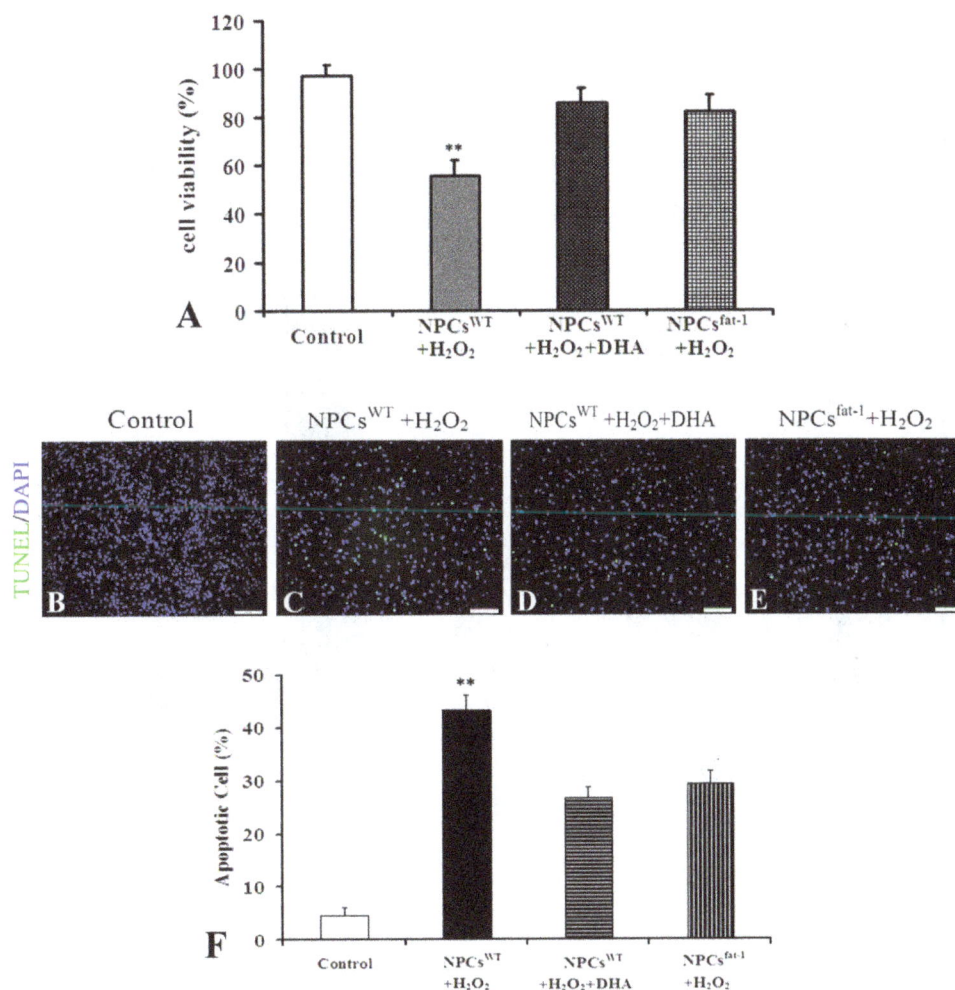

2.4. Expression Analyses of Nrf2-ARE Pathway Genes

To study the anti-oxidative mechanisms of ω-3 PUFAs against H$_2$O$_2$-induced apoptosis in NPCs, we first investigated whether Nrf2, the principal transcription factor that regulates the basal and inducible expression of a battery of antioxidant genes, was up-regulated after pretreatment with DHA and in NPCs^{fat-1}. Real-time RT-PCR assays showed that both DHA pretreatment and NPCs^{fat-1} induced a nearly 2.5-fold increase in the transcript level of Nrf2 when compared with the controls (Figure 4). Furthermore, significant increases in the expression level of the downstream gene and phase II detoxification gene transcripts (HO-1, NQO-1, GCLC, GCLM) were found in NPCs^{fat-1} and DHA-pretreated NPCs^{WT} (Figure 4).

Figure 4. Expression analyses of Nrf2-ARE pathway genes in DHA-pretreated NPCs^{WT} and NPCs^{fat-1}. Real-time RT-PCR assays showed that both DHA pretreatment and NPCs^{fat-1} induced significant increases in the transcript level of Nrf2 and its downstream gene and phase II detoxification gene transcripts HO-1, NQO-1, GCLC, GCLM when compared with the controls. Data are shown as mean ± SD ($n = 3$); * $p < 0.05$ *versus* control; ** $p < 0.01$ *versus* control.

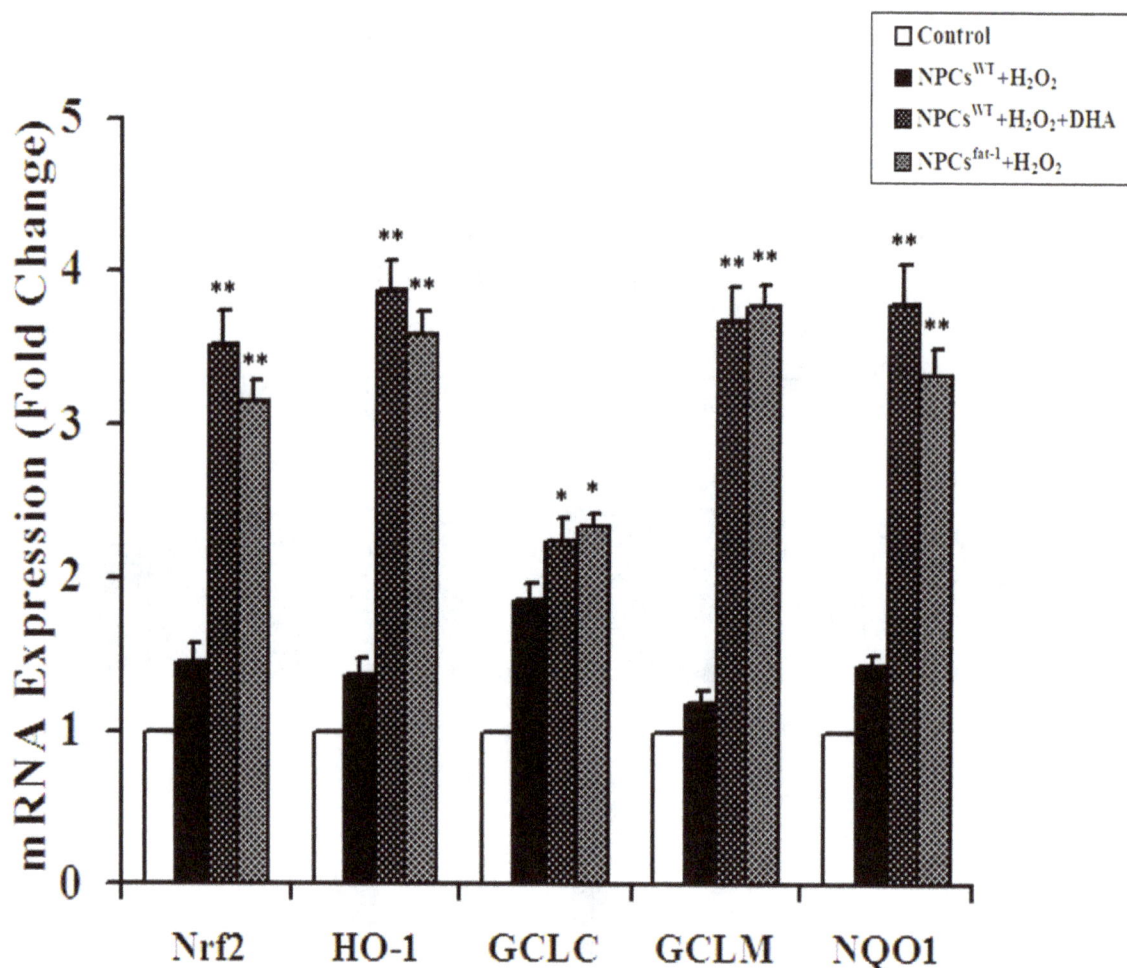

2.5. Expression Profiles of Nuclear and Cytosolic Nrf2 by Western Blot Analysis

Real-time RT-PCR assays demonstrated that both DHA pretreatment and NPCs^{fat-1} induced significant increases in the transcript level of Nrf2. As nuclear translocation of protein Nrf-2 is very critical in inducing gene expression of anti-oxidant genes, we then investigated the expression profiles of cytosolic and nuclear fraction Nrf2 by Western blot analysis. Consistent with real-time RT-PCR results, Western blot analysis demonstrated that both DHA pretreatment and NPCs^{fat-1} significantly increased the protein expression of nuclear Nrf2 when compared with the controls (Figure 5A,C). However, the expression of cytosolic Nrf2 in DHA pretreatment and NPCs^{fat-1} was significantly decreased when compared with the control (Figure 5B,D). These results demonstrated an obvious translocation of Nrf2 from the cytoplasm to the nucleus in DHA pretreatment and NPCs^{fat-1}.

Figure 5. DHA pretreatment and NPCs[fat-1] induced a significant increase in nuclear Nrf2 expression and a significant decrease in cytosolic Nrf2 expression. DHA-pretreated NPCs[WT] and NPCs[fat-1] were treated with 200 μM H_2O_2 for the indicated time points. Cells were lysed and fractionated to isolate nuclear and cytosolic fractions as indicated. Fractions were confirmed using Western blot with histone H3 for nuclear fractions (**A**) and β-actin as a marker for cytosolic fractions (**B**); Densitometry analysis showed that DHA pretreatment and NPCs[fat-1] induced a significant increase in nuclear Nrf2 expression (**C**) and a significant decrease in cytosolic Nrf2 expression (**D**). Data are shown as mean ± SD ($n = 3$); * $p < 0.05$ *versus* control; ** $p < 0.01$ *versus* control.

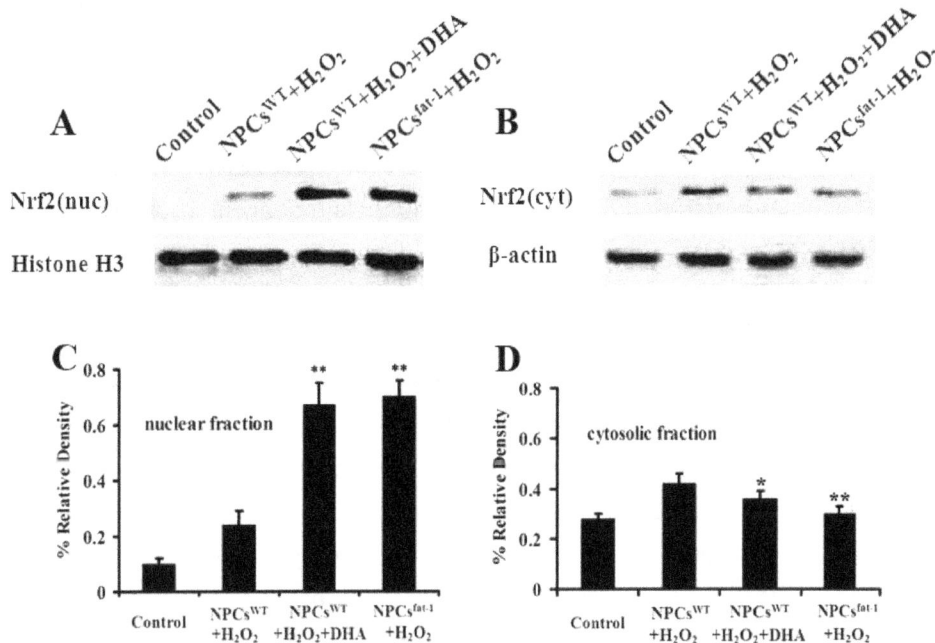

3. Discussion

Neurodegenerative diseases are characterized by the progressive loss of neurons and usually influence the cognitive function, movement control or muscle strength [1,2]. Neurodegenerative diseases are commonly late-onset disorders, including Alzheimer's disease (AD), Parkinson's disease (PD), Huntington's disease (HD), and Amyotrophic lateral sclerosis (ALS). Oxidative stress has a significant role in the pathogenesis and/or progression of several neurodegenerative diseases and aging diseases, which are closely associated with disease-specific proteins aggregation, inflammation, mitochondrial dysfunction, and neurotoxicity [1,2,18]. Effective antioxidants have promising potential for therapeutic application. A prospective strategy in disease control has been focused on development of antioxidants as preventive and therapeutic medicine.

NPCs are largely undifferentiated cells originating in the central nervous system. They have the potential to give rise to offspring cells and efficiently differentiate into neurons, astrocytes and oligodendrocytes [19,20]. The inherent biological properties of NPCs may provide multiple strategies to treat CNS dysfunction and enable them to be an optimal model to screen antioxidants, which have therapeutic potentials for the treatment of neurological diseases.

DHA is an *n*-3 long chain PUFA, highly enriched in the central nervous system, and is critical for brain development and function. DHA is reported to play a neuroprotective role against oxidative stress in astrocytes [14], oligodendroglia cells [21], retinal ganglion cells [22], and human lymphocytes [23]. The animal with DHA diet or transgenic *fat-1* mice rich in endogenous *n*-3 PUFA showed a better behavior performance [11–13,24]. H_2O_2 is a common oxidant to cause oxidative damage to cells and widely used in establishment of oxidative injury models [25]. Our present study reported that both exogenous and endogenous ω-3 PUFAs significantly protected NPCs against H_2O_2-induced oxidative injury, suggesting that ω-3 PUFAs might be an effective supplement for the prevention of neurodegenerative diseases which are associated with oxidative stress. DHA has been reported to scavenge the intracellular radical productions induced by hydrogen peroxide (H_2O_2), superoxide anion ($O_2^{\bullet-}$), and hydroxyl radical (•OH) [22]. Many previous studies reported that DHA treatment could significantly reduce ROS production, which is a possible mechanism underlying DHA's protective effects [14]. However, no significant differences in intracellular ROS levels were found between treatments in our study using 2′,7′-dichlorofluorescin diacetate (DCFH-DA) to measure intracellular ROS levels. A previous study reported that DHA at some certain concentrations showed no effects on the fluorescence change by use of ROS probe [26]. It could be a reason why we were unable to detect the differences in intracellular ROS levels between the DHA treatment and control group. Another possible reason may be the cell type used in our study. Our study investigated whether DHA could protect NPCs against oxidative injury. NPCs are capable of self-renewal and they grow and proliferate rapidly in the culture, which may make it difficult to accumulate ROS to sufficient levels for measurement inside cells. To detect ROS changes in NPCs may require more sensitive methods or probes. Regarding protective effects of DHA on attenuating oxidative stress/damage induced by H_2O_2, our study suggests that DHA exert its antioxidative effects possibly via initiating a translocation of Nrf2 from the cytoplasm to the nucleus and subsequently stimulating the expression of a battery of antioxidant and phase II detoxification molecules as a response to oxidative injury.

The nuclear factor erythroid 2-related factor 2 (Nrf2) is an emerging regulator of cellular resistance to oxidants [4–6]. Nrf2 is localized mainly in the cytoplasm bound to its specific repressor Keap1. Oxidative stress will cause the liberation of Nrf2 and allow it to translocate into the nucleus. Nrf2 will induce transcriptional upregulation of numerous antioxidant and phase II detoxification genes to provide efficient cytoprotection [27]. The activated Nrf2 has shown protective effects in animal models of many neurodegenerative disorders [28–30]. In the present investigation, we demonstrated that both exogenous and endogenous DHA enhanced Nrf2 translocation from the cytoplasm to the nucleus of cultured NPCs when exposed to the oxidative stress and subsequently stimulated the mRNA levels of Nrf2, GCLC, GCLM, NQO-1, and HO-1. These results confirm previous findings that treatment of DHA can induce antioxidant and detoxifying genes [31,32].

Although our study demonstrated a similar antioxidative effect between exogenous and endogenous DHA on cultured NPCs, a difference in mechanisms underlying protective effects of exogenous and endogenous DHA may exist. The fat-1 gene, which is absent in mammals, encodes omeg-3 polyunsaturated fatty acids (ω-3 PUFAs) that convert ω-6 to ω-3 PUFAs, leading to an elevated amount of ω-3 PUFAs such as DHA and higher ω-3 PUFAs/ω-6 PUFAs ratio in cells and tissues from fat-1 mice. PUFAs are essential components of membrane phospholipids and have a specific influence on membrane properties. Membranes enriched in ω-3 PUFAs show increased membrane fluidity [33] and can directly or indirectly affect the function of a number of membrane proteins such as receptors since receptors and their affinity to their respective hormones/growth factors/proteins depend on the fluidity of the cell membrane [34,35]. The increased membrane fluidity can be involved in antioxidative effects of endogenous DHA in addition to its direct action on initiating Nrf2 translocation from the cytoplasm to the nucleus, while exogenous DHA is considered to exert its antioxidative effects possibly via a direct action on initiating a translocation of Nrf2 from the cytoplasm to the nucleus and subsequently stimulating the expression of a battery of antioxidant and phase II detoxification molecules as a response to oxidative injury.

4. Experimental Section

4.1. Animals

We obtained *fat-1* breeders on a C57BL/6 background from Dr. Jing X. Kang (Harvard Medical School, Boston, MA, USA) and arose in the Laboratory Animal Center, University of Macau (Macau, China). Mice were housed in a temperature-controlled, 12:12 light/dark room and were allowed free access to water and food. The F1 progeny were obtained by mating C57BL/6 × C3H *fat-1* breeders with C57BL/6 WT mice. Generations of heterozygous fat-1 mice were then mated with WT littermates to obtain WT and heterozygous *fat-1* mice. All animals were treated in accordance with prevailing laws on animal experiments that were approved by the ethical committee of the University of Macau (Macau, China).

4.2. Genomic DNA Extractions and PCR Amplification

The *fat-1* C57BL6 mice (fat-1) and *fat-1* negative C57BL6 mice (WT) were identified by genotyping using PCR. Genomic DNA was prepared from 1 to 2 mm sections of tail tip using DNA Isolation Kit for Cells and Tissues (Roche, Mannheim, Germany). The DNA was used running polymerase chain reactions (PCR) using oligonucleotide primers that are specific for the transgene. Primer pair sets for the *fat-1* gene were constructed from Invitrogen (Carlsbad, CA, USA) and were as follows: Fat-1 forward: 5′-TGTTCATGCCTTCTT-CTTTTTCC-3′; reverse: 5′-GCGACCATACCTCAAACTTGGA-3′. PCR was carried out using rTaq (Takara, Otsu, Japa) with the following conditions: 95 °C 60 s (1 cycle); 95 °C 20 s, 58 °C 30 s, 72 °C 40 s (34 cycles). Amplified fragments were separated by 1.5% agarose gel electrophoresis.

4.3. Cell Isolation and Culture

Under sterile conditions, cerebral cortex from E13.5 *fat-1* mice and WT littermates were dissected out and prepared for NPCs culture following procedures described previously with minor modifications [25]. Briefly, the cortex was separated from surrounding tissues. After peeling off the meninges, the cortex was transferred into a 15 mL centrifuge tube containing culture medium (described below) and dissociated to a single-cell suspension by gentle mechanical trituration through a fire polished Pasteur pipette. The dissociated cells were filtered through a cell strainer (BD Falcon, Franklin Lakes, NJ, USA) and then cultured in T25 flask in suspension. The culture medium consisted of DMEM-F12, BSA (1 mg/mL), B27 (20 IU/mL), N2 (10 IU/mL), EGF (20 ng/mL), and bFGF (20 ng/mL). Cells were maintained in an incubator with a humidified atmosphere containing 5% CO_2 at 37 °C. NPCs isolated from fat-1 mice and their WT littermates were confirmed by genomic DNA analyses and designated as NPCs^{fat-1} and NPCs^{WT} respectively. The medium was changed every two days. After five to six days, cells grew in neurospheres with the diameter of approximately 150 μm. Cells in the neurospheres were passaged at the ratio of 1:6 after initial plating. These subcultured cells were designated as "first passage" (P1). The third passage (P3) cells were used for all the following experiments. For differentiation studies, growth factors were removed from the culture medium and 1% fetal bovine serum (FBS, Gibco, Life Technologies Inc., Grand Island, NY, USA) was added. The cultures were allowed to differentiate for up to five days.

4.4. Exposure to H_2O_2 and Pretreatment with DHA

Dilutions of H_2O_2 (Sigma-Aldrich, St. Louis., MO, USA) were made fresh from a 30% stock solution into cell culture medium to the different terminal concentrations. The NPCs^{WT} were seeded at a density of 1×10^4 cells per well into 96-well plates, then incubated in a humidified atmosphere of 95% air and 5% CO_2 at 37 °C. A 200 μM H_2O_2 concentration in NPCs was determined to be optimal for this study (data not shown).

DHA (Sigma-Aldrich, St. Louis., MO, USA) was dissolved in 100% ethanol and kept at −20 °C in the dark as described in a previous study [24]. Immediately before use, the DHA stock solution was diluted in the bath solution and adjusted to the final concentrations needed. To examine the protective effects of DHA on H_2O_2-mediated apoptosis, NPCs^{WT} at a confluence of around 75% was pretreated with DHA (0, 0.1, 1, 10, and 50 μM) for 2 h and followed by oxidative injury induced by H_2O_2 treatment. NPCs^{fat-1} were exposed to H_2O_2 directly to investigate the protective effects of endogenous ω-3 PUFAs against oxidative injury. These cultures were then proceeded to cell viability analysis and TUNEL staining.

4.5. Analysis of Cell Viability and TUNEL

Cell viability was assessed using the WST-8 dye (Beyotime Inst Biotech, Haimen, China) according to the manufacturer's instructions. After 10 μL WST-8 dye was add to each well, cells were incubated at 37 °C for 2 h and the absorbance was finally determined at 450 nm using a microplate reader (Molecular Devices, Sunnyvale, CA, USA). The results were expressed as relative cell viability (%). The apoptotic cell death of NPCs was estimated using TUNEL staining (Roche Applied Science, Indianapolis, IN, USA) according to the manufacturer's protocol. Cell cultures were counterstained with DAPI (5 μg/mL), which stained the nuclei of all cells.

4.6. Real-Time RT-PCR

Total RNA was extracted from NPCs^{WT}, NPCs^{fat-1} and pretreated NPCs^{WT} with 10 μM DHA after exposed to 200 μM H_2O_2 for 6 h using TRIzol reagent (Invitrogen, Carlsbad, CA, USA) according to the manufacturer's instructions. cDNAs were amplified and quantified in ABI Prism 7500 Sequence Detection System (Applied Biosystems, Foster City, CA, USA) using dye SYBR Green I (Takara, Otsu, Japan). The fold change in the levels of Nrf2, HO-1, GCLC, GCLM, and NQO-1 between the NPCs^{WT} and NPCs^{fat-1}, normalized by the level of β-actin, was determined using the following equation: Fold change = $2^{-\Delta(\Delta Ct)}$, where $\Delta Ct = Ct(target) - Ct(\beta\text{-actin})$ and $\Delta(\Delta Ct) = \Delta Ct(treated) - \Delta Ct(untreated)$. The primer sequences are listed in Table 1.

Table 1. Primers for real-time PCR assay.

Gene	Primer (5′-3′)
Nrf2	F: TTCTTTCAGCAGCATCCTCTCCAC
	R: ACAGCCTTCAATAGTCCCGTCCAG
NQO1	F: GCGAGAAGAGCCCTGATTGTACTG
	R: TCTCAAACCAGCCTTTCAGAATGG
HO-1	F: CAAGCCGAGAATGCTGAGTTCATG
	R: GCAAGGGATGATTTCCTGCCAG
GCLM	F: GCCACCAGATTTGACTGCCTTTG
	R: TGCTCTTCACGATGACCGAGTACC
GCLC	F: ACATCTACCACGCAGTCAAGGACC
	R: CTCAAGAACATCGCCTCCATTCAG
β-actin	F: TCGTGCGTGACATTAAGGAGAAG
	R: GTTGAAGGTAGTTTCGTGGATGC

4.7. Immunofluorescence

Immunocytochemistry was performed to characterize NPCs^WT and NPCs^fat-1. Briefly, cells were fixed with 4% paraformaldehyde, blocked with 5% goat serum, and then incubated with primary antibodies overnight at 4 °C, including rabbit anti-nestin (1:1000, Millipore, Billerica, MA, USA), mouse anti-Rip (1:50, kindly gift from Dr. XM Xu, University of Louisville, Louisville, KY, USA), rabbit anti-GFAP (1:1000, Sigma-Aldrich, St. Louis., MO, USA), mouse anti-Tuj1 (1:1000, Sigma-Aldrich, St. Louis., MO, USA). The cells were then rinsed three times with PBS and incubated for 30 min with species-specific secondary antibody conjugated to the fluorescent labels Alexa 568 or 488 (1:400, Invitrogen, Carlsbad, CA, USA). Cell cultures were counterstained with DAPI (5 μg/mL) to stained the nuclei of all cells. Finally, the cells were visualized under a fluorescent laser microscope (IX73, Olympus Corp., Tokyo, Japan).

4.8. Western Blotting Analysis

Cells were washed twice with ice-cold PBS and lysed using a Nuclear and Cytoplasmic Protein Extraction Kit (Beyotime, Haimen, China) according to the protocol described by the manufacturer. The protein concentrations were determined using Bradford method. The protein extracts were separated in sodium dodecylsulfate polyacrylamide gel electrophoresis (SDS-PAGE) gel and then transferred to a poly-vinylidene difluoride (PVDF) membrane. They were then incubated overnight at 4 °C with primary monoclonal antibodies against Nrf2 (1:1000; R & D Systems, Minneapolis, MN, USA). Histone H3 (1:1000; Cell Signaling Technology, Beverly, MA, USA) and β-actin (1:1000; Cell Signaling Technology, Beverly, MA, USA) used as controls were detected in the nuclear fraction and cytosolic fraction, respectively. The blots were washed thoroughly in TBST buffer and incubated for 1 h with appropriate HRP-linked secondary antibodies (1:1000; Cell Signaling Technology, Beverly, MA, USA). Immunoreactive proteins were visualized with the ECL Western blotting detection reagent (Amersham

Biosciences, GE Healthcare, Piscataway, NJ, USA). Relative band intensities were determined by Quality-one 1-D analysis software (Bio-Rad, Hercules, CA, USA).

4.9. Statistical Analysis

The results were expressed as the mean ± S.D. of triplicate measurements representative of three independent experiments. The one-way analysis of variance and Tukey test were used for the multiple comparisons. Statistical significance was defined as $P < 0.05$.

5. Conclusions

Both exogenous and endogenous DHA showed protective effects on NPCs against oxidative injury possibly via Nrf-ARE pathway, suggesting that DHA might be an effective supplement for the prevention of neurodegenerative diseases which are associated with oxidative stress.

Acknowledgments

This study was supported by Macao Science and Technology Development Fund (003/2012/A and 018/2013/A1) and multi-year research grant, university of Macau, MYRG122 (Y1-L3)-ICMS12-SHX and MYRG110 (Y1-L2)-ICMS13-SHX.

Author Contributions

Conceived and designed the experiments: HS, JK, JW; Performed the experiments: QL, DW, NN; Analyzed the data: CH, HR, CL; Wrote the paper: HS, QL.

References

1. Barnham, K.J.; Masters, C.L.; Bush, A.I. Neurodegenerative diseases and oxidative stress. *Nat. Rev. Drug Discov.* **2004**, *3*, 205–214.

2. Xu, J.; Kao, S.Y.; Lee, F.J.; Song, W.; Jin, L.W.; Yankner, B.A. Dopamine-dependent neurotoxicity of alpha-synuclein: A mechanism for selective neurodegeneration in Parkinson disease. *Nat. Med.* **2002**, *8*, 600–606.

3. Zahler, S.; Kupatt, C.; Becker, B.F. Endothelial preconditioning by transient oxidative stress reduces inflammatory responses of cultured endothelial cells to TNF-alpha. *FASEB J.* **2000**, *14*, 555–564.

4. Johnson, J.A.; Johnson, D.A.; Kraft, A.D.; Calkins, M.J.; Jakel, R.J.; Vargas, M.R.; Chen, P.C. The Nrf2-ARE pathway: An indicator and modulator of oxidative stress in neurodegeneration. *Ann. N. Y. Acad. Sci.* **2008**, *1147*, 61–69.

5. Kraft, A.D.; Johnson, D.A.; Johnson, J.A. Nuclear factor E2-related factor 2-dependent antioxidant response element activation by tert-butylhydroquinone and sulforaphane occurring preferentially in astrocytes conditions neurons against oxidative insult. *J. Neurosci.* **2004**, *24*, 1101–1112.

6. Ma, Q. Role of nrf2 in oxidative stress and toxicity. *Annu. Rev. Pharmacol. Toxicol.* **2013**, *53*, 401–426.

7. Chang, Y.L.; Chen, S.J.; Kao, C.L.; Hung, S.C.; Ding, D.C.; Yu, C.C.; Chen, Y.J.; Ku, H.H.; Lin, C.P.; Lee, K.H.; *et al.* Docosahexaenoic acid promotes dopaminergic differentiation in induced pluripotent stem cells and inhibits teratoma formation in rats with Parkinson-like pathology. *Cell Transplant.* **2012**, *21*, 313–332.

8. Russell, F.D.; Bürgin-Maunder, C.S. Distinguishing health benefits of eicosapentaenoic and docosahexaenoic acids. *Mar. Drugs.* **2012**, *10*, 2535–2559.

9. Tixier-Vidal, A.; Picart, R.; Loudes, C.; Bauman, A.F. Effects of polyunsaturated fatty acids and hormones on synaptogenesis in serum-free medium cultures of mouse fetal hypothalamic cells. *Neuroscience* **1986**, *17*, 115–132.

10. Greiner, R.S.; Moriguchi, T.; Hutton, A.; Slotnick, B.M.; Salem, N., Jr. Rats with low levels of brain docosahexaenoic acid show impaired performance in olfactory-based and spatial learning tasks. *Lipids* **1999**, *34*, S239–S243.

11. Gamoh, S.; Hashimoto, M.; Sugioka, K.; Hossain, M.S.; Hata, N.; Misawa, Y.; Masumura, S. Chronic administration of docosahexaenoic acid improves reference memory-related learning ability in young rats. *Neuroscience* **1999**, *93*, 237–241.

12. Gamoh, S.; Hashimoto, M.; Hossain, S.; Masumura, S. Chronic administration of docosahexaenoic cid improves the performance of radial arm maze task in aged rats. *Clin. Exp. Pharmacol. Physiol.* **2001**, *28*, 266–270.

13. Lim, G.P.; Calon, F.; Morihara, T.; Yang, F.; Teter, B.; Ubeda, O.; Salem, N., Jr.; Frautschy, S.A.; Cole, G.M. A diet enriched with the omega-3 fatty acid docosahexaenoic acid reduces amyloid burden in an aged Alzheimer mouse model. *J. Neurosci.* **2005**, *25*, 3032–3040.

14. Kim, E.J.; Park, Y.G.; Baik, E.J.; Jung, S.J.; Won, R.; Nahm, T.S.; Lee, B.H. Dehydroascorbic acid prevents oxidative cell death through a glutathione pathway in primary astrocytes. *J. Neurosci. Res.* **2005**, *79*, 670–679.

15. Kang, Z.B.; Ge, Y.; Chen, Z.; Cluette-Brown, J.; Laposata, M.; Leaf, A.; Kang, J.X. Adenoviral gene transfer of Caenorhabditis elegans *n*-3 fatty acid desaturase optimizes fatty acid composition in mammalian cells. *Proc. Natl. Acad. Sci. USA* **2001**, *98*, 4050–4054.

16. Kang, J.X.; Wang, J.; Wu, L.; Kang, Z.B. Transgenic mice: Fat-1 mice convert *n*-6 to *n*-3 fatty acids. *Nature* **2004**, *427*, doi:10.1038/427504a.

17. Xia, S.; Lu, Y.; Wang, J.; He, C.; Hong, S.; Serhan, C.N.; Kang, J.X. Melanoma growth is reduced in fat-1 transgenic mice: Impact of omega-6/omega-3 essential fatty acids. *Proc. Natl. Acad. Sci. USA* **2006**, *103*, 12499–12504.

18. Agar, J.; Durham, H. Relevance of oxidative injury in the pathogenesis of motor neuron diseases. *Amyotroph. Lateral Scler. Other Motor Neuron Disord.* **2003**, *4*, 232–242.

19. Reynolds, B.A.; Weiss, S. Generation of neurons and astrocytes from isolated cells of the adult mammalian central nervous system. *Science* **1992**, *255*, 1707–1710.

20. Lin, H.J.; Wang, X.; Shaffer, K.M.; Sasaki, C.Y.; Ma, W. Characterization of H_2O_2-induced acute apoptosis in cultured neural stem/progenitor cells. *FEBS Lett.* **2004**, *570*, 102–106.

21. Brand, A.; Schonfeld, E.; Isharel, I.; Yavin, E. Docosahexaenoic acid-dependent iron accumulation in oligodendroglia cells protects from hydrogen peroxide-induced damage. *J. Neurochem.* **2008**, *105*, 1325–1335.

22. Shimazawa, M.; Nakajima, Y.; Mashima, Y.; Hara, H. Docosahexaenoic acid (DHA) has neuroprotective effects against oxidative stress in retinal ganglion cells. *Brain Res.* **2009**, *1251*, 269–275.

23. Bechoua, S.; Dubois, M.; Dominguez, Z.; Goncalves, A.; Némoz, G.; Lagarde, M.; Prigent, A.F. Protective effect of docosahexaenoic acid against hydrogen peroxide-induced oxidative stress in human lymphocytes. *Biochem. Pharmacol.* **1999**, *57*, 1021–1030.

24. He, C.; Qu, X.; Cui, L.; Wang, J.; Kang, J.X. Improved spatial learning performance of fat-1 mice is associated with enhanced neurogenesis and neuritogenesis by docosahexaenoic acid. *Proc. Natl. Acad. Sci. USA* **2009**, *106*, 11370–11375.

25. Su, H.X.; Zhang, W.M.; Guo, J.S.; Guo, A.C.; Yuan, Q.J.; Wu, W.T. Neural Progenitor Cells Enhance the Survival and Axonal Regeneration of Injured Motoneurons after Transplantation into the Avulsed Ventral Horn of Adult Rats. *J. Neurotrauma* **2009**, *26*, 67–80.

26. Zhao, Z.; Wen, H.; Fefelova, N.; Allen, C.; Guillaume, N.; Xiao, D.; Huang, C.; Zang, W.; Gwathmey, J.K.; Xie, L.H. Docosahexaenoic Acid reduces the incidence of early afterdepolarizations caused by oxidative stress in rabbit ventricular myocytes. *Front. Physiol.* **2012**, *3*, doi:10.3389/fphys.2012.00252.

27. Lee, J.M.; Johnson, J.A. An important role of Nrf2-ARE pathway in the cellular defense mechanism. *J. Biochem. Mol. Biol.* **2004**, *37*, 139–143.

28. Ryu, J.; Zhang, R.; Hong, B.H.; Yang, E.J.; Kang, K.A.; Choi, M.; Kim, K.C.; Noh, S.J.; Kim, H.S.; Lee, N.H.; *et al.* Phloroglucinol attenuates motor functional deficits in an animal model of Parkinson's disease by enhancing Nrf2 activity. *PLoS One* **2013**, *8*, e71178.

29. Kanninen, K.; Heikkinen, R.; Malm, T.; Rolova, T.; Kuhmonen, S.; Leinonen, H.; Ylä-Herttuala, S.; Tanila, H.; Levonen, A.L.; Koistinaho, M.; *et al.* Intrahippocampal injection of a lentiviral vector expressing Nrf2 improves spatial learning in a mouse model of Alzheimer's disease. *Proc. Natl. Acad. Sci. USA* **2009**, *106*, 16505–16510.

30. Nanou, A.; Higginbottom, A.; Valori, C.F.; Wyles, M.; Ning, K.; Shaw, P.; Azzouz, M. Viral delivery of antioxidant genes as a therapeutic strategy in experimental models of amyotrophic lateral sclerosis. *Mol. Ther.* **2013**, *21*, 1486–1496.

31. Yang, Y.C.; Lii, C.K.; Wei, Y.L.; Li, C.C.; Lu, C.Y.; Liu, K.L.; Chen, H.W. Docosahexaenoic acid inhibition of inflammation is partially via cross-talk between Nrf2/heme oxygenase 1 and IKK/NF-κB pathways. *J. Nutr. Biochem.* **2013**, *24*, 204–212.

32. Stulnig, G.; Frisch, M.T.; Crnkovic, S.; Stiegler, P.; Sereinigg, M.; Stacher, E.; Olschewski, H.; Olschewski, A.; Frank, S. Docosahexaenoic acid (DHA)-induced heme oxygenase-1 attenuates cytotoxic effects of DHA in vascular smooth muscle cells. *Atherosclerosis* **2013**, *230*, 406–413.

33. Stillwell, W.; Wassall, S.R. Docosahexaenoic acid: Membrane properties of a unique fatty acid. *Chem. Phys. Lipids* **2003**, *126*, 1–27.

34. Yamashima, T. A putative link of PUFA, GPR40 and adult-born hippocampal neurons for memory. *Prog. Neurobiol.* **2008**, *84*, 105–115.

35. Lafourcade, M.; Larrieu, T.; Mato, S.; Duffaud, A.; Sepers, M.; Matias, I.; De Smedt-Peyrusse, V.; Labrousse, V.F.; Bretillon, L.; Matute, C.; *et al.* Nutritional omega-3 deficiency abolishes endocannabinoid-mediated neuronal functions. *Nat. Neurosci.* **2011**, *14*, 345–350.

5

Antifouling Activity of Synthetic Alkylpyridinium Polymers using the Barnacle Model

Veronica Piazza [1], Ivanka Dragić [2], Kristina Sepčić [2], Marco Faimali [1], Francesca Garaventa [1], Tom Turk [2] and Sabina Berne [2,*]

[1] ISMAR—CNR Institute of Marine Science, U.O.S. Genova, Via De Marini 6, 16149 Genova, Italy;
E-Mails: veronica.piazza@ge.ismar.cnr.it (V.P.); marco.faimali@ismar.cnr.it (M.F.);
francesca.garaventa@ismar.cnr.it (F.G.)

[2] Department of Biology, Biotechnical Faculty, University of Ljubljana, Večna pot 111,
Ljubljana 1000, Slovenia; E-Mails: ivana84@gmail.com (I.D.); kristina.sepcic@bf.uni-lj.si (K.S.);
tom.turk@bf.uni-lj.si (T.T.)

* Author to whom correspondence should be addressed; E-Mail: sabina.berne@bf.uni-lj.si

Abstract: Polymeric alkylpyridinium salts (poly-APS) isolated from the Mediterranean marine sponge, *Haliclona (Rhizoniera) sarai*, effectively inhibit barnacle larva settlement and natural marine biofilm formation through a non-toxic and reversible mechanism. Potential use of poly-APS-like compounds as antifouling agents led to the chemical synthesis of monomeric and oligomeric 3-alkylpyridinium analogues. However, these are less efficient in settlement assays and have greater toxicity than the natural polymers. Recently, a new chemical synthesis method enabled the production of poly-APS analogues with antibacterial, antifungal and anti-acetylcholinesterase activities. The present study examines the antifouling properties and toxicity of six of these synthetic poly-APS using the barnacle (*Amphibalanus amphitrite*) as a model (cyprids and II stage nauplii larvae) in settlement, acute and sub-acute toxicity assays. Two compounds, APS8 and APS12-3, show antifouling effects very similar to natural poly-APS, with an anti-settlement effective concentration that inhibits 50% of the cyprid population settlement (EC_{50}) after 24 h of 0.32 mg/L and 0.89 mg/L, respectively. The toxicity of APS8 is negligible, while APS12-3 is three-fold more toxic (24-h LC_{50}: nauplii, 11.60 mg/L; cyprids, 61.13 mg/L) than natural poly-APS. This toxicity of APS12-3 towards nauplii is, however, 60-fold and 1200-fold lower than that of the common co-biocides, Zn- and Cu-pyrithione, respectively.

Additionally, exposure to APS12-3 for 24 and 48 h inhibits the naupliar swimming ability with respective IC_{50} of 4.83 and 1.86 mg/L.

Keywords: *Amphibalanus amphitrite*; antifouling; natural product antifoulants; alkylpyridinium polymers; *Haliclona (Rhizoniera) sarai*; barnacle; settlement assay; toxicity assay; swimming inhibition assay

---

1. Introduction

Marine biofouling is a dynamic natural process that occurs on ocean-submerged surfaces and leads to the undesired accumulation of organic polymers and of microbial, plant and animal communities and their by-products [1]. Although there is a wide diversity of fouling organisms and a variety of contributing environmental factors, a general sequence of fouling events is frequently observed [2]. The initial conditioning of the submerged surfaces by the adsorption of organic macromolecules is followed by the attachment of marine bacteria, which form a complex multi-species biofilm [3,4]. The complexity of "microfoulers" is further increased when fungi, diatoms and protozoa colonize this microbial slime layer [5]. Within hours to days, "soft macrofouling" is observed, as algal spores and various invertebrate larvae begin to attach and develop [6,7]. Shelled invertebrates, like barnacles, mussels and tubeworms, represent the "hard macrofouling" phase, which results in the formation of a mature fouling community [8].

The settlement and accumulation of marine organisms is a severe problem on engineered structures that are submerged in the sea, and it incurs substantial economic costs in the shipping [9,10], desalination [11] and offshore oil and gas [12] industries and in marine aquaculture [13]. Traditionally, the most effective strategy for controlling biofouling has been achieved using paints and antifouling coatings that contain toxic constituents (e.g., Cu, Zn) or biocides (e.g., tributyltin, bis(tributyltin) oxide) [14,15]. However, the accumulation of these compounds in harbors and ports led to massive pollution problems [14] and had detrimental effects on non-target marine organisms [16]. Consequently, the use of organotin-based antifouling coatings was prohibited by the Antifouling System Convention of the International Maritime Organization (effective from 17 September 2008).

Modern antifouling approaches are investigating behavioral, chemical and physical defense mechanisms that have evolved in living organisms, to translate these into novel antifouling applications [17]. The creation of self-polishing and foul-release coatings has been inspired by the skin of marine mammals and fish, which respond to environmental stimuli, like temperature and pH [1]. Bio-inspired physical antifouling strategies have exploited the physical properties for fouling prevention, such as the surface energy and microtopography [1,15,18–20]. Antifoulants based on natural products have been proposed as one of the best ecologically relevant antifouling solutions [21,22], due to their lower toxicity, reversible effects at lower effective concentrations and biodegradability, compared with conventional biocides [23]. Natural antifouling compounds are produced as secondary metabolites by a wide range of organisms, and they include terpenoids, steroids, carotenoids, phenolics, furanones, alkaloids, peptides and lactones [24–26].

To date, over 80 different bioactive 3-alkylpyridinium and 3-alkylpyridine compounds have been identified in marine sponges of the order Haplosclerida [27,28]. One of the most studied of these compounds are water-soluble polymeric 3-alkylpyridinium salts (poly-APS), secondary metabolites produced by the Mediterranean marine sponge, *Haliclona (Rhizoniera) sarai* [29]. Poly-APS (5.52 kDa; [30]) have been chemically defined as polymers that are composed of 29 monomeric *N*-butyl-3-butyl pyridinium units, with 3-octyl chains linked to the nitrogen of the adjacent unit in a head-to-tail organization (see Figure 1). In aqueous solutions, poly-APS behave as cationic detergents at concentrations >0.23 g/L and form large supramolecular structures of a 23 nm mean hydrodynamic radius [31]. At concentrations >1 g/L, poly-APS have toxic and lethal effects in rodents upon intravenous administration [23].

Figure 1. Chemical structures of natural polymeric 3-alkylpyridinium salts (poly-APS) and the synthetic poly-APS investigated. Molecular weights: natural poly-APS, 5.52 kDa; APS3, 1.46 kDa; APS7, 2.33 kDa; APS8, 11.9 kDa; APS12, 12.5 kDa; APS12-2, 14.7 kDa; APS12-3, 6.08 kDa.

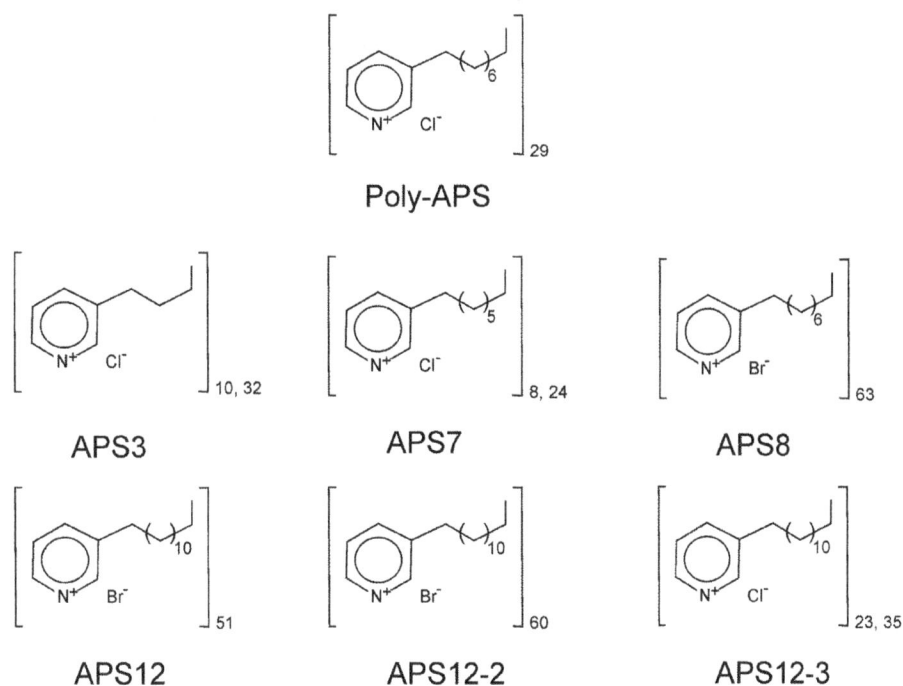

Among the numerous biological activities of poly-APS that were detailed by [27], their antifouling activity is of particular interest for the current study. In a static laboratory bioassay [2] using the most common hard macrofouler species, the barnacle, *Amphibalanus amphitrite*, poly-APS effectively deterred the settlement of cyprids (effective concentration that inhibits 50% of the cyprid population settlement (EC_{50}), 0.27 mg/L), with low toxicity towards nauplii (lethal concentration that kills 50% of the cyprid population (LC_{50}), 30.01 mg/L) and the non-target organisms, the alga, *Tetraselmis suecica*, and the mussel, *Mytilus galloprovincialis* [32]. Additionally, in a laboratory anti-microfouling activity assay and in the range of 0.1 mg/L to 1.0 mg/L, poly-APS prevented the formation of the natural marine biofilm, through inhibition of the growth of certain marine bacteria [33]. The molecular mechanisms behind poly-APS antifouling activity presumably involve neurotransmission blockade/modulation [34] through acetylcholinesterase inhibition, combined with their surfactant-like properties that decrease the surface tension [28].

Despite extensive studies of such natural antifoulants over the past 20 years, their incorporation into antifouling paints has been hampered by their limited supply [24]. Sufficient amounts of target poly-APS for broad biological screening and of poly-APS analogues for structure-activity relationship studies can be supplied through synthetic organic chemistry [35]. Using such an approach, dimers and tetramers of linear 3-alkylpyridinium salts have been produced that have high antibacterial and moderate anti-acetylcholinesterase activities [36]. In the barnacle anti-settlement assay, these oligomeric compounds did not reach the antifouling potential of natural poly-APS and were considerably more toxic [37]. In contrast, screening for novel antimicrobial agents has revealed that some of these APS analogues have considerable antibacterial activity towards biofilm-forming marine bacteria and has indicated that their charged pyridinium moiety and bromine atom, and the length of their alkyl chain, are decisive factors in this bioactivity [38]. Recently, new microwave-assisted polymerization has allowed the production of high-molecular-weight analogues of poly-APS [39,40]. Certain poly-APS analogues have been seen to behave similarly to natural poly-APS when their antifungal, antibacterial, anti-acetylcholinesterase, antitumoral and membrane-damaging activities have been assessed [39–41].

In the present study, we investigated the antifouling activity of six synthetic poly-APS analogues using *A. amphitrite* cyprids as the macrofouling model in a static laboratory settlement assay. In parallel, we examined the toxicity towards *A. amphitrite* cyprids after 24 h, 48 h and 72 h of exposure to these synthetic poly-APS and calculated their therapeutic ratios. We monitored the acute and sub-acute toxicity towards II stage *A. amphitrite* nauplii after 24 h and 48 h exposure to these poly-APS analogues. Finally, the behavior of II stage nauplii was studied for one of the most promising analogues, APS12-3, using swimming speed alteration assay as a measure of sub-lethal toxicity.

2. Results and Discussion

Marine biofouling is a complex and dynamic natural process that is very difficult to reproduce under laboratory conditions. As direct evaluation of antifouling coatings *in situ* is expensive and time-consuming, several bioassays have been developed to estimate the antifouling potential of novel natural products [2]. Due to the large diversity of organisms implicated in the marine biofouling process, it has been recommended that as many target species are used as possible, taken from both the microfouling and macrofouling communities [25]. Generally, bacteria, diatoms and fungi isolated from marine biofilms are studied in microfouling bioassays, while sessile hard-foulers (e.g., barnacles, tube worms, mussels) and soft-foulers (e.g., the bryozoan, *Bugula neritina*, the polychaete, *Hydroides elegans*) or seaweed (e.g., *Ulva*) are frequently used as representative macrofouling organisms. In our previous studies [32,33,42], we demonstrated that natural poly-APS can effectively inhibit the settlement and/or growth of different target fouling organisms. Using synthetic monomeric and oligomeric analogues of natural poly-APS, we also demonstrated their considerable antimicrobial activity [38]. However, in the anti-settlement assay against *A. amphitrite* cyprids, we identified only one synthetic poly-APS analogue (1,8-di(3-pyridyl)octane) that showed similar efficacy to natural poly-APS, although with a noticeably different toxic mechanism of action [37].

In the present study, we focused on the antifouling activities of synthetic poly-APS using *A. amphitrite* as the primary invertebrate model for biofouling [43]. The structures of the synthetic poly-APS evaluated in this study are illustrated in Figure 1.

2.1. Anti-Settlement Assay

We investigated the settlement of laboratory-reared cyprids of *A. amphitrite* after 24 h, 48 h and 72 h of exposure to these synthetic poly-APS in 24-well plates (see the Experimental section). The results of the 72 h anti-settlement assays are shown in Figure 2a.

APS8 was the most effective of the synthetic poly-APS for the inhibition of the settlement of cyprids ($p < 0.01$; Figure 2a). With an EC_{50} of 0.32 mg/L after 24 h (Table 1), APS8 matches the anti-settlement activity of natural poly-APS (EC_{50}, 0.27 mg/L). Moreover, when these are expressed as molar concentrations, APS8 is almost two-fold more effective than natural poly-APS (Table 1). The biphasic dose response seen for APS8 (Figure 2a) is characterized by a significant inhibition of settlement at 0.5 and 1 mg/L, followed by the loss of anti-settlement activity at concentrations from 5 to 10 mg/L and then by the return of inhibition at 50 and 100 mg/L, thus suggesting the phenomenon of hormesis [44]. The underlying mechanism of such adaptive responses might be mediated via specific receptor and/or cell-signaling pathways [45,46]. For natural poly-APS, one of the possible molecular mechanisms of this anti-settlement activity is interference with the cyprid cholinergic system [28]. A recent novel hypothesis proposed that the cellular quality control systems that are involved in the recognition, repair and prevention of cell stress represent the underlying molecular mechanisms that account for the benefits of hormesis [47].

We observed a hormetic-like response also for APS7, even though this is not supported by *a posteriori* comparison of the means and less clearly for APS12. However, for both analogues, 50% inhibition of the cyprid settlement occurred at concentrations that are an order of magnitude higher than APS8 (Table 1). For all of these tested synthetic poly-APS, the settlement inhibition decreased with the time of exposure. This effect is particularly evident for polymer APS3, which shows no significant effect on the cyprid settlement ($p = 0.45$; F = 1.08) and is thus less interesting for further antifouling research.

The synthetic polymers, APS12-2 and APS12-3, inhibit the settlement of cyprids in a concentration-dependent manner, with the respective EC_{50} of 8.78 and 0.89 mg/L at 24 h (Table 1). As reported by Zovko *et al.* [40], APS12-3 has the highest antibacterial and antifungal activity among all of these synthetic poly-APS. By comparing the data for the inhibition of cyprid settlement with those for the toxicity of APS12-3 (Figure 2b), for the longer incubation times of 48 h and 72 h, it can be concluded that the inhibitory effects probably occur due to the toxic action of APS12-3. Interestingly, however, the potent inhibition of cyprid settlement by APS12-3 is reversible at concentrations up to 1.6 mg/L (Figure 3). The other synthetic poly-APS tested here did not show any significant toxic effects against these barnacle cyprids at concentrations of up to 100 mg/L (Figure 2b).

Figure 2. Anti-settlement activity (**a**) and toxicity (**b**) of synthetic poly-APS on *A. amphitrite* cyprids after 72-h exposure to the different poly-APS concentrations (as indicated). Data are expressed as the means ± standard error. * $p < 0.05$; ** $p < 0.001$.

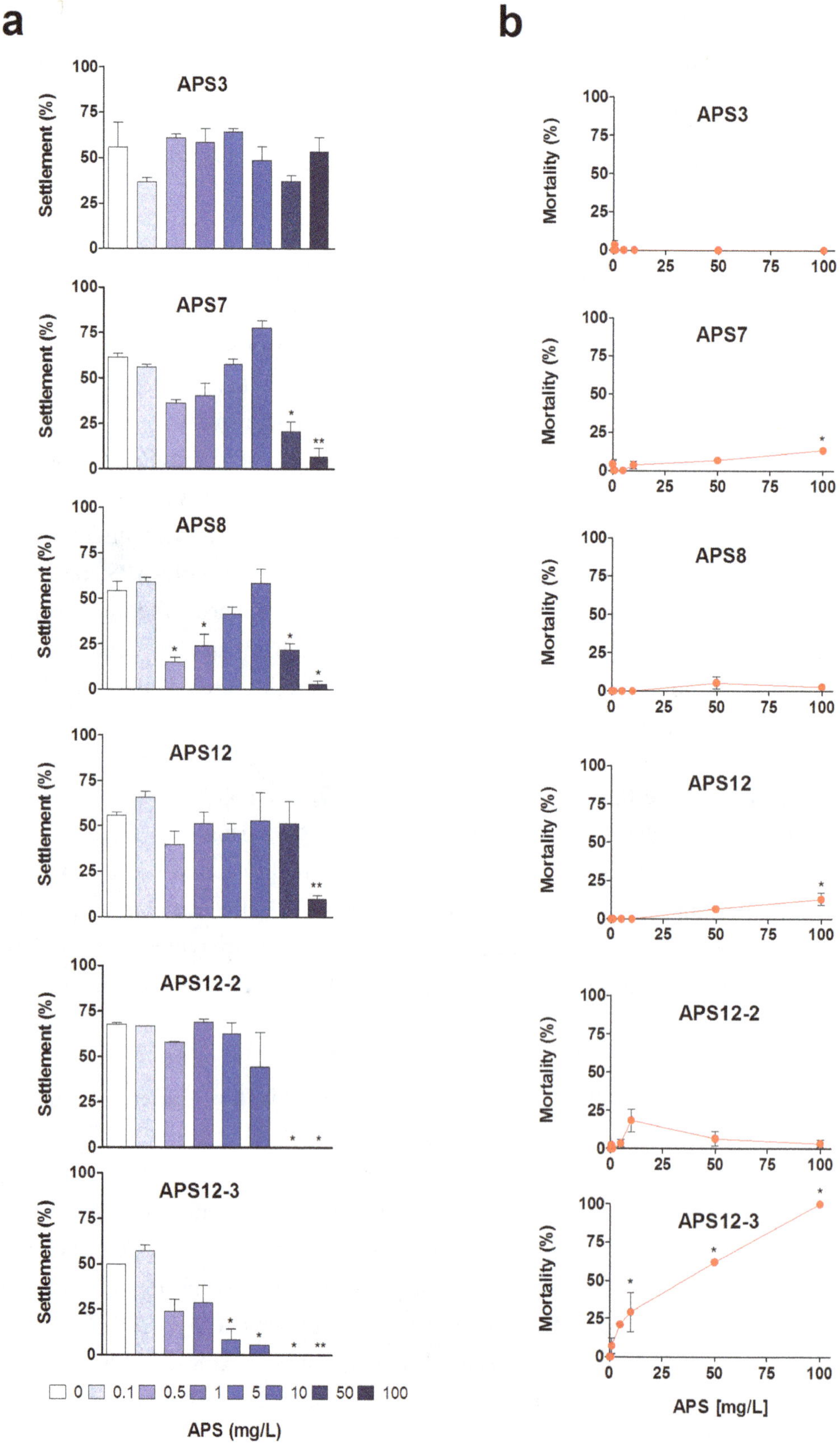

Table 1. Antifouling activities of natural poly-APS, Zn and Cu pyrithiones, and the synthetic poly-APS, assessed as the settlement (EC_{50}) and mortality (LC_{50}) of *A. amphitrite* cyprids.

Compound	Treatment (h)	EC_{50} (mg/L)	LC_{50} (mg/L)	EC_{50} (µM)	LC_{50} (µM)
Poly-APS [1]	24	0.27 (0.15–0.47)		0.049	
Zn pyrithione [1]	24	0.02		0.063	
Cu pyrithione [1]	24	<0.01		<0.032	
APS3	24	5.72 (4.24–7.72)	>100	3.9	
	48	>100	>100		
	72	>100	>100		
APS7	24	10.50 (8.47–13.01)	>100	4.5	
	48	25.86 (23.29–28.71)	>100	11.1	
	72	29.38 (26.17–32.99)	>100	12.6	
APS8 [2]	24	0.32 (0.26–0.39)	>100	0.026	
	48	0.50 (0.36–0.70)	>100	0.042	
	72	2.33 (1.78–3.04)	>100	0.195	
APS12	24	*nc*	>100		
	48	*nc*	>100		
	72	49.82 (37.18–66.76)	>100	4.0	
APS12-2	24	8.78 (8.37–9.20)	>100	0.597	
	48	9.38 (8.76–10.05)	>100	0.638	
	72	11.13 (10.38–11.94)	>100	0.757	
APS12-3 [2]	24	0.89 (0.48–1.65)	61.13 (51.65–72.36)	0.146	10.0
	48	4.03 (3.49–4.65)	24.24 (20.09–29.24)	0.661	4.0
	72	4.76 (4.44–5.11)	17.97 (14.88–21.70)	0.781	2.9

Data are expressed as EC_{50} or LC_{50} (95% confidence interval); EC_{50}, effective concentration that inhibits 50% of the cyprid population settlement; LC_{50}, lethal concentration that kills 50% of the cyprid population; *nc*, not calculable; [1] data measured after a 24 h-treatment with poly-APS or commercial co-biocide [32]; [2] EC_{50} calculated from an additional experiment (see the Experimental Section).

Figure 3. Recovery of *A. amphitrite* cyprid settlement in fresh filtered natural seawater after 72-h treatment with different concentrations of APS12-3 monitored for three consecutive days. Data are the means ± standard error.

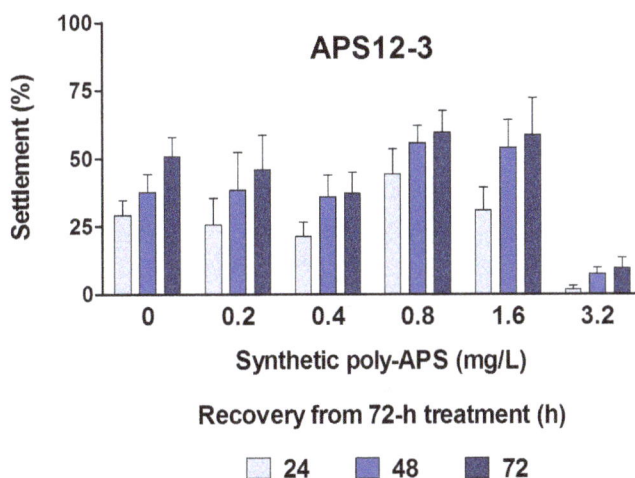

2.2. Naupliar Mortality, Immobility and Swimming Speed Alteration Assays

Amphibalanus amphitrite nauplii normally swim continuously, and their inability to move is evidence of toxicity [48]. On this basis, we measured the acute (mortality) and sub-acute (immobility) toxicity end-points of II stage nauplii exposed to the different solutions of the synthetic poly-APS for 24 h and 48 h. These immobility and naupliar mortality results are shown in Figure 4.

In ecotoxicological studies, the behavioral changes are more sensitive, short-term indicators of chemical toxicity than the assessment of lethal effects. Therefore, we investigated the sub-lethal toxicity of one of the synthetic analogues with the highest anti-settlement activity, APS12-3, using a swimming speed alteration test (Figure 5). Due to a lack of compound, the same evaluation has not been performed also for APS8. The synthetic poly-APS APS12-3 inhibited the mobility of nauplii in a concentration-dependent manner. The EC_{50} for immobility was 9.43 mg/L (after 24 h), and the ability of the nauplii to swim was completely lost at 50 mg/L APS12-3, and above. Thus, APS12-3 is moderately toxic to these nauplii, and indeed, the mortality (LC_{50} of 11.60 mg/L after 24 h) is three-fold greater compared to that for natural poly-APS (24-h LC_{50} of 30.01 mg/L; Table 2). The greater toxicity of APS12-3 towards this model organism, compared to natural poly-APS, is evidenced also by their 24-h swimming speed inhibition IC_{50} of 4.83 and >10 mg/L, respectively.

Increasing concentrations of APS12-2 also progressively inhibited the naupliar mobility and resulted in increased mortality. Prolonging the time of exposure to APS12-2 increased this immobility of nauplii (EC_{50}, 36.92 mg/L at 24 h; 2.28 mg/L at 48 h). The toxicity of APS12-2 towards these nauplii is comparable to that of APS12-3.

Table 2. Lethal (mortality) and sub-lethal (immobility) toxicity of natural poly-APS, Zn and Cu pyrithiones and the synthetic poly-APS, as assessed with *A. amphitrite* II stage nauplii.

Compound	Treatment (h)	EC_{50} (mg/L)	LC_{50} (mg/L)	EC_{50} (µM)	LC_{50} (µM)
Poly-APS [1]	24	>10	30.01 (21.71–41.49)	>1.81	5.43
Zn pyrithione [1]	24	0.23 (0.16–0.33)	0.19 (0.13–0.30)	0.725	0.6
Cu pyrithione [1]	24	0.03 (0.03–0.04)	<0.01	0.095	<0.032
APS3	24	>100	>100		
	48	>100	>100		
APS7	24	>100	>100		
	48	30.64 (25.43–36.9)	94.02 (81.73–108.17)	13.1	40.3
APS8	24	>100	>100		
	48	>100	79.37 (68.29–92.25)		6.7
APS12	24	>100	>100		
	48	>100	>100		
APS12-2	24	36.92 (29.33–46.46)	>100	2.5	
	48	2.28 (1.95–2.67)	4.80 (4.21–5.46)	0.15	0.32
APS12-3	24	9.43 (8.10–10.97)	11.60 (10.06–13.38)	1.5	1.9
	48	3.61 (3.14–4.16)	5.44 (4.64–6.37)	0.59	0.89

Data are expressed as EC_{50} or LC_{50} (95% confidence interval); EC_{50}, effective concentration that inhibits mobility of 50% naupliar population; LC_{50}, lethal concentration that kills 50% of the naupliar population; [1] data measured after a 24 h-treatment with poly-APS or commercial co-biocide [32].

Figure 4. Immobility (white bars) and mortality (blue bars) of the synthetic poly-APS for *A. amphitrite* II stage nauplii after 24 h and 48 h exposure. Data are the means ± standard error. * $p < 0.05$; ** $p < 0.001$.

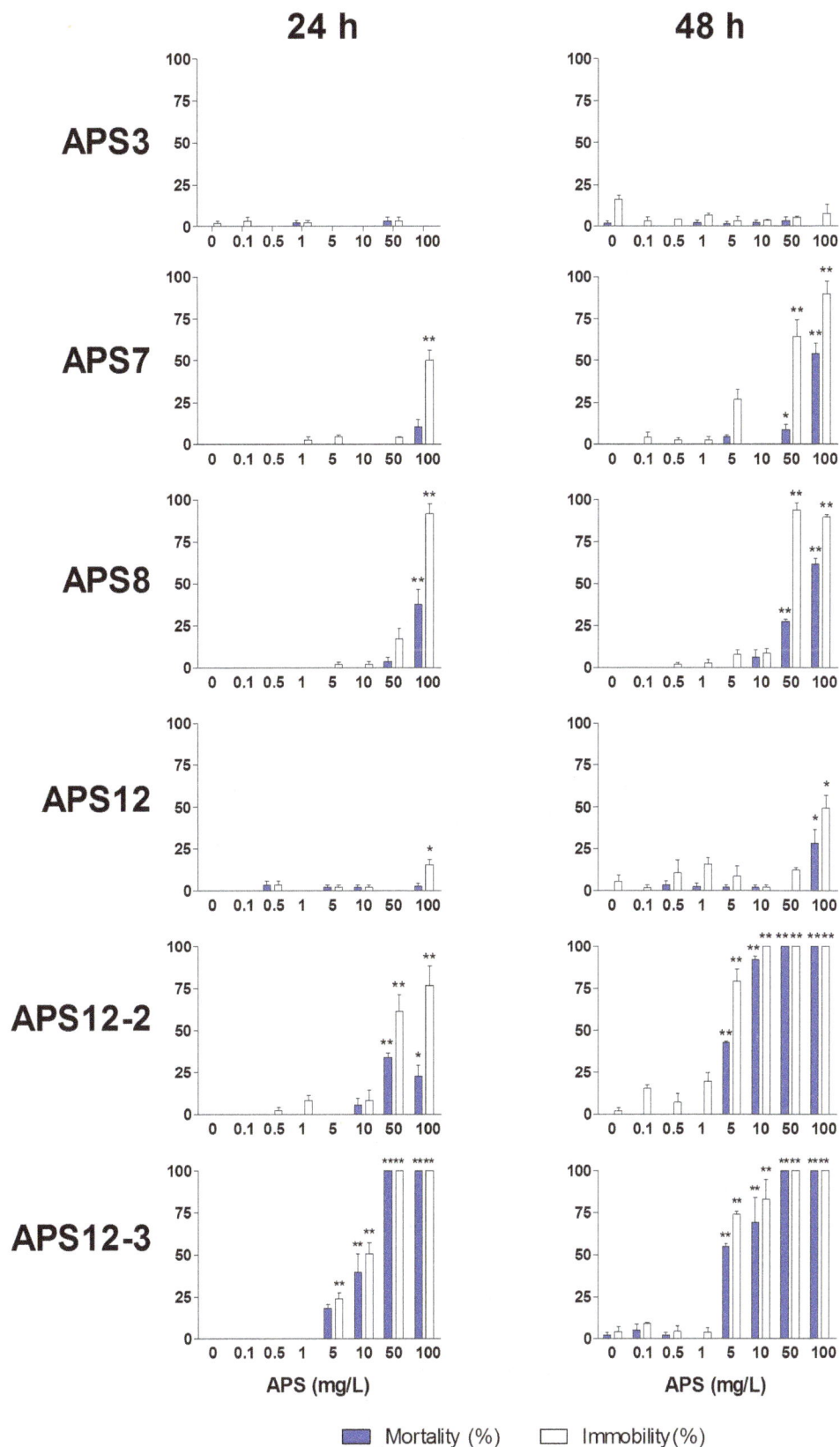

Figure 5. Swimming speed alteration (green bars), immobility (blue bars) and mortality (red bars) of the APS12-3 for *A. amphitrite* II stage nauplii after 24-h (**a**) and 48-h (**b**) incubations. Data are the means ± standard error. * $p < 0.05$; ** $p < 0.001$.

a

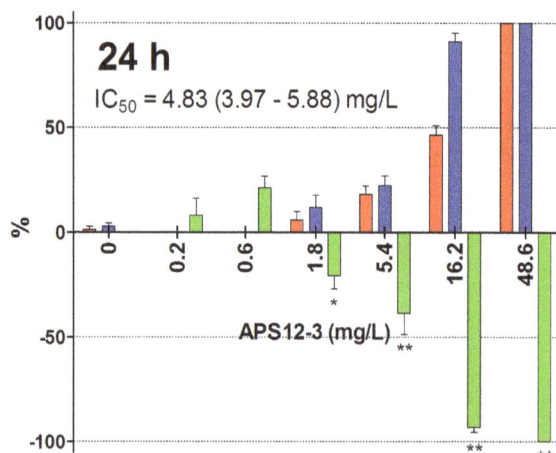

b

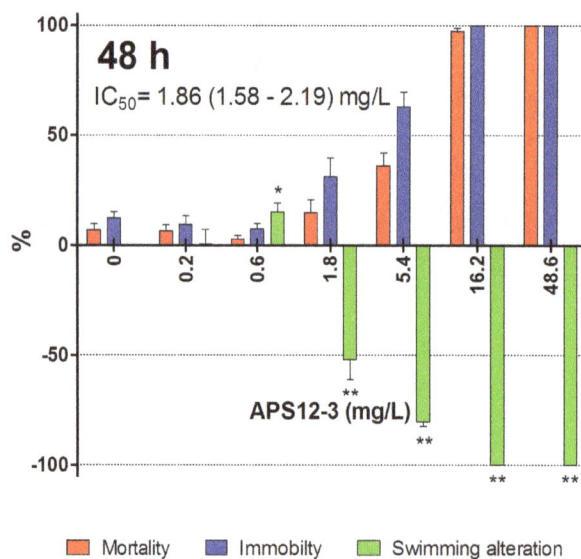

■ Mortality   ■ Immobilty   ■ Swimming alteration

At concentrations above 10 mg/L, APS7 and APS8 inhibited the naupliar mobility. This effect was more prominent after 48-h exposure at the higher doses of 50 mg/L and 100 mg/L APS7 and APS8. Both APS7 and APS8 showed a low toxicity towards these nauplii, with 48-h LC_{50} values of 94.02 mg/L and 79.37 mg/L, respectively.

The naupliar immobility was only minor upon exposure to APS3 and APS12.

A therapeutic ratio (TR; lethal concentration for 50% mortality (LC_{50}) divided by the effective concentration for 50% inhibition of settlement (EC_{50})) is commonly used to evaluate compounds' potential [2,26]. Generally, compounds with a TR > 50 and $EC_{50} < 5$ mg/L are considered to be promising non-toxic antifouling candidates [25]. We calculated TR values both for naupliar (TR_N) and cyprids (TR_C) toxicity (Table 3). TR_C denotes an antifouling mechanism (toxic or non-toxic), whereas TR_N indicates an environmental impact of a compound, since it is calculated by using LC_{50} towards non-target organisms (e.g., nauplii as representatives of the plankton) [49].

After a 24-h treatment of cyprids with synthetic analogue APS12-3, this analogue has a very promising antifouling activity (EC_{50} = 0.89 mg/L), and it seems to act through a non-toxic mechanism with a TR_C of 68.68. However, it is potentially hazardous for the environment (TR_N = 13.3). After 72 h, the antifouling activity of APS12-3 is less significant, and the compound displays a high toxicity towards target and non-target organisms (TR_C = 3.77 and TR_N = 1.35, respectively).

Another synthetic analogue, APS8, shows an anti-settlement activity (EC_{50}, 0.32 mg/L) similar to natural poly-APS (EC_{50}, 0.27 mg/L). Its TR_N is higher than that of the natural compound (158.74 and 111.15, respectively) suggesting an even lower environmental risk. Based on this, we propose its use as one of the pharmacophores in the prospective studies of non-toxic antifouling compounds.

Table 3. Therapeutic ratio values calculated using both data of naupliar (TR_N) and cyprids (TR_C) toxicity.

Compounds	Treatment (h)	EC_{50} (mg/L)	$LC_{50(N)}$ (mg/L)	$LC_{50(C)}$ (mg/L)	TR_C	TR_N
Poly-APS [1]	24	0.27	30.01	/	/	111.15
Zn pyrithione [1]	24	0.02	0.19	/	/	9.50
Cu pyrithione [1]	24	<0.01	<0.01	/	/	/
APS3	24	5.72	>100	>100	nc	nc
	48	nc	>100	>100	nc	nc
	72	nc	/	>100	nc	/
APS7	24	10.50	>100	>100	nc	nc
	48	25.86	94.02	>100	nc	3.64
	72	29.38	/	>100	nc	/
APS8	24	0.32	>100	>100	nc	
	48	0.50	79.37	>100	nc	158.74
	72	2.33	/	>100	nc	/
APS12	24	nc	>100	>100	nc	nc
	48	nc	>100	>100	nc	nc
	72	49.82	/	>100	nc	/
APS12-2	24	8.78	>100	>100	nc	11.39
	48	9.38	4.8	>100	nc	0.51
	72	11.13	/	>100	nc	/
APS12-3	24	0.89	11.6	61.13	68.68	13.3
	48	4.03	5.44	24.24	6.01	1.35
	72	4.76	/	17.97	3.77	/

[1] Data measured after a 24 h-treatment with poly-APS or commercial co-biocide [32]; nc, not calculable.

3. Experimental Section

3.1. Chemicals

Stock solutions (100 mg/L) of each synthetic poly-APS were prepared by dissolving them in filtered (0.22 μm) natural seawater. The information on the synthesis and NMR analyses of APS8,

APS12 and APS12-2 is available in [39] and of APS3, APS7 and APS12-3 in [40]. The chemical structures of the synthetic poly-APS are illustrated in Figure 1. The Zn pyrithione (Zinc Omadine®) and Cu pyrithione (Copper Omadine®) were from Arch Chemicals Incorporated, Atlanta, USA.

3.2. Rearing of Amphibalanus amphitrite Larvae

Adult barnacles were maintained in aerated, filtered (0.45 μm) natural seawater at 20 °C, on a 16-h:8-h light-dark cycle. They were fed every two days with *Artemia salina* (50 to 100 mL; 200 larvae/mL) and *Tetraselmis suecica* (100 to 200 mL; 2×10^6 cells/mL). The seawater was changed three times per week, and the barnacles were periodically rinsed with fresh water to remove epibionts or debris.

To obtain the nauplii for cyprid cultures, the adults were left to dry for 30 min to 40 min and then immersed in fresh seawater. The hatched nauplii were attracted to a light source and collected using a Pasteur pipette. They were reared to the cyprid stage as described in [50], by keeping them at 28 °C in natural filtered (0.22 μm) seawater and feeding them three times a week with *Tetraselmis suecica* (2×10^6 cells/mL). In these conditions nauplii reach the cyprids stage in 5–6 days. The cyprids were harvested by filtration and aged for 4 days prior to use, in filtered (0.45 μm) natural seawater at 4 °C in the dark [48].

3.3. Settlement Assay

The effects of poly-APS on the barnacle cyprids settlement were tested using these *A. amphitrite* cyprids. In the preliminary settlement assays, all poly-APS-like compounds were tested within a wide range of concentrations (from 0 to 100 mg/L). The settlement assays were conducted in 24-well microplates (Greiner Bio-One International AG, Austria) by adding 20 to 25 cyprids per well, with each well containing 2 mL of the relevant poly-APS solution (*i.e.*, 0.1, 0.5, 1, 5, 10, 50, 100 mg/L), or control seawater (*i.e.*, 0 mg/L poly-APS). For each compound, the experiment was performed in duplicate (two wells per concentration).

The test plates were sealed to prevent evaporation and incubated at 28 °C in the dark. The settlement was evaluated after 24 h, 48 h and 72 h of incubation. The larvae were examined under a dissecting microscope, to record the number of dead and permanently attached and metamorphosed individuals. The experiments were terminated by the addition of three droplets of 40% formaldehyde into each test well and the counting of the settled and non-settled larvae. The results were expressed as the percentages (±standard error) of the settlement of the total number of larvae incubated (20–25). The EC_{50} was determined as the concentration of poly-APS causing 50% inhibition of the cyprids' population settlement.

The settlement assay was repeated only for those synthetic poly-APS-like compounds that showed the best antifouling activities: APS8 and APS12-3. To better define the EC_{50}, the assays were slightly modified in terms of the range of concentrations of these poly-APS solutions (0, 0.3, 0.6, 1.2, 2.4, 4.8, 9.6, 19.2 mg/L for APS8; 0, 0.2, 0.4, 0.8, 1.6, 3.2, 6.4, 12.8 mg/L for APS12-3), and the experiments were performed with three replicates (three wells per concentration). The data for the inhibition of cyprid settlement are not illustrated, but EC_{50} values are reported in Table 1.

As APS12-3 caused significant mortality of these cyprids at the higher concentrations, another experiment was designed to study the recovery of the settlement ability of these cyprid larvae. Briefly, after the 72 h treatment with APS12-3, the unsettled cyprids were collected from the wells, rinsed with filtered seawater and transferred into new microplates with clean fresh filtered (0.45 μm) natural seawater. The percentages of settled cyprids were then determined after 24 h, 48 h and 72 h at 28 °C.

3.4. Toxicity Assay

Acute (mortality) and sub-acute (immobility) toxicities of the synthetic poly-APS were assessed using II stage nauplii or cyprids of *A. amphitrite*. All of these tests were performed in duplicate (two wells per concentration) in 24-well microplates (Greiner Bio-One International AG, Kremsmünster, Austria), each well containing 2 mL of the relevant poly-APS solution (*i.e.*, 0.1, 0.5, 1, 5, 10, 50 and 100 mg/L), or filtered seawater as control (*i.e.*, 0 mg/L poly-APS), with 20 to 25 larvae per well.

The naupliar mortality and immobility were evaluated after 24 h and 48 h of incubation at 20 °C in the dark, while the cyprids' mortality was evaluated after 24 h, 48 h and 72 h of incubation at 28 °C. After the exposure to the relevant poly-APS or control filtered seawater, the larvae were examined under a dissecting microscope, and the number of dead larvae was recorded. The data are presented as the percentage of mortality ± standard error, and the LC_{50} is expressed as the concentration of the poly-APS that induced death in 50% of the tested organisms. When assessing naupliar immobility, the number of immobile larvae is expressed as the sum of dead larvae (not swimming and moving appendages for 10 s of observation) and non-swimming larvae (not shifting their barycenter, but moving their appendages). The EC_{50} was calculated as the concentration of the toxicant that caused 50% immobility of the exposed organisms after 24 and 48 h.

3.5. Naupliar Swimming Speed Test

A swimming speed assay [51,52] was used to measure the behavioral effects of the synthetic analogue APS12-3 on the *A. amphitrite* larvae. Briefly, II stage nauplii (15–20 per well) were exposed to the APS12-3 test solutions in 24-well microplates for 48 h, at 20 °C in the dark, without aeration and feeding. All of these experiments were performed in duplicate (two wells per concentration).

A Swimming Behavioral Recorder System (e-magine IT, Genova, Italy) was used to track the paths of the swimming nauplii. Prior to the video recording under infrared light, the nauplii were adapted to the dark for 2 min, to gain their steady swimming speeds and to reach a uniform spatial distribution. The swimming behavior was monitored in the dark at 20 °C, for about three seconds, at 25 frames/s. The resulting digital images were analyzed using advanced image processing software to reconstruct the individual naupliar swimming paths and to measure the average swimming speed (mm/s) for each of the nauplii (15–20 nauplii). Finally, the data are expressed as percentages in terms of the swimming alteration, after normalization to the mean swimming speed (v) of the control (filtered natural seawater), as follows (Equation 1):

$$Swimming\ alteration\ (\%) = \frac{v_{treated} - v_{control}}{v_{control}} \times 100\% \tag{1}$$

The IC_{50} value was determined as the concentration of APS12-3 that caused changes in swimming behavior in 50% of the test nauplii.

3.6. Statistical Analyses

The EC_{50} for the cyprid settlement inhibition after 24, 48 and 72 h, the EC_{50} for the naupliar immobility after 24 and 48 h, the LC_{50} for the larval mortality after 24, 48 and 72 h and the IC_{50} for naupliar swimming speed alteration were calculated using a trimmed Spearman–Karber analysis [53]. One-way analysis of variance (ANOVA) was performed, with the level of significance set at $p < 0.05$ or $p < 0.001$, followed by Student–Newman–Keuls (SNK) tests to compare the treatment means [54].

4. Conclusions

The present study was designed to determine the antifouling potential of six synthetic poly-APS that differ in the lengths of their alkyl chains (3–12 carbon atoms), their degree of polymerization (eight to 63 monomeric subunits) and the nature of their counter ions (chloride or bromide). Although several studies have reported that these structural features influence the biological activities of amphiphilic compounds, particularly in terms of their membrane-damaging potential and antimicrobial activities [55–59], we were not able to observe such a relationship.

The antifouling activity of natural poly-APS is believed to derive from its surfactant-like properties, or is potentially due to the inhibition of the cholinergic system, which is involved in cyprid settlement [28]. The underlying antifouling mechanism of these synthetic poly-APS remains unsolved; however, the hormetic responses observed imply the involvement of receptor system(s). Recently, a strong interaction between the synthetic poly-APS APS8 and human α7 nicotinic acetylcholine receptors was reported [60]. Acetylcholine serves as a neurotransmitter/neuromodulator during the settlement of *A. amphitrite* larvae [50], and therefore, the antifouling activity of APS8 and possibly also of these other synthetic poly-APS might derive from the binding of these synthetic poly-APS to these receptors and the subsequent competition with acetylcholine.

Although numerous natural products with anti-settlement activities have been reported to date, only a few have shown potential for commercialization [24]. Among these, natural poly-APS has been suggested as a non-toxic natural antifoulant [22,26]. With the recent chemical synthesis of various poly-APS [39,40] now ensuring an adequate supply of these compounds, we focused our current research on the finding of synthetic poly-APS with comparable or improved antifouling activities to natural poly-APS. Among the polymers tested, APS8 prevented the settlement of *A. amphitrite* cyprids via a non-toxic mechanism and with similar potency to natural poly-APS. On the other hand, APS12-3 showed an anti-settlement efficacy similar to the natural poly-APS, but with higher toxicity.

Acknowledgments

Financial support from the Slovenian Research Agency through grant J1-4044 (Tom Turk), from the Italian National Research Council (CNR)—Institute of Marine Science and from the Erasmus Programme (Ivanka Dragić) is gratefully acknowledged. We sincerely thank Christopher Berrie for critical reading and editing of the manuscript.

Author Contributions

Veronica Piazza, Francesca Garaventa, Marco Faimali, Kristina Sepčić, Tom Turk and Sabina Berne contributed to conception and design of the study. Veronica Piazza and Ivanka Dragić performed the experiments. Veronica Piazza, Francesca Garaventa, Sabina Berne and Kristina Sepčić analyzed and interpreted the data. All authors participated in drafting the article and critically revising it.

References

1. Kirschner, C.M.; Brennan, A.B. Bio-Inspired Antifouling Strategies. *Annu. Rev. Mater. Res.* **2012**, *42*, 211–229.

2. Briand, J.-F. Marine antifouling laboratory bioassays: An overview of their diversity. *Biofouling* **2009**, *25*, 297–311.

3. Jain, A.; Bhosle, N.B. Biochemical composition of the marine conditioning film: Implications for bacterial adhesion. *Biofouling* **2009**, *25*, 13–19.

4. Dobretsov, S. Marine Biofilms. In *Biofouling*; Dürr, S., Thomason, J.C., Eds.; Wiley-Blackwell: Oxford, UK, 2009; pp. 123–136.

5. Railkin, A.I.; Ganf, T.A.; Manylov, O.G. Biofouling as a Process. In *Marine Biofouling: Colonization Processes and Defenses*; CRC Press: Boca Raton, FL, USA, 2003; pp. 25–39.

6. Joint, I.; Tait, K.; Callow, M.E.; Callow, J.A.; Milton, D.; Williams, P.; Cámara, M. Cell-to-cell communication across the prokaryote-eukaryote boundary. *Science* **2002**, *298*, 1207.

7. Hadfield, M.G.; Paul, V.J. Natural chemical cues for settlement and metamorphosis of marine-invertebrate larvae. In *Marine Chemical Ecology*; McClintock, J.B., Baker, B.J., Eds.; CRC Press: Boca Raton, FL, USA, 2001; pp. 431–461.

8. Hadfield, M.G. Biofilms and marine invertebrate larvae: What bacteria produce that larvae use to choose settlement sites. *Ann. Rev. Mar. Sci.* **2011**, *3*, 453–470.

9. Schultz, M.P. Effects of coating roughness and biofouling on ship resistance and powering. *Biofouling* **2007**, *23*, 331–341.

10. Schultz, M.P.; Bendick, J.A.; Holm, E.R.; Hertel, W.M. Economic impact of biofouling on a naval surface ship. *Biofouling* **2011**, *27*, 87–98.

11. Elimelech, M.; Phillip, W.A. The future of seawater desalination: Energy, technology, and the environment. *Science* **2011**, *333*, 712–717.

12. Page, H.M.; Dugan, J.E.; Piltz, F. Fouling and antifouling in oil and other offshore industries. In *Biofouling*; Dürr, S., Thomason, J.C., Eds.; Wiley-Blackwell: Oxford, UK, 2009; pp. 252–266.

13. Fitridge, I.; Dempster, T.; Guenther, J.; de Nys, R. The impact and control of biofouling in marine aquaculture: A review. *Biofouling* **2012**, *28*, 649–669.

14. Dafforn, K.A.; Lewis, J.A.; Johnston, E.L. Antifouling strategies: History and regulation, ecological impacts and mitigation. *Mar. Pollut. Bull.* **2011**, *62*, 453–465.

15. Rosenhahn, A.; Schilp, S.; Kreuzer, H.J.; Grunze, M. The role of "inert" surface chemistry in marine biofouling prevention. *Phys. Chem. Chem. Phys.* **2010**, *12*, 4275–4286.

16. Sonak, S.; Pangam, P.; Giriyan, A.; Hawaldar, K. Implications of the ban on organotins for protection of global coastal and marine ecology. *J. Environ. Manag.* **2009**, *90*, S96–S108.

17. Ralston, E.; Swain, G. Bioinspiration—The solution for biofouling control? *Bioinspir. Biomim.* **2009**, *4*, 015007.

18. Salta, M.; Wharton, J.A.; Stoodley, P.; Dennington, S.P.; Goodes, L.R.; Werwinski, S.; Mart, U.; Wood, R.J.K.; Stokes, K.R. Designing biomimetic antifouling surfaces. *Philos. Trans. A Math. Phys. Eng. Sci.* **2010**, *368*, 4729–4754.

19. Scardino, A.J.; de Nys, R. Mini review: Biomimetic models and bioinspired surfaces for fouling control. *Biofouling* **2011**, *27*, 73–86.

20. Banerjee, I.; Pangule, R.C.; Kane, R.S. Antifouling coatings: Recent developments in the design of surfaces that prevent fouling by proteins, bacteria, and marine organisms. *Adv. Mater.* **2011**, *23*, 690–718.

21. Rittschof, D. Natural product antifoulants: One perspective on the challenges related to coatings development. *Biofouling* **2000**, *15*, 119–127.

22. Fusetani, N. Biofouling and antifouling. *Nat. Prod. Rep.* **2004**, *21*, 94–104.

23. Turk, T.; Frangež, R.; Sepčić, K. Mechanisms of toxicity of 3-alkylpyridinium polymers from marine sponge *Reniera sarai*. *Mar. Drugs* **2007**, *5*, 157–167.

24. Raveendran, T.V.; Limna Mol, V.P. Natural product antifoulants. *Curr. Sci.* **2009**, *97*, 508–520.

25. Qian, P.-Y.; Xu, Y.; Fusetani, N. Natural products as antifouling compounds: Recent progress and future perspectives. *Biofouling* **2010**, *26*, 223–234.

26. Fusetani, N. Antifouling marine natural products. *Nat. Prod. Rep.* **2011**, *28*, 400–410.

27. Turk, T.; Sepčić, K.; Mancini, I.; Guella, G. 3-Akylpyridinium and 3-alkylpyridine compounds from marine sponges, their synthesis, biological activities and potential use. In *Studies in Natural Products Chemistry*; Elsevier: Amsterdam, The Netherlands, 2008; Volume 35, pp. 355–397.

28. Sepčić, K.; Turk, T. 3-Alkylpyridinium compounds as potential non-toxic antifouling agents. In *Antifouling Compounds SE-4*; Fusetani, N., Clare, A., Eds.; Springer: Berlin/Heidelberg, Germany, 2006; Volume 42, pp. 105–124.

29. Sepčić, K.; Guella, G.; Mancini, I.; Pietra, F.; Serra, M.D.; Menestrina, G.; Tubbs, K.; Maček, P.; Turk, T. Characterization of anticholinesterase-active 3-alkylpyridinium polymers from the marine sponge *Reniera sarai* in aqueous solutions. *J. Nat. Prod.* **1997**, *60*, 991–996.

30. Grandič, M.; Sepčić, K.; Turk, T.; Juntes, P.; Frangež, R. *In vivo* toxic and lethal cardiovascular effects of a synthetic polymeric 1,3-dodecylpyridinium salt in rodents. *Toxicol. Appl. Pharmacol.* **2011**, *255*, 86–93.

31. Malovrh, P.; Sepčić, K.; Turk, T.; Maček, P. Characterization of hemolytic activity of 3-alkylpyridinium polymers from the marine sponge *Reniera sarai*. *Comp. Biochem. Physiol. Part C Pharmacol. Toxicol. Endocrinol.* **1999**, *124*, 221–226.

32. Faimali, M.; Sepčić, K.; Turk, T.; Geraci, S. Non-toxic antifouling activity of polymeric 3-alkylpyridinium salts from the Mediterranean sponge *Reniera sarai* (Pulitzer-Finali). *Biofouling* **2003**, *19*, 47–56.

33. Garaventa, F.; Faimali, M.; Sepčić, K.; Geraci, S. Laboratory analysis of antimicrofouling activity of Poly-APS extracted from *Reniera sarai* (Porifera: Demospongiae). *Biol. Mar. Mediterr.* **2003**, *10*, 565–567.

34. Qian, P.-Y.; Chen, L.; Xu, Y. Mini-review: Molecular mechanisms of antifouling compounds. *Biofouling* **2013**, *29*, 381–400.

35. Mancini, I.; Defant, A.; Guella, G. Recent synthesis of marine natural products with antibacterial activities. *Anti-Infect. Agents Med. Chem.* **2007**, *6*, 17–48.

36. Mancini, I.; Sicurelli, A.; Guella, G.; Turk, T.; Maček, P.; Sepčić, K. Synthesis and bioactivity of linear oligomers related to polymeric alkylpyridinium metabolites from the Mediterranean sponge Reniera sarai. *Org. Biomol. Chem.* **2004**, *2*, 1368–1375.

37. Faimali, M.; Garaventa, F.; Mancini, I.; Sicurelli, A.; Guella, G.; Piazza, V.; Greco, G. Antisettlement activity of synthetic analogues of polymeric 3-alkylpyridinium salts isolated from the sponge Reniera sarai. *Biofouling* **2005**, *21*, 49–57.

38. Chelossi, E.; Mancini, I.; Sepčić, K.; Turk, T.; Faimali, M. Comparative antibacterial activity of polymeric 3-alkylpyridinium salts isolated from the Mediterranean sponge Reniera sarai and their synthetic analogues. *Biomol. Eng.* **2006**, *23*, 317–323.

39. Houssen, W.E.; Lu, Z.; Edrada-Ebel, R.; Chatzi, C.; Tucker, S.J.; Sepčić, K.; Turk, T.; Zovko, A.; Shen, S.; Mancini, I.; *et al.* Chemical synthesis and biological activities of 3-alkyl pyridinium polymeric analogues of marine toxins. *J. Chem. Biol.* **2010**, *3*, 113–125.

40. Zovko, A.; Vaukner Gabrič, M.; Sepčić, K.; Pohleven, F.; Jaklič, D.; Gunde-Cimerman, N.; Lu, Z.; Edrada-Ebel, R.; Houssen, W.E.; Mancini, I.; *et al.* Antifungal and antibacterial activity of 3-alkylpyridinium polymeric analogs of marine toxins. *Int. Biodeterior. Biodegrad.* **2012**, *68*, 71–77.

41. Grandič, M.; Aráoz, R.; Molgó, J.; Turk, T.; Sepčić, K.; Benoit, E.; Frangež, R. Toxicity of the synthetic polymeric 3-alkylpyridinium salt (APS3) is due to specific block of nicotinic acetylcholine receptors. *Toxicology* **2013**, *303*, 25–33.

42. Eleršek, T.; Kosi, G.; Turk, T.; Pohleven, F.; Sepčić, K. Influence of polymeric 3-alkylpyridinium salts from the marine sponge Reniera sarai on the growth of algae and wood decay fungi. *Biofouling* **2008**, *24*, 137–143.

43. Holm, E.R. Barnacles and biofouling. *Integr. Comp. Biol.* **2012**, *52*, 348–355.

44. Calabrese, E.J. Biphasic dose responses in biology, toxicology and medicine: Accounting for their generalizability and quantitative features. *Environ. Pollut.* **2013**, *182*, 452–460.

45. Calabrese, E.J.; Mattson, M.P. Hormesis provides a generalized quantitative estimate of biological plasticity. *J. Cell Commun. Signal.* **2011**, *5*, 25–38.

46. Calabrese, E.J. Hormetic mechanisms. *Crit. Rev. Toxicol.* **2013**, *43*, 580–606.

47. Wiegant, F.A.C.; de Poot, S.A.H.; Boers-Trilles, V.E.; Schreij, A.M.A. Hormesis and cellular quality control: A possible explanation for the molecular mechanisms that underlie the benefits of mild stress. *Dose-Response* **2012**, *11*, 413–430.

48. Rittschof, D.; Clare, A.S.; Gerhart, D.J.; Mary, S.A.; Bonaventura, J. Barnacle *in vitro* assays for biologically active substances: Toxicity and settlement inhibition assays using mass cultured *Balanus amphitrite amphitrite* Darwin. *Biofouling* **1992**, *6*, 115–122.

49. Piazza, V.; Roussis, V.; Garaventa, F.; Greco, G.; Smyrniotopoulos, V.; Vagias, C.; Faimali, M. Terpenes from the red alga *Sphaerococcus coronopifolius* inhibit the settlement of barnacles. *Mar. Biotechnol.* **2011**, *13*, 764–772.

50. Faimali, M.; Falugi, C.; Gallus, L.; Piazza, V.; Tagliafierro, G. Involvement of acetyl choline in settlement of *Balanus amphitrite*. *Biofouling* **2003**, *S19*, 213–220.

51. Faimali, M.; Garaventa, F.; Piazza, V.; Greco, G.; Corrà, C.; Magillo, F.; Pittore, M.; Giacco, E.; Gallus, L.; Falugi, C.; *et al.* Swimming speed alteration of larvae of *Balanus amphitrite* as a behavioural end-point for laboratory toxicological bioassays. *Mar. Biol.* **2006**, *149*, 87–96.

52. Garaventa, F.; Gambardella, C.; di Fino, A.; Pittore, M.; Faimali, M. Swimming speed alteration of *Artemia* sp. and *Brachionus plicatilis* as a sub-lethal behavioural end-point for ecotoxicological surveys. *Ecotoxicology* **2010**, *19*, 512–519.

53. Hamilton, M.A.; Russo, R.C.; Thurston, R.V. Trimmed Spearman-Karber method for estimating median lethal concentrations in toxicity bioassays. *Environ. Sci. Technol.* **1977**, *11*, 714–719.

54. De Muth, J.E. *Basic Statistics and Pharmaceutical Statistical Applications*, 2nd ed.; Chapman and Hall/CRC: Boca Raton, FL, USA, 2006; p. 744.

55. Kleszczyńska, H.; Bielecki, K.; Sarapuk, J.; Bonarska-Kujawa, D.; Pruchnik, H.; Trela, Z.; Łuczyński, J. Biological activity of new *N*-oxides of tertiary amines. *Z. Naturforsch. C J. Biosci.* **2009**, *61*, 715–720.

56. Zarif, L.; Riess, J.G.; Pucci, B.; Pavia, A.A. Biocompatibility of alkyl and perfluoroalkyl telomeric surfactants derived from THAM. *Biomater. Artif. Cells Immobil. Biotechnol.* **1993**, *21*, 597–608.

57. Kuroda, K.; DeGrado, W.F. Amphiphilic polymethacrylate derivatives as antimicrobial agents. *J. Am. Chem. Soc.* **2005**, *127*, 4128–4129.

58. Sarapuk, J.; Kleszczyńska, H.; Pernak, J.; Kalewska, J.; Rózycka-Roszak, B. Influence of counterions on the interaction of pyridinium salts with model membranes. *Z. Naturforsch. C* **1999**, *54*, 952–955.

59. Kleszczynska, H.; Sarapuk, J.; Rozycka-Roszak, B. The role of counterions in the interaction of some cationic surfactants with model membranes. *Pol. J. Environ. Stud.* **1998**, *7*, 327–329.

60. Zovko, A.; Viktorsson, K.; Lewensohn, R.; Kološa, K.; Filipič, M.; Xing, H.; Kem, W.R.; Paleari, L.; Turk, T. APS8, a polymeric alkylpyridinium salt blocks α7 nAChR and induces apoptosis in non-small cell lung carcinoma. *Mar. Drugs* **2013**, *11*, 2574–2594.

6

Type II Collagen and Gelatin from Silvertip Shark (*Carcharhinus albimarginatus*) Cartilage: Isolation, Purification, Physicochemical and Antioxidant Properties

Elango Jeevithan, Bin Bao *, Yongshi Bu, Yu Zhou, Qingbo Zhao and Wenhui Wu *

Department of Marine Pharmacology, College of Food Science and Technology, Shanghai Ocean University, Shanghai 201306, China; E-Mails: srijeevithan@gmail.com (E.J.); bruce.bu@unilever.com (Y.B.); yzhou0724@163.com (Y.Z.); bobozhao123@163.com (Q.Z.)

* Authors to whom correspondence should be addressed; E-Mails: bbao@shou.edu.cn (B.B.); whwu@shou.edu.cn (W.W.)

Abstract: Type II acid soluble collagen (CIIA), pepsin soluble collagen (CIIP) and type II gelatin (GII) were isolated from silvertip shark (*Carcharhinus albimarginatus*) cartilage and examined for their physicochemical and antioxidant properties. GII had a higher hydroxyproline content (173 mg/g) than the collagens and cartilage. CIIA, CIIP and GII were composed of two identical α_1 and β chains and were characterized as type II. Amino acid analysis of CIIA, CIIP and GII indicated imino acid contents of 150, 156 and 153 amino acid residues per 1000 residues, respectively. Differing Fourier transform infrared (FTIR) spectra of CIIA, CIIP and GII were observed, which suggested that the isolation process affected the secondary structure and molecular order of collagen, particularly the triple-helical structure. The denaturation temperature of GII (32.5 °C) was higher than that of CIIA and CIIP. The antioxidant activity against 1,1-diphenyl-2-picrylhydrazyl radicals and the reducing power of CIIP was greater than that of CIIA and GII. SEM microstructure of the collagens depicted a porous, fibrillary and multi-layered structure. Accordingly, the physicochemical and antioxidant properties of type II collagens (CIIA, CIIP) and GII isolated from shark cartilage were found to be suitable for biomedical applications.

Keywords: shark cartilage; type II collagens; denaturation temperature; FTIR; SEM; antioxidant activity

1. Introduction

Type II collagen is a principal component of the extracellular matrix of articular cartilage and constitutes 90%–95% of the total protein content in the cartilage [1]. In the current commercial market, most of the type II collagen is isolated from terrestrial mammalian cartilage due to its high biocompatibility, and it is widely used in the pharmaceutical, food, healthcare, and cosmetic industries [2]. However, the incidences of diseases such as bovine spongiform encephalopathy (BSE) and foot and mouth disease (FMD) have raised concerns about its safety. Alternatively, collagen from fish processing waste may be a suitable substitute. Production of fish collagen and gelatin not only adds significant value to the fish processing sector but also to other pharmacological industries.

Gelatin, a denatured form of collagen, has been extensively studied [3,4]; however, studies related to fish collagen are limited to isolation and characterization [2,5,6]. The functional properties of collagen and gelatin are highly influenced by their molecular structure, amino acid composition and internal linkages, which are affected by the processing conditions.

Recently, several researchers have focused on the oral tolerance of type II collagen to treat autoimmune diseases [1,7]. Oral tolerance is an immunological mechanism by which external agents that enter the body through the digestive system are recognized and ignored by the immune system. The effects of oral administration of type II collagen obtained from bovine, chicken and sheep sources have been evaluated for the treatment of rheumatoid arthritis (RA) [8–10].

The antioxidant activity of collagen is an essential property for the oral tolerance mechanism in autoimmune diseases [11]. Recently, an interest in natural antioxidants has increased because they are widely distributed and safer than synthetic antioxidants. Few studies have been conducted on the antioxidant activity of collagen and gelatin isolated from marine animals, and include sea cucumber skin [12,13], jellyfish skin [14]; squid skin [15] and tuna skin and bone [16]. Recently, Merly and Smith [17] studied the immunomodulatory properties of type II collagen from commercially available shark cartilage capsules. To our knowledge, the physicochemical and antioxidant activities of silvertip shark (*Carcharhinus albimarginatus*) cartilage type II collagen and gelatin have not been examined. Therefore, the present study investigated the physicochemical and antioxidant properties of acid- and pepsin-soluble type II collagens (ASC, PSC) and type II gelatin from silvertip shark cartilage and their potential for further applications.

2. Results and Discussion

2.1. Biochemical Composition of Shark Cartilage

The cartilage ash content was high (25.41%) compared with the protein content (8.95%) (data not shown). This high level of ash was also reported in the cartilage of Amur sturgeon and Nile perch (26.7% and 39.1%, respectively) [18,19]. The generally high ash content of shark cartilage is attributed to its high mineral content [3]. A protein content of 19.2% was reported for another shark species (*Isurus oxyrinchus*) cartilage [20] and was higher than our finding. The moisture and fat content of the silvertip shark cartilage was 63.56% and 1.69%, respectively. Cho *et al.* [20] reported that the fat content of *I. oxyrinchus* shark cartilage was 1.4%, which was lower than in the present report. The moisture, protein, ash and fat contents of cartilage of several species of shark are between 66.84%–78.3%,

14.01%–19.20%, 1.10%–12.09% and 0.21%–1.40%, respectively [20,21]. Fish cartilage generally contains a lower protein content than in skin due to higher ash and fat contents [3].

The hydroxyproline content of shark cartilage was 30.28 mg/g (data not shown), which was lower than that found in CIIA, CIIP and GII. A lower hydroxyproline content of 11.37–13.44 mg/g for shark cartilage was reported by Kittiphattanabawon *et al.* [21]. Variations in the hydroxyproline content among fish species occur because of the differences in their body temperature and seasonal variations. The hydroxyproline content of CIIA, CIIP and GII were 92.7, 113 and 173 mg/g, respectively. Kittiphattanabawon *et al.* [21] reported that hydroxyproline content of PSC and ASC from brownbanded bamboo shark (*Chiloscyllium punctatum*) cartilage was between 103.71 and 104.49 mg/g, which is similar to our findings. Accordingly, silvertip shark cartilage may be a better resource for the isolation of collagen.

2.2. Protein Pattern

The degree of purification by gel filtration chromatography and the protein pattern of CIIA, CIIP and GII were evaluated by SDS-PAGE. The purified CIIA and CIIP showed two distinct bands, α_1 and β, with molecular weights of approximately 120 and 210 kDa, respectively (Figure 1a).

Figure 1. (**a**) Electrophoretic pattern of type II collagens and type II gelatin isolated form shark cartilage; (**b**) Purification of type II collagen and type II gelatin by gel filtration chromatography. The horizontal lines in the chromatograms represent pooled fractions that were analyzed by SDS-PAGE. GII: type II gelatin, CIIA: type II acid soluble collagen, CIIP: type II pepsin soluble collagen.

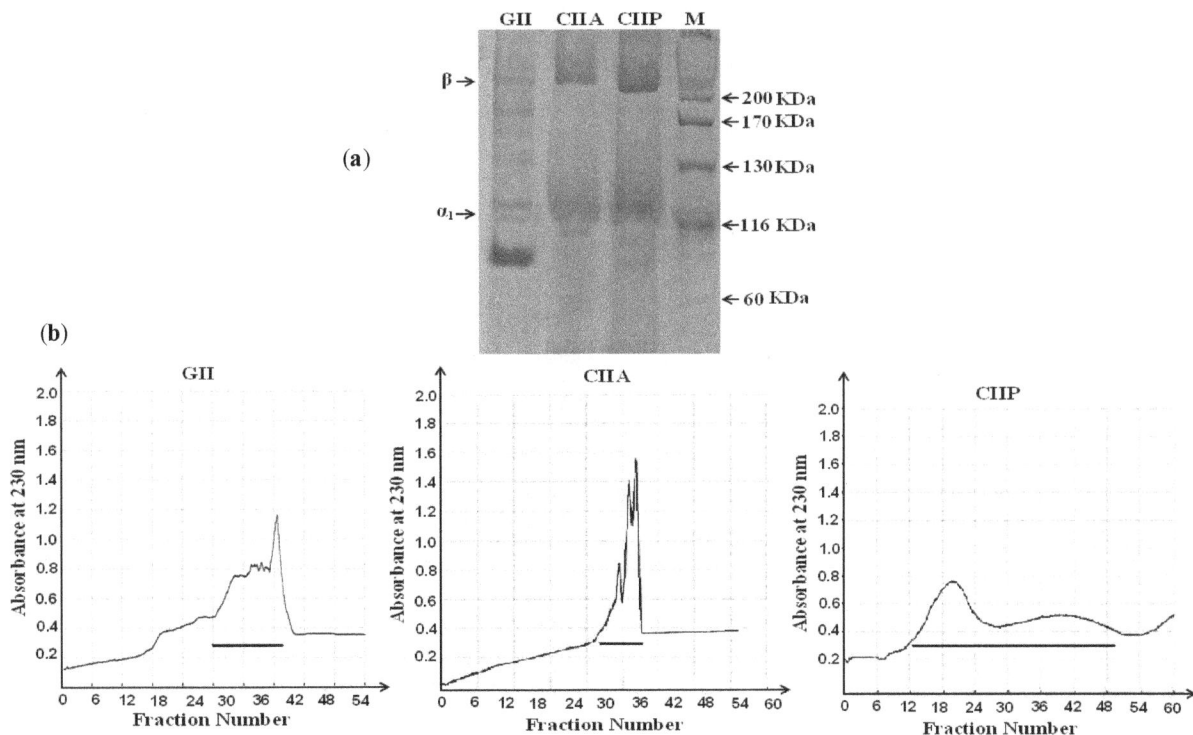

The band intensity of the α_1-chain was not greater than the β chain in CIIA and CIIP. This may be due to the formation of a β dimer resulting from the inter- and intramolecular cross-linking of the α component in the collagen structure. A similar protein pattern for type II collagen isolated from the cartilage of sharks and chicks has been reported in several studies [21–23]. When compared with CIIA and CIIP, the mobility of the GII α_1-chain was lower (approximately 100 kDa) and the band intensity was higher, which may indicate a conversion of the β dimer into the α component in the heat extraction process. In addition, smaller MW fractions were observed in GII in between the α component and β dimer, which was attributed to the denaturation of collagen during the isolation process. Therefore, type II collagen isolated from shark cartilage is pure and is composed of two chains, such as α_1 and β, and the secondary structure of collagen was altered by heat extraction.

2.3. Peptide Mapping

Shark cartilage type II collagens (CIIA and CIIP) digested by the V8 protease from *Staphylococcus aureus* V8 (EC 3.4.21.19, Sigma–Aldrich, Shanghai, China) exhibited a similar pattern. After hydrolysis, the α chain and β component of the collagens were degraded into small molecular weight peptides ranging from 180 to 40 kDa (Figure 2). This was in accordance with the peptide pattern for brownbanded bamboo shark cartilage collagen reported by Kittiphattanabawon *et al.* [21]. Peptide mapping revealed that CIIA and CIIP had similar secondary structures. The V8 protease exhibits a high degree of specificity for glutamic acid and aspartic acid residues in proteins [24]. The contents of glutamic acid and aspartic acid residues were higher in CIIP (76.20 and 45.76 residues per 1000 residues, respectively) than CIIA (71.57 and 42.91 residues per 1000 residues, respectively). Therefore, CIIP was more susceptible to hydrolysis by the V8 protease than CIIA, although, a similar pattern of peptide fragments were observed. This result was in accordance with the peptide mapping of ASC and PSC extracted from the skin of the brownstripe red snapper [25].

Figure 2. Peptide map of collagen and gelatin isolated form shark cartilage. CIIP: type II pepsin soluble collagen, CIIA: type II acid soluble collagen, GII: type II gelatin, M: Marker.

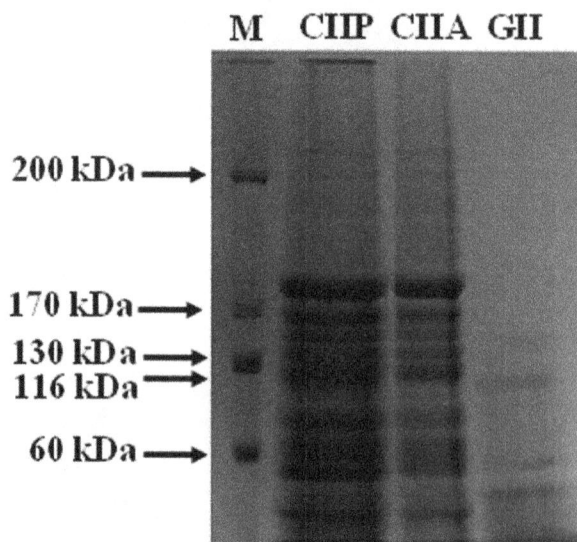

GII exhibited a different peptide pattern than CIIA and CIIP. The α_1-chain, as well as high MW cross-linked β components, of GII were completely degraded by the V8 protease and thus more small peptides were observed. This suggests that the α_1-chain and β components of GII were more susceptible to the V8 protease than CIIA and CIIP. Although GII had lower glutamic acid and aspartic acid contents than the collagens, the degree of hydrolysis was higher in GII.

This may be due to differences in the accessibility of susceptible bonds to the proteinase and to differences in the secondary structure between the collagens and gelatin, which were observed in the FTIR spectra. Hence, the peptide pattern of type II collagens (PSC, ASC) and type II gelatin from shark cartilage may be influenced by internal cross-linking, secondary structure and the composition of amino acids.

2.4. Maximum Absorption

The maximum absorptions of CIIA, CIIP and GII were 238.6, 237.7 and 237.6 nm, respectively (data not shown). The maximum absorption reported for eel skin collagen was similar to our findings [26]. In the present study, shark type II collagens and type II gelatin did not exhibit an absorption peak at 280 nm. This was due to a low tyrosine content (2–8 residues per 1000 residues), which absorbs UV light at 280 nm [27]. Kittiphattanabawon *et al.* [21] stated that no absorption peak of collagen at 280 nm indicates the purity of the collagen and efficacy of non-collagenous protein removal. This indicates adequate efficiency of the collagen and gelatin isolation processes used in this study.

2.5. Amino Acid Composition

CIIA, CIIP and GII extracted from shark cartilage had similar amino acid profiles (Table 1). In general, CIIA, CIIP and GII had high glycine contents (326, 319 and 353 residues per 1000 residues, respectively), which was the primary amino acid followed by alanine and proline. This result was in accordance with other fish collagen studies [18,20,25]. Every third amino acid residue was glycine in the α chain of collagen, except in the first 14 amino acid residues of the *N*-terminus and the first 10 amino acid residues of the *C*-terminus [28]. Higher contents of glycine, alanine and hydroxyproline were observed in GII compared with CIIA and CIIP. Type II collagens and type II gelatin had low contents of Met, His, Cys and Tyr, which is similar to other fish collagens [2,29,30]. The difference in the amino acid composition between type II collagens and type II gelatin was due to the different isolation processes.

Compared with type II collagens and type II gelatin, shark cartilage contained a lower content of glycine, alanine and proline (310, 122 and 93 residues per 1000 residues, respectively), but a higher amount of glutamic acid, aspartic acid, threonine, tyrosine and cysteine. The removal of non-collagenous protein during the isolation process likely resulted in the changes in the amino acid composition between the raw material and collagens. Imino acid contents of shark cartilage, type II collagens and type II gelatin were from 15.6% to 14.3%, which was relatively lower than that of ASC (21%) and PSC (22%) isolated from brownstripe red snapper skin [25]. CIIA had a lower imino acid content than CIIP and GII. The higher imino acid content of PSC compared with ASC was due to the removal of telopeptides by pepsin digestion [25].

Table 1. Amino acid compositions of ASC, PSC and gelatin from shark cartilage (residues/1000 residues). CIIP: type II pepsin soluble collagen, CIIA: type II acid soluble collagen, GII: type II gelatin.

Amino Acids	Shark Cartilage	CIIA	CIIP	GII
Hyp	49.96	47.50	49.25	51.28
Asp	51.84	42.91	45.76	39.84
Thr	27.26	23.54	25.57	21.85
Ser	35.62	36.37	38.27	32.62
Glu	86.45	71.57	76.20	71.70
Pro	93.61	103.30	106.78	102.31
Gly	310.47	326.90	319.69	353.12
Ala	122.87	133.92	132.63	140.45
Cys	5.88	3.93	4.26	3.74
Val	25.63	25.14	25.43	22.65
Met	13.69	10.68	13.54	12.34
Ile	18.55	21.19	21.81	16.94
Leu	40.16	30.01	29.95	25.16
Tyr	15.13	8.75	7.19	2.73
Phe	19.51	18.95	14.97	14.66
Lys	22.34	30.93	29.37	27.35
His	9.98	9.33	9.38	8.15
Arg	50.96	55.05	49.87	53.14
Total	1000	1000	1000	1000
Imino acid	143.58	150.81	156.03	153.59

2.6. Relative Viscosity

Relative viscosity is a physicochemical property of collagen. The viscosity of collagen and gelatin decreased continuously with increasing temperature, and this trend was similar for both collagens and gelatin. A sharp decline in the relative viscosity of collagen and gelatin was found above 30°C (Figure 3). The viscosity observed in this study was similar to that reported for other fish collagens [2,26,31]. The increasing temperature breaks hydrogen bonds between adjacent polypeptide chains of the collagen molecules and transforms intact trimers into individual chains or dimers, and, ultimately, causes the denaturation of the collagen structure. As a result, the helical structure of collagen converts into a random coil with a reduced viscosity. Hence, the viscosity is primarily influenced by the secondary structure of type II collagens and type II gelatin.

Viscosity of GII was lower than type II collagens, and this is due to the lower molecular weight and proportion of β chains, as well as to the loss of the triple helical structure of gelatin during the denaturation process [2].

Figure 3. Relative viscosity of type II collagens and type II gelatin. The viscosity obtained at 5 °C was considered as 100%. CIIP: type II pepsin soluble collagen, CIIA: type II acid soluble collagen, GII: type II gelatin.

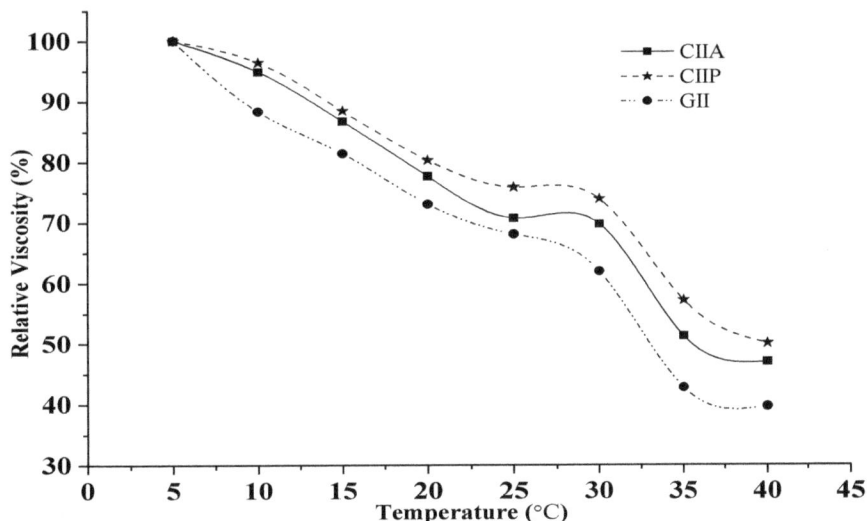

2.7. Effect of pH and NaCl Concentration on Solubility

The solubility of CIIA, CIIP and GII increased with increasing pH up to 5 and 6, and above this pH, the solubility decreased (Figure 4A).

Figure 4. Effect of pH (**A**) and salt concentration (**B**) on the solubility of collagens and gelatin. The highest solubility of collagens and gelatin was considered as 100% solubility. CIIP: type II pepsin soluble collagen, CIIA: type II acid soluble collagen, GII: type II gelatin.

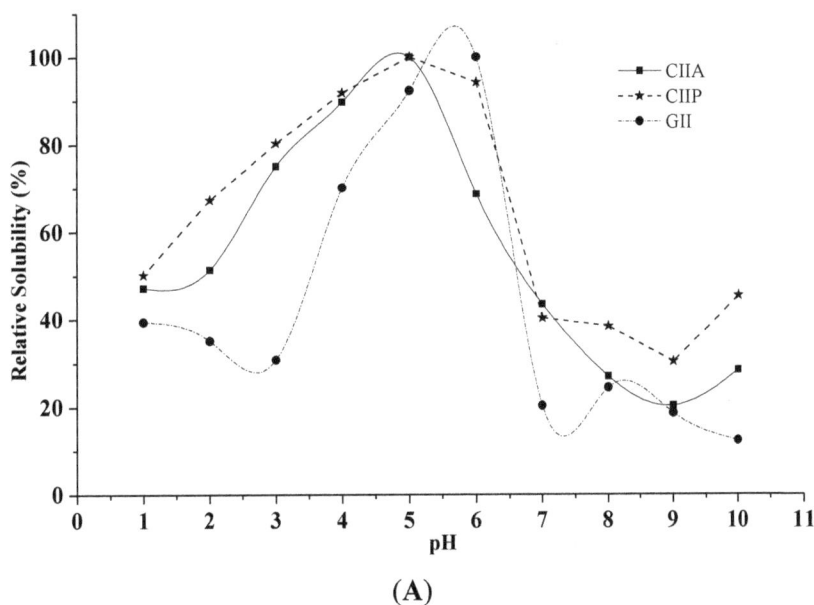

(A)

Figure 4. *Cont.*

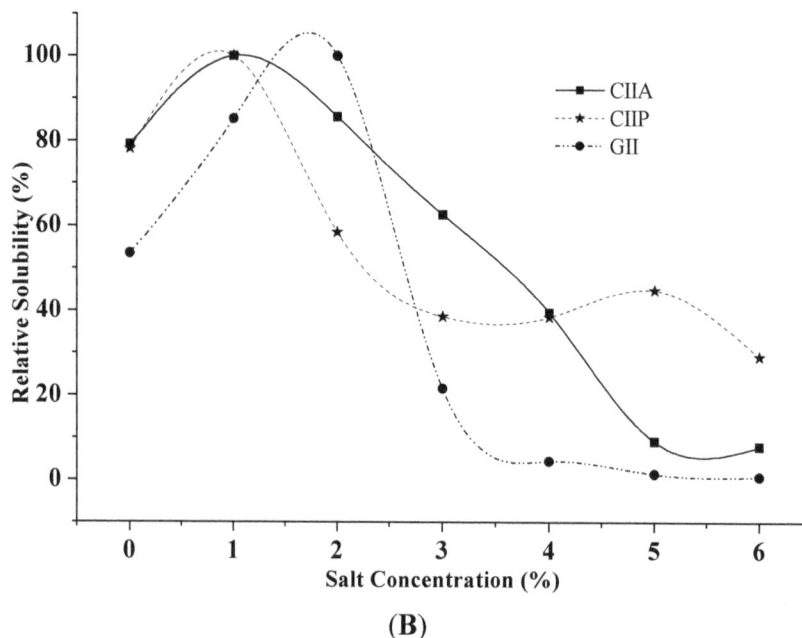

(B)

Low solubilization of type II collagens and type II gelatin was observed in an alkaline pH from 8 to 9. Bae *et al.* [29] reported that decreased solubility between pH 7 and 9 was due to protein precipitation and the increased relative viscosity of collagen. Moreover, Kittiphattanabawon *et al.* [28] suggested that the variation in solubility at varying pH is attributed to the differences in the molecular conformations of collagen.

Maximum solubility was observed at the salt concentration of 1% for type II collagens and 2% for type II gelatin (Figure 4B). A drastic decrease in solubility was observed between 3% and 4% salt concentration for CIIA and GII and between 1% and 2% for CIIP. The maximum solubility of ASC and PSC from eel skin was reported at 3%–4% salt concentration [26], which is slightly higher than our findings. Jongjareonrak *et al.* [25] explained that the addition of salt decreased the solubility of collagens by increasing the ionic strength and enhancing hydrophobic interactions between protein chains, which leads to protein precipitation. In the present study, CIIP was more soluble than CIIA at 1% salt concentration, which was due to the partial hydrolysis of high molecular weight cross-linked CIIP by pepsin. Variations in the solubility may also be due to differing hydrophobic amino acid contents and the isoelectric point of collagen and gelatin [25].

2.8. Thermal Stability

The denaturation (Td) and melting temperatures (Tm) of type II collagens and type II gelatin was determined by DSC. Thermal denaturation denotes unfolding of the triple helix to a random coil and leads to a loss of the unique characteristics of collagens. Denaturation and melting temperature of type II collagens and gelatin are shown in Table 2 and Figure 5. Type II collagens and type II gelatin had two endothermic peaks: the first peak occurred with the thermal denaturation of collagen and the second peak with the breakage of peptide chains and the destruction of the material (melting temperature). According to the DSC spectra, the melting temperature of CIIA, CIIP and GII was observed at 52.04, 58.07 and 66.67 °C, respectively. Wang *et al.* [31] suggested that the difference in thermal stability of ASC and PSC from Amur sturgeon skin may be due to the level of hydration and the number and nature of covalent cross-linkages.

Table 2. Antioxidant and thermal transition temperature of type II collagens and type II gelatin. CIIP; type II pepsin soluble collagen, CIIA: type II acid soluble collagen, GII: type II gelatin.

Sample	DPPH Radical Scavenging (%)	Reducing Power (Absorbance at 700 nm)	DSC	
			Denaturation Temp (°C)	Melting Temp (°C)
CIIA	20.08	0.22	30.00	52.04
CIIP	24.77	0.24	31.25	58.07
GII	16.56	0.20	32.50	66.67
BHT	65.72	0.28	-	-

The denaturation temperature (Td) of GII (32.50 °C) was higher than that of CIIP (31.25 °C) and CIIA (30.00 °C). The Td of other fish and calf collagen ranged from 19.4 to 40.8 °C [19,23,26,30]. The denaturation temperature may be affected by the degree of hydroxylation of the Pro and the Gly-Pro-Hyp sequence in collagen and gelatin [24]. The pyrrolidine rings of imino acids restrict the polypeptide chain conformation and strengthen the triple helix. Imino acid content is primarily responsible for the thermal stability and the triple stranded helix formation [32]. However, GII had a lower imino acid content than CIIP and a higher denaturation temperature. Similarly, Li *et al.* [6] reported that porcine collagen had higher thermal stability than Amur sturgeon collagen but a lower imino acid content; these authors proposed that the molecular conformation, amino acid sequence and stable covalent intra- and intermolecular cross-linkages may also influenced the thermal stability of collagen. In the present study, we have determined that the changes in the denaturation temperature are due to the different molecular conformations of type II collagens and type II gelatin.

Figure 5. DSC thermograms of type II collagens and type II gelatin. CIIP: type II pepsin soluble collagen, CIIA: type II acid soluble collagen, GII: type II gelatin.

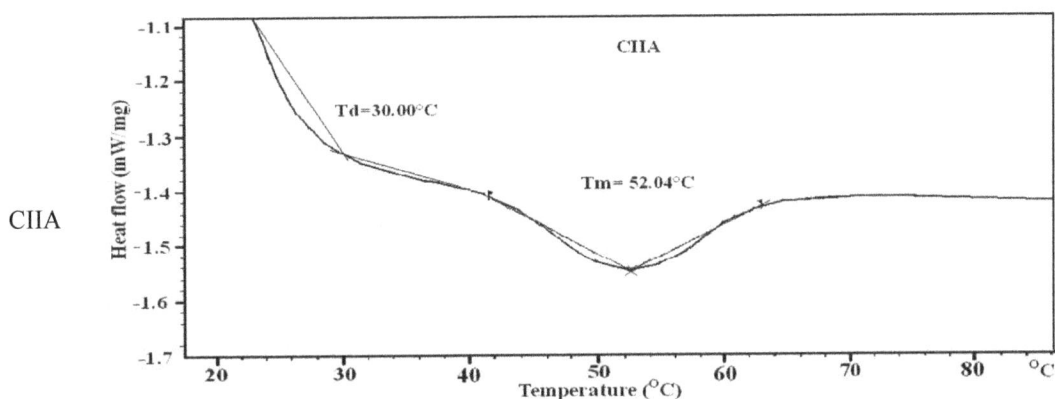

Figure 5. *Cont.*

CIIP

GII

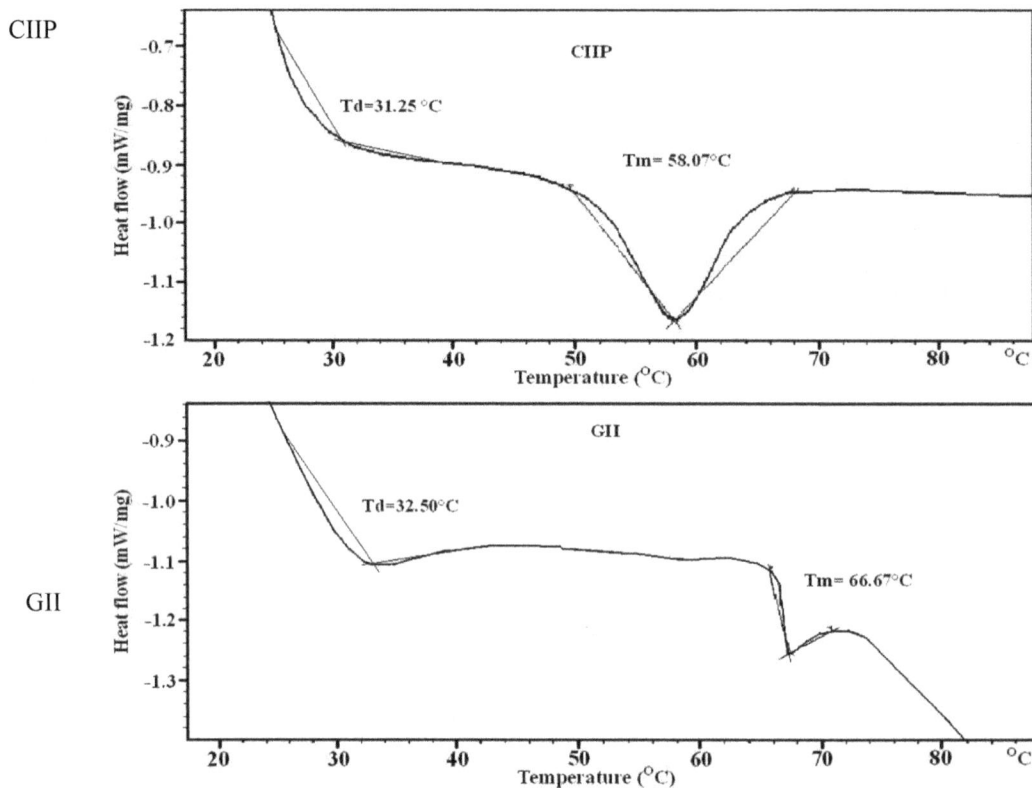

Bae *et al.* [29] stated that the Td of fish collagen above 33 °C was considered as high heat resistance. In the present study, the Td of shark type II collagens and type II gelatin indicated higher stability and heat resistance than other reported fish collagens. Although, fish collagen has several important biochemical properties, the lower denaturation temperature of fish collagen than mammalian collagen is a major drawback in practical applications. However, the Td of type II collagens and type II gelatin from silvertip shark is similar to mammalian collagen and therefore may be a suitable alternative source of mammalian collagen.

2.9. FTIR Spectra

FTIR spectroscopy was employed to monitor the functional groups and secondary structure of type II collagens and type II gelatin. The frequencies at which major peaks occurred for type II collagens and gelatin are given in Table 3. The amide-A peak of CIIA, CIIP and GII appeared at 3340.38, 3331.13 and 3350.73 cm^{-1}, respectively (Figure 6). This peak is generally associated with N–H stretching coupled with the hydrogen bond of a carbonyl group in a peptide chain.

The amide-B peak was observed at 2927.42, 2932.28 and 2948.13 cm^{-1} for CIIA, CIIP and GII, respectively. This peak represents the asymmetric stretching vibration of alkenyl C–H, as well as NH$_3^+$. The amide-A and amide-B peaks of type II gelatin appeared at higher wavenumbers than that of type II collagens. The higher wavenumbers of GII are likely due to the increased protein-protein intermolecular cross-linkages through hydrogen bonds of low molecular weight peptides, rather than high molecular weight collagens [19].

Table 3. General peak assignments of the FTIR spectra of type II collagens and type II gelatin isolated from shark cartilage. CIIP: type II pepsin soluble collagen, CIIA: type II acid soluble collagen, GII: type II gelatin.

Peak Wavenumber (cm^{-1})			
CIIP	**CIIA**	**GII**	**Assignment**
3331.13	3340.38	3350.73	Amide A: NH stretch coupled with a hydrogen bond
2932.28	2927.42	2948.13	Amide B: CH$_2$ asymmetrical stretch
2882.11	2855.34	-	CH$_3$-symmetric stretch: mainly proteins
1796.47	1734.76	-	Carbonyl C=O stretch: lipids
1659.84	1659.77	1657.77	Amide I: C=O stretch/hydrogen bond coupled with COO–
1550.67	1554.23	1551.26	Amide II: NH bend coupled with an CN stretch
1452.09	1452.16	1453.79	CH$_2$ bend
1338.83	1338.22	1338.47	CH$_2$ wag of proline
1239.69	1240.40	1237.81	Amide III: NH bend coupled with an CN stretch
1157.44	1161.1	1129.96	CO-O-C asymmetric stretch: glycogen and nucleic acids
1079.18	1079.52	-	C–O stretch
876.16	874.26	874.30	Skeletal stretch
616.96	611.58	597.11	Skeletal stretch

Figure 6. FTIR spectra of type II collagens and type II gelatin. CIIP: type II pepsin soluble collagen, CIIA: type II acid soluble collagen, GII: type II gelatin.

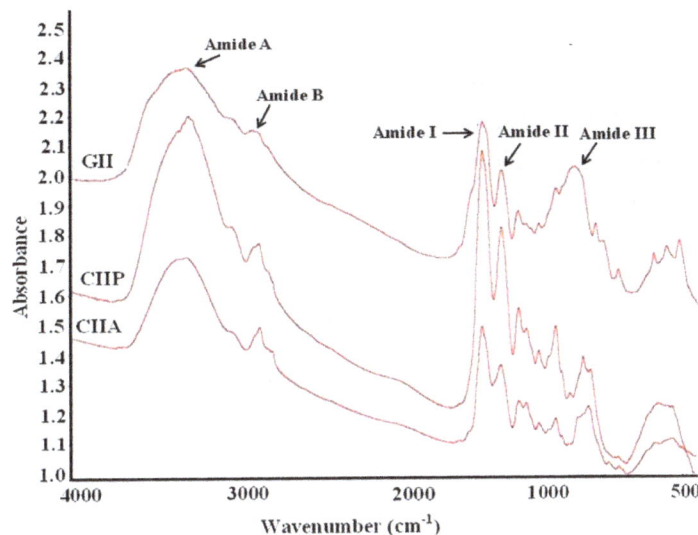

The amide-I peak was observed at 1659.77, 1659.84 and 1657.77 cm^{-1} for CIIA, CIIP and GII, respectively, which was similar to that of other fish collagens [19,27,31]. The amide-I region is mainly used for the analysis of the secondary structure of proteins [19]. The amide-I vibration mode is primarily due to the C=O stretching vibration of the peptide linkages (approximately 80%). The shift of the amide-I peak to a lower wavenumber is associated with the coiled structure of gelatin resulting from the heat denaturation during the isolation process [29].

The characteristic peak of the amide-II region of CIIA, CIIP and GII was observed at 1554.23, 1550.67 and 1551.26 cm^{-1}, respectively. The amide-II vibration modes are attributed to the N–H in-plane bend (40%–60%) and the C–N stretching vibration (18%–40%). Muyonga *et al.* [19] also observed the amide II peak at 1540–1558 cm^{-1} for Nile perch skin collagen. The intensity of the amide peak is associated with the triple helical structure of collagen. In the present study, the intensity of the amide I peak of GII was lower than type II collagens and this indicated the loss of the triple helix structure of collagen during the isolation process. The protein structure of gelatin also confirmed the above statement (Figure 1). Therefore, FTIR spectra clearly indicated that the triple helix structure, molecular order and intermolecular cross-linkages of collagen varied from gelatin.

2.10. CD Spectra

The triple helix of collagen consisted of three molecular strands. The prolines are arranged in a left handed polyproline-II-helical conformation, and these helices coil together to form a right handed super helix. Circular dichroism spectroscopy (CD) is used to investigate the molecular order of collagen.

CD spectra of type II collagens and type II gelatin exhibited a rotatory maximum at 221–221.5 nm, a minimum at 196–197 nm and a crossover point at 213.5–215.5 nm (Figure 7), which is characteristic of triple helical protein conformation [33]. The maximum and minimum rotatory CD spectra reported for eel skin collagen were 230 and 204 nm, respectively [26], and those of Adult black drum and sheepshead seabream bone were 220–221 and 197–199 nm, respectively [2]. However, for gelatin that was isolated by heat treatment, the CD spectra was similar to that of collagen. Aewsiri *et al.* [34] reported that after the complete denaturation of collagen, the characteristic triple helical positive peak at 220–230 nm disappeared and only a negative peak at 200 nm remained for gelatin. The triple helical structure of collagen can transform into a random coil when it is heated above its denaturation temperature [19]. The present study showed that the triple helical structures of type II gelatin were not completely destroyed during the isolation process. This is in agreement with the CD spectra of gelatin isolated from Amur sturgeon skin [35].

Figure 7. Circular dichroism spectra of type II collagens and type II gelatin. _____ type II pepsin soluble collagen; -------- type II acid soluble collagen; _____ GII-type II gelatin.

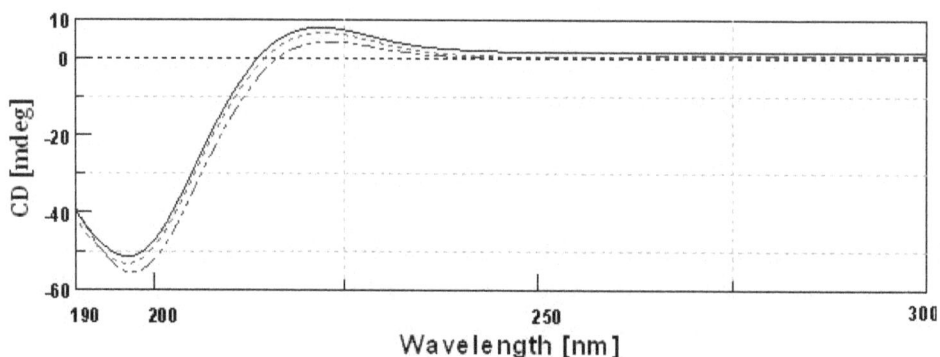

2.11. Antioxidant Activity

In the present study, the protective abilities of type II collagens and type II gelatin against oxidation was examined with the DPPH radical scavenging capacity and the reducing power. Natural or artificial antioxidants scavenge the DPPH radicals by donating a proton to the system. Type II collagens and type II gelatin were capable of scavenging hydroxyl radicals. The DPPH radical scavenging rate is 24%–16% for type II collagens and type II gelatin, which is lower than that of BHT (65%) (Table 2). CIIP exhibited higher scavenging activity than CIIA and GII. Zhu *et al.* [13] reported that the DPPH radical scavenging activity of PSC from sea cucumber was 45.58%, which was higher than our findings. It is thought that collagen can inactivate reactive oxygen species, reduce hydroperoxides, enzymatically eliminate specific oxidants, chelate pro-oxidative transition metals and scavenge free radicals, which may contribute to their antioxidant activities [13]. Moreover, certain specific amino acid residues and their sequences are thought to be responsible for antioxidant activities [15]. Zhong *et al.* [12] suggested that the antioxidant activities of the body wall of sea cucumber may be attributed to collagens. Zhuang *et al.* [14] reported that collagen from jellyfish exhibited improved antioxidant activity in animal mice skin by protecting the endogenous antioxidant enzymes and suggested that the high contents of glycine, proline, and hydrophobic amino acids of the collagen were responsible for the activity. In the present study, higher content of proline and hydrophobic amino acids (valine, leucine, isoleucine, cysteine and methionine) in CIIP than that of CIIA and GII also supports the superior antioxidant activity of CIIP.

The reducing power is determined by the ability of collagen to reduce ferric into ferrous ions. Similar to the DDPH activity, the reducing power was higher in CIIP than in CIIA and GII, with a value of 0.24 at 700 nm (Table 2). The present result was comparable to the porcine protein reducing power of 0.25 [36] and lower than that of the buckwheat protein (0.77–1.24) [37]. The reducing power of BHT was higher than the isolated collagen and gelatin. Li *et al.* [6] reported that the reducing ability increased with increasing concentrations of ASC from 1 to 5 mg/mL. The antioxidant property of proteins is primarily influenced by their size, configuration and amino acid composition [6]. Antioxidant activity of type II collagens and gelatin is one of the important properties for its use as a drug, especially for the treatment of RA. Some researchers have reported the effective inhibitory activity of cartilage type II collagen on rheumatoid arthritis [1,7,22]. There is substantial evidence that the CII-deficiency in transgenic mice results in embryos that die at birth and have skeletal malformations, thus indicating the importance of this protein in bone development [38]. Therefore, further investigations are required to identify specific antioxidant peptides from silvertip shark cartilage collagen and gelatin.

2.12. Microscopic Structure

The SEM microscopic structure showed a homogenous, porous, fibrillary and multi-layered structure of collagen and gelatin (Figure 8). The microstructure of CIIA and CIIP was a rough, fibril-dominated network with a coarse structural integrity. Conversely, the surface of gelatin was a smooth, less fibril-dominated network and was partially degraded. This similar structure was reported for collagen isolated from Amur sturgeon skin [31]. The microstructural changes in the collagen and gelatin are related to the functional properties. At a higher magnification, the collagen and gelatin had a fiber-like structure with an average size of 0.4 to 0.7 μm for CIIP and 0.75 to 1.5 μm for CIIA. An inter-linkage between CIIP and glycoprotein was observed at higher magnification. This structure was similar to the eel skin ASC and PSC structures reported by Veeruraj *et al.* [26]. Accordingly, the surface morphology of collagen greatly differs from that of gelatin due to isolation process.

Figure 8. Scanning electron microscopic structure of type II collagens and type II gelatin isolated from silvertip shark cartilage. A1, A2, A3: type II pepsin soluble collagen; B1, B2, B3: type II acid soluble collagen; C1, C2, C3: type II gelatin.

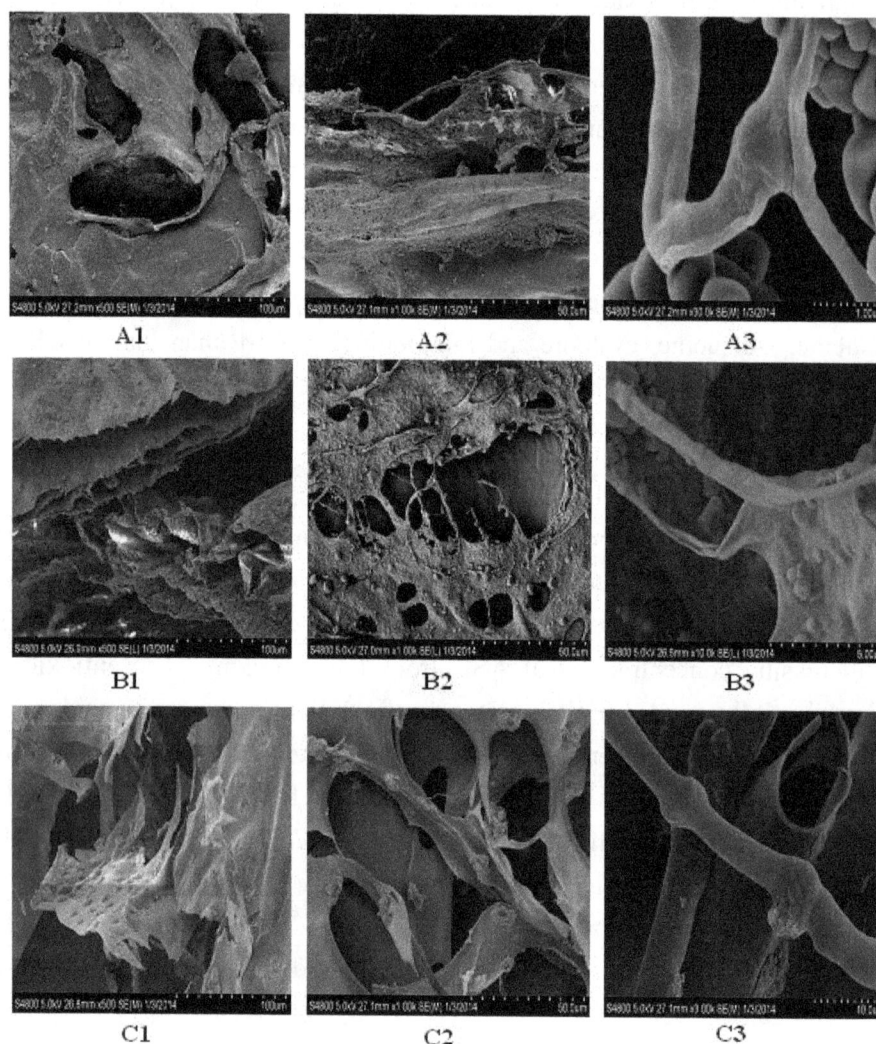

The microstructural changes in collagen are related to the functional properties. The present SEM image showed that the shark collagens and gelatin had a cross-section with an inter-connected network pore configuration. Joseph *et al.* [39] suggested that the moderate pore size of collagen was suitable for *in vivo* studies and that the pore size of collagen was influenced by the water content during preparation. Earlier, Jansson *et al.* [40] reported that the collagen extracted from the cartilage of horse mackerel and croaker could be used as a biofilm or scaffold for wound healing purposes. In addition, other architectural features, such as pore shape, pore wall morphology and interconnectivity of collagen, have also been suggested for use in cell seeding, growth, gene expression, migration, mass transport, and new tissue formation. Generally, the uniform and regular network structure of collagen as a drug carrier is propitious for a well-proportioned drug distribution [41]. The microscopic structure of type II collagens and gelatin isolated from shark cartilage may provide a suitable biomaterial for a drug carrier system.

3. Experimental Section

3.1. Raw Materials

The cartilage (cartilaginous skeleton) of the silvertip shark was used as the raw material for the isolation of fish collagen and gelatin and was obtained from a private fish processing plant, M/s. Yueqing Ocean Biological Health Care Product Co., Ltd. Zhejiang province, Shanghai, China. The cartilage was washed with potable water and cut into small pieces prior to the isolation of collagen and gelatin. Moisture, ash, fat and protein content of the silvertip shark cartilage was determined according to AOAC [42] methods: No. 950.46, 928.08, 960.39 and 920.153, respectively. The sample hydroxyproline content was determined according to the method of Bergman and Loxley [43].

3.2. Isolation of Collagen and Gelatin

3.2.1. Pretreatment of Shark Cartilage

The cartilage was homogenized in a tissue homogenizer using phosphate buffer (pH 6.5; 0.2 mol/L sodium dihydrogen phosphate and 0.2 mol/L disodium hydrogen phosphate heptahydrate). The entire process of collagen isolation was performed at 4 °C. The homogenate was treated with double-distilled water at a ratio of 1:6 (w/v) for 24 h to remove water soluble substances. The cartilage was decalcified with 0.5 Methylenediaminetetraacetic acid (EDTA) (pH 7.4) at a ratio of 1:10 (w/v) for 48 h. The solution was replaced every 12 h. This pretreated shark cartilage was used for the further isolation of acid soluble collagen (ASC) and pepsin soluble collagen (PSC).

3.2.2. Isolation of ASC and PSC

Type II collagen (CII) was isolated from the shark cartilage according to the method of Kittiphattanabawon et al. [21] with modification. The pretreated shark cartilage was soaked into 0.5 M acetic acid (1:6, w/v) for 4 days with continuous shaking and the extracts were centrifuged at 10,000 rpm for 30 min at 4 °C. The supernatant was collected and salted out by adding 2 M NaCl. The precipitates were re-dissolved in a minimum volume of 0.5 M acetic acid and dialyzed against distilled water for 2 days until a neutral pH was obtained. The dialyzed sample was lyophilized (Labconco Freezone 2.5, Kansas City, MI, USA) and further referred to as ASC. For the isolation of PSC, the pretreated shark cartilage was soaked in 0.5 M acetic acid containing 1% pepsin (1:6, w/v) and then followed the above procedure.

3.2.3. Isolation of Gelatin

Type II gelatin (GII) was isolated according to Jeevithan et al. [4]. Briefly, the shark cartilage was treated twice with 0.2% NaOH at a ratio of 1:6 w/v for 45 min to remove the non-collagenous protein. After thorough washing, cartilage was treated twice with 0.2% H_2SO_4 at a ratio of 1:6 w/v for 45 min. Samples were then treated with 1% citric acid twice at a ratio of 1:6 w/v for 45 min. The final isolation was carried out with distilled water at a ratio of 1:1 *w/v* at 45°C for 24 h. The extract was filtered through Whatman No. 4 filter paper under vacuum, lyophilized and used for further characterization.

3.3. Purification by Gel Filtration Chromatography

Gel filtration chromatography was performed to purify CIIA, CIIP and GII. In brief, samples (100 mg) were dissolved in 10 mL of a running buffer (20 mM sodium acetate buffer, pH 4.8) and were applied to a Sephadex G-100 (Sigma, Shanghai, China) column (25.0 × 3.0 cm). The column was previously equilibrated with the same buffer until A_{230} was less than 0.05 to achieve baseline correction. The elution volume was 200 mL, and the flow rate was 1 mL/min. The protein fractions were identified by UV absorbance spectroscopy and 5 mL fractions were collected. As shown in Figure 1b, the collagen peak was pooled and salted out by addition of 2 M NaCl. The precipitate was dialyzed against distilled water and freeze dried.

3.4. Sodium Dodecyl Sulfate-Polyacrylamide Gel Electrophoresis (SDS-PAGE)

The protein pattern was analyzed using sodium dodecyl sulfate poly-acrylamide gel electrophoresis (SDS-PAGE) according to the method of Laemmli [44] with modification. Briefly, the purified samples were dissolved in 5% SDS, kept in a water bath at 60 °C for 20 min and centrifuged at 3500 rpm. The supernatant was mixed with a sample buffer (1:1) containing Tris HCl (pH 6.8; 12.1 g Tris base was dissolved in 80 mL distilled water and the pH was adjusted with concentrated HCl), 1% 2-mercaptoethanol, 40% sucrose, 20% glycerol, 0.02% bromophenol blue and 1% SDS. The mixtures were loaded onto a polyacrylamide gel, composed of 7.5% separating gel and 4% stacking gel, and were subjected to electrophoresis at a constant current of 50 mA. After electrophoresis, gels were fixed with a mixture of 5:1 methanol:acetic acid would be composed of 83.3% and 16.7% methanol and acetic acid, respectively for 1 h. This was followed by staining with 0.5% Coomassie blue R-250 in 150% methanol and 50% acetic acid for 30 min. Finally, the gels were destained with a mixture of 300% methanol and 100% acetic acid for 2 h.

3.5. Peptide Mapping

Peptide mapping was performed according to the method of Kittiphattanabawon *et al.* [28] with modification. Samples (5 mg) were dissolved in 1 mL of 0.1 M Tris-HCL (pH 6.8) buffer containing 0.5% SDS and heated at 100 °C for 2 h. To initiate digestion, 20 µL of the enzyme solution, V8-protease (EC 3.4.21.19, Sigma-Aldrich, Shanghai, China), with a concentration of 5 µg/mL was added to each mixture. The reaction mixtures were then incubated at 37 °C for 1 h and the proteolysis was halted by increasing the temperature to 100 °C for 5 min. Peptides generated by the protease digestion were separated by SDS-PAGE using 10% separating gel and 4% stacking gel as previously described.

3.6. Viscosity

Samples (0.03%) were dissolved in 0.1 M acetic acid and the viscosity was measured using a Brookfield LVDV-II+P viscometer (Brookfield Engineering Laboratories Ltd., Middleboro, MA,

USA) equipped with an ultra-low viscosity adapter at a speed of 90 rpm. Samples were heated using a rotary water bath from 5 to 40 °C. The sample solutions (20 mL) were incubated for 30 min at each temperature prior to measurement. The relative viscosity was calculated compared with that obtained at 5 °C.

3.7. Collagen Solubility Test

Optimum solubility at different pH and salt concentrations was determined according to the method of Jongjareonrak et al. [24]. Collagen samples were dissolved in 0.5 M acetic acid with gentle stirring at 4 °C for 12 h to obtain a final concentration of 6 mg/mL.

3.7.1. Effect of pH

The collagen solution (5 mL) was transferred into a series of centrifuge tubes, adjusted to pH values ranging from 1 to 10 by addition of the appropriate amount of 6 M NaOH or 6 M HCl. The resulting sample solution totaled 10 mL with distilled water. The solution was stirred gently for 30 min at 4 °C and centrifuged at 5000× g for 30 min. An aliquot (1 mL) of the supernatant was collected from each tube and the protein content was measured by the Lowry method [45]. The relative solubility of collagen was calculated compared with the pH rendering the highest solubility.

3.7.2. Effect of NaCl

The collagen solution (5 mL) was mixed with 5 mL of cold NaCl in acetic acid of various concentrations (0%–12%, w/v) to obtain final concentrations of 1%–6% (w/v). The mixture was stirred gently at 4 °C for 30 min and centrifuged at 10,000× g for 30 min at 4 °C. The relative solubility was calculated compared with that of the salt concentration exhibiting the highest solubility.

3.8. UV Absorption Spectrum

UV absorption spectrum of type II collagen samples was measured using a UV spectrophotometer. Samples were dissolved in 0.5 M acetic acid and UV spectra were measured between 190 and 400 nm at a scan speed of 2 nm/s with an interval of 1 nm.

3.9. Amino Acid Profiling

The collagen samples were hydrolyzed under reduced pressure in 6 M HCl at 110 °C for 24 h. Amino acid composition was analyzed using an amino acid analyzer (Hitachi L-8800, Tokyo, Japan). The amino acid content is expressed as the number of residues/1000 residues.

3.10. Fourier Transform Infrared Spectroscopy (FTIR)

FTIR spectra of the samples were obtained using a Nicolet 6700-Fourier transform infrared spectrometer (Thermofisher Scientific Inc., Waltham, MA, USA) equipped with a DLaTGS detector. The lyophilized samples (5 mg) were mixed with dried KBr (100 mg), ground in a mortar and pestle and subjected to a pressure of approximately 5×10^6 Pa in an evacuated die to produce a 13×1 mm clear transparent disk. The absorption intensity of the peaks was calculated using the base-line method. The resultant spectra were analyzed using ORIGIN 8.0 software (Thermo Nicolet, Madison, WI, USA).

3.11. Circular Dichroism (CD)

The molecular conformations of collagen and gelatin were assessed by CD using a spectropolarimeter (Jasco J-810, Shanghai, China) according to the method of Cao et al. [22] with modification. Briefly, the samples were dissolved in 0.1 M acetic acid to obtain a final concentration of 0.1 mg/mL and were stirred for 6 h. Then, the sample solutions were placed in a quartz cell with a path length of 10 mm. The spectra were recorded between 190 and 300 nm at 25 °C. The acetic acid spectrum was used as a reference and CD spectra of the samples were obtained after subtracting the reference spectrum.

3.12. Scanning Electron Microscopy (SEM)

Morphological characteristics of the isolated collagens and gelatin were visualized by SEM-S4800 (Hitachi, Tokyo, Japan). Collagen samples were mounted on specimen stubs with two-sided carbon tape. The sample surface was sputter-coated with a thin layer (~8–10 nm) of gold ions using a sputter-coater. The samples were then introduced into the specimen chamber and examined for surface morphology at a 20 kV accelerating voltage.

3.13. Thermal Stability

Differential scanning calorimetry (DSC) was conducted following the method of Rochdi et al. [46]. The samples were rehydrated with deionized water at a solid/solution ratio of 1:10 (w/v). DSC was performed using a differential scanning calorimeter (Model DSC822e, Mettler-Toledo GmbH, Greifensee, Switzerland). The temperature calibration was conducted using an indium standard. Samples were weighed into aluminum pans and sealed. Subsequently, samples were scanned at 5 °C/min from 20 to 120 °C using ice water as the cooling medium. An empty pan was used as the reference. The denaturation and melting temperatures were estimated from the DSC thermogram.

3.14. Antioxidant Activity

3.14.1. DPPH Radical Scavenging Assay

DPPH radical scavenging activity was acquired based on the method of Shimada et al. [47]. Briefly, the test sample (500 µL) was added to 500 µL of 99.5% ethanol and 125 µL of 99.5% ethanol containing 0.02% DPPH. The mixture was kept in the dark at room temperature for 60 min before measuring absorbance at 517 nm. In the blank, the sample was replaced with distilled water (500 µL) and the above procedure was followed. Butylated hydroxytoluenum (BHT) was used as a positive control. A lower absorbance indicated higher DPPH scavenging activity. Radical scavenging activity was calculated as follows:

Radical scavenging activity = [(blank absorbance-sample absorbance)/blank absorbance] × 100%.

3.14.2. Reducing Power

Protein solutions (1 mg) were mixed with 2.5 mL of phosphate buffer (0.2 M, pH 6.6) and 2.5 mL of 1.0% potassium ferricyanide and the mixture was incubated at 50°C for 20 min. Trichloroacetic acid (2.5 mL of a 10% solution) was added, and the mixture was centrifuged. The supernatant (2.5 mL) was mixed with water (2.5 mL) and 0.1% ferric chloride (0.5 mL), and the absorbance was measured at 700 nm. Higher absorbance indicated higher reducing power. BHT was used as a reference antioxidant.

4. Conclusions

Collagen and gelatin was isolated from shark cartilage using an improved isolation method with a high denaturation temperature. Compared with type II gelatin, CIIP was richer in proline and alanine and lower in hydroxyproline. The homogenous, porous, fibrillary and multi-layered structure of type II collagens and type II gelatin are good characteristics of biomaterials used in cell attachment. In conclusion, the solubility, susceptibility to proteolytic enzymes, high denaturation temperature and efficient antioxidant activities suggest that silvertip shark cartilage type II collagens and type II gelatin have promising potential for use in various practical applications in biomedical industries.

Acknowledgments

This work received financial support from the National High Technology Research and Development Program of China (No. 2011AA09070109) and the National Natural Science Foundation of China (No. 81341082).

References

1. Poole, A.R. Cartilage in Health and Disease. In *Arthritis and Allied Conditions: A Textbook of Rheumatology*, 15th ed.; Koopman, W.J., Moreland, L.W., Eds.; Williams and Wilkins: Baltimore, MD, USA, 2005; pp. 223–269.

2. Ogawa, M.; Moody, M.W.; Portier, R.J.; Bell, J.; Schexnayder, M.A.; Losso, J.N. Biochemical properties of black drum and sheepshead seabream skin collagen. *J. Agric. Food Chem.* **2003**, *51*, 8088–8092.

3. Jeya Shakila, R.; Jeevithan, E.; Varatharajakumar, A.; Jeyasekaran, G.; Sukumar, D. Functional characterization of gelatin extracted from bones of red snapper and grouper in comparison with mammalian gelatin. *LWT Food. Sci. Technol.* **2012**, *48*, 30–36.

4. Jeevithan, E.; Jeya Shakila, R.; Varatharajakumar, A.; Jeyasekaran, G.; Sukumar, D. Physico- functional and mechanical properties of chitosan and calcium salts incorporated fish gelatin scaffolds. *Int. J. Biol. Macromol.* **2013**, *60*, 262–267.

5. Chirita, M. Mechanical properties of collagen biomimetic films formed in the presence of calcium, silica and chitosan. *J. Bionic Eng.* **2008**, *5*, 149–158.

6. Li, Z.R.; Wang, B.; Chi, C.F.; Zhang, Q.H.; Gong, Y.D.; Tang, J.J.; Luo, H.Y.; Ding, G.F. Isolation and characterization of acid soluble collagens and pepsin soluble collagens from the skin and bone of Spanish mackerel (*Scomberomorous niphonius*). *Food Hydrocoll.* **2013**, *31*, 103–113.

7. Chen, L.; Bao, B.; Wang, N.; Xie, J.; Wu, W.H. Oral Administration of Shark Type II Collagen Suppresses Complete Freund's Adjuvant-Induced Rheumatoid Arthritis in Rats. *Pharmaceuticals* **2012**, *5*, 339–352.

8. Zhao, W.; Tong, T.; Wang, L.; Li, P.P.; Chang, Y.; Zhang, L.L.; Wei, W. Chicken type II collagen induced immune tolerance of mesenteric lymph node lymphocytes by enhancing beta2-adrenergic receptor desensitization in rats with collagen-induced arthritis. *Int. Immunopharmacol.* **2011**, *11*, 12–18.

9. Zhu, P.; Li, X.Y.; Wang, H.K.; Jia, J.F.; Zheng, Z.H.; Ding, J.; Fan, C.M. Oral administration of type-II collagen peptide 250–270 suppresses specific cellular and humoral immune response in collagen-induced arthritis. *Clin. Immunol.* **2007**, *122*, 75–84.

10. Garcia, G.; Komagata, Y.; Slavin, A.J.; Maron, R.; Weiner, H.L. Suppression of collagen-induced arthritis by oral or nasal administration of type II collagen. *J. Autoimmun.* **1999**, *13*, 315–324.

11. Youn, J.; Hwang, S.H.; Ryoo, Z.Y.; Lynes, M.A.; Paik, D.J.; Chung, H.S.; Kim, H.Y. Metallothionein suppresses collagen-induced arthritis via induction of TGF-b and down-regulation of proinflammatory mediators. *Clin. Exp. Immunol.* **2002**, *129*, 232–239.

12. Zhong, Y.; Khan, M.A.; Shahidi, F. Compositional characteristics and antioxidant properties of Fresh and Processed Sea Cucumber (*Cucumaria frondosa*). *J. Agric. Food Chem.* **2007**, *55*, 1188–1192.

13. Zhu, B.; Dong, X.; Zhou, D.; Gao, Y.; Yang, J.; Li, D.; Zhao, X.; Rena, T.; Yea, W.; Tana, H.; Wua, H.; Yu, C. Physicochemical properties and radical scavenging capacities of pepsin-solubilized collagen from sea cucumber (*Stichopus japonicas*). *Food Hydrocoll.* **2012**, *28*, 182–188.

14. Zhuang, Y.; Zhao, X.; Li, B. Optimization of antioxidant activity by response surface methodology in hydrolysates of jellyfish (*Rhopilema esculentum*) umbrella collagen. *J. Zhejiang Univ. Sci. B* **2009**, *10*, 572–579.

15. Mendis, E.; Rajapakse, N.; Byun, H.G.; Kim, S.K. Investigation of jumbo squid (*Dosidicus gigas*) skin gelatin peptides for their *in vitro* antioxidant effects. *Life Sci.* **2005**, *77*, 2166–2178.

16. Je, J.Y.; Qian, Z.J.; Byun, H.G.; Kim, S.K. Purification and characterization of an antioxidant peptide obtained from tuna backbone protein by enzymatic hydrolysis. *Process Biochem.* **2007**, *42*, 840–846.

17. Merly, L.; Smith, S.L. Collagen type II, alpha 1 protein: a bioactive component of shark cartilage. *Int. Immunopharmacol.* **2013**, *15*, 309–315.

18. Liang, Q.; Wang, L.; Sun, W.; Wang, Z.; Xu, J.; Ma, H. Isolation and characterization of collagen from the cartilage of Amur sturgeon (*Acipenser schrenckii*). *Process Biochem.* **2014**, *49*, 318–323.

19. Muyonga, J.H.; Cole, C.G.B.; Duodu, K.G. Characterisation of acid soluble collagen from skins of young and adult Nile perch (*Lates niloticus*). *Food Chem.* **2004**, *85*, 81–89.

20. Cho, S.M.; Kwak, K.S.; Park, D.C.; Gu, Y.S.; Ji, C.I.; Jang, D.H.; Lee, Y.B.; Kim, S.B. Processing optimization and functional properties of gelatin from shark (*Isurus oxyrinchus*) cartilage. *Food Hydrocoll.* **2004**, *18*, 573–579.

21. Kittiphattanabawon, P.; Benjakul, S.; Visessanguan, W.; Shahidi, F. Isolation and characterization of collagen from the cartilages of brownbanded bamboo shark (*Chiloscyllium punctatum*) and blacktip shark (*Carcharhinus limbatus*). *LWT Food Sci. Technol.* **2010**, *43*, 792–800.

22. Cao, H.; Shi, F.X.; Xu, F.; Yu, J.S. Molecular structure and physicochemical properties of pepsin-solubilized type II collagen from the chick sternal cartilage. *Eur. Rev. Med. Pharmacol. Sci.* **2013**, *17*, 1427–1437.

23. Helen, E.; Cahir, A. Spatial organization of type I and II collagen in the canine meniscus. *J. Orthop. Res.* **2005**, *23*, 142–149.

24. Vercaigne-Marko, D.; Kosciarz, E.; Nedjar-Arroume, N.; Guillochon, D. Improvement of *Staphylococcus aureus*-V8-protease hydrolysis of bovine haemoglobin by its adsorption on to a solid phase in the presence of SDS: peptide mapping and obtention of two haemopoietic peptides. *Biotechnol. Appl. Biochem.* **2000**, *31*, 127–134.

25. Jongjareonrak, A.; Benjakul, S.; Visessanguan, W.; Nagai, T.; Tanaka, M. Isolation and characterisation of acid and pepsin-solubilised collagens from the skin of brownstripe red snapper (*Lutjanus vitta*). *Food Chem.* **2005**, *93*, 475–484.

26. Veeruraj, A.; Arumugam, M.; Balasubramanian, T. Isolation and characterization of thermostable collagen from the marine eel-fish (*Evenchelys macrura*). *Process Biochem.* **2013**, *48*, 1592–1602.

27. Duan, R.; Zhang, J.; Du, X.; Yao, X.; Konno, K. Properties of collagen from skin, scale and bone of carp (*Cyprinus carpio*). *Food Chem.* **2009**, *112*, 702–706.

28. Kittiphattanabawon, P.; Benjakul, S.; Visessanguan, W.; Nagai, T.; Tanaka, M. Characterisation of acid-soluble collagen from skin and bone of bigeye snapper (*Priacanthus tayenus*). *Food Chem.* **2005**, *89*, 363–372.

29. Bae, I.; Osatomi, K.; Yoshida, A.; Osako, K.; Yamaguchi, A.; Hara, K. Biochemical properties of acid-soluble collagens extracted from the skins of underutilized fishes. *Food Chem.* **2008**, *108*, 49–54.

30. Woo, J.W.; Yu, S.J.; Cho, S.M.; Lee, Y.B.; Kim, S.B. Extraction optimization and properties of collagen from yellowfin tuna (*Thunnus albacares*) dorsal skin. *Food Hydrocoll.* **2008**, *22*, 879–887.

31. Wang, L.; Liang, Q.; Chen, T.; Wang, Z.; Xu, J.; Ma, H. Characterization of collagen from the skin of Amur sturgeon (*Acipenser schrenckii*). *Food Hydrocoll.* **2014**, *38*, 104–109.

32. Ikoma, T.; Kobayashi, H.; Tanaka, J.; Walsh, D.; Mann, S. Physical properties of type I collagen extracted from fish scales of *Pagrus major* and *Oreochromis niloticas*. *Int. J. Biol. Macromol.* **2003**, *32*, 199–204.

33. Usha, R.; Ramasami, T. Structure and conformation of intramolecularly cross-linked collagen. *Colloids Surf. B* **2005**, *41*, 21–24.

34. Aewsiri, T.; Benjakul, S.; Visessanguan, W.; Wierenga, P.A.; Gruppen, H. Improvement of foaming properties of cuttlefish skin gelatin by modification with *N*-hydroxysuccinimide esters of fatty acid. *Food Hydrocoll.* **2011**, *25*, 1277–1284.

35. Nikoo, M.; Benjakul, S.; Bashari, M.; Alekhorshied, M.; Cissouma, A.I.; Yanga, N.; Xu, X. Physicochemical properties of skin gelatin from farmed Amur sturgeon (*Acipenser schrenckii*) as influenced by acid pretreatment. *Food Biosci.* **2014**, *5*, 19–26.

36. Chang, C.Y.; Wu, K.C.; Chiang, S.H. Antioxidant properties and protein compositions of porcine haemoglobin hydrolysates. *Food Chem.* **2007**, *100*, 1537–1543.

37. Tang, C.H.; Peng, J.; Zhen, D.W.; Chen, Z. Physicochemical and antioxidant properties of buckwheat (*Fagopyrum esculentum* Moench) protein hydrolysates. *Food Chem.* **2009**, *115*, 672–678.

38. Aszodi, A.; Hunziker, E.B.; Olsen, B.R.; Fassler, R. The role of collagen II and cartilage fibril-associated molecules in skeletal development. *Osteoarthr. Cartil.* **2001**, *9*, 150–159.

39. Joseph, G.; Jun, O.; Teruo, M. Biodegradable honeycomb collagen scaffold for dermal tissue engineering. *J. Biomed. Mater. Res. Part A* **2008**, *87*, 1103–1111.

40. Jansson, K.; Haegerstrand, A.; Kratz, G. A biodegradable bovine collagen membrane as a dermal template for human *in vivo* wound healing. *Scandinavian J. Plast. Reconstr. Surg. Hand Surg.* **2001**, *35*, 369–375.

41. Zhang, Y.; Liu, W.T.; Li, G.Y.; Shi, B.; Miao, Y.Q.; Wu, X.H. Isolation and partial characterization of pepsin-soluble collagen from the skin of grass carp (*Ctenopharyngodon idella*). *Food Chem.* **2007**, *103*, 906–912.

42. AOAC. *Official Methods of Analysis*; Association of Official Ananlytical Chemists Inc.: Arlington, VA, USA, 2000.

43. Bergman, I.; Loxley, R. Two improved and simplified methods for the spectrophotometric determination of hydroxyproline. *Anal. Chem.* **1963**, *35*, 1961–1965.

44. Laemmli, U.K. Cleavage of structural proteins during the assembly of the head of bacteriophage T4. *Nature* **1970**, *227*, 680–685.

45. Lowry, O.H.; Rosebrough, N.J.; Farr, A.L.; Randall, R.J. Protein measurement with Folin phenol reagent. *J. Biol. Chem.* **1951**, *193*, 256–275.

46. Rochdi, A.; Foucat, L.; Renou, J.P. NMR and DSC studies during thermal denaturation of collagen. *Food Chem.* **2000**, *69*, 295–299.

47. Shimada, K.; Fujikawa, K.; Yahara, K.; Nakamura, T. Antioxidative properties of xanthan on the autoxidation of soybean oil in cycloextrin emulsion. *J. Agric. Food Chem.* **1992**, *40*, 945–948.

7

The Marine Sponge-Derived Inorganic Polymers, Biosilica and Polyphosphate, as Morphogenetically Active Matrices/ Scaffolds for the Differentiation of Human Multipotent Stromal Cells: Potential Application in 3D Printing and Distraction Osteogenesis

Xiaohong Wang [1,*], **Heinz C. Schröder** [1], **Vladislav Grebenjuk** [1], **Bärbel Diehl-Seifert** [2], **Volker Mailänder** [3,4], **Renate Steffen** [1], **Ute Schloßmacher** [1] and **Werner E. G. Müller** [1,*]

[1] ERC Advanced Investigator Grant Research Group, Institute for Physiological Chemistry, University Medical Center, Johannes Gutenberg University, Duesbergweg 6, D-55128 Mainz, Germany; E-Mails: hschroed@uni-mainz.de (H.C.S.); grebenyu@uni-mainz.de (V.G.); steffen@uni-mainz.de (R.S.); schlossm@uni-mainz.de (U.S.)

[2] NanotecMARIN GmbH, 55128 Mainz, Germany; E-Mail: helmut-baerbel.Seifert@t-online.de

[3] Max Planck Institute for Polymer Research, Ackermannweg 10, 55129 Mainz, Germany; E-Mail: volker.mailaender@mpip-mainz.mpg.de

[4] Medical Clinic, University Medical Center, Johannes Gutenberg University, Langenbeckstr. 1, D-55131 Mainz, Germany

* Authors to whom correspondence should be addressed; E-Mails: wang013@uni-mainz.de (X.W.); wmueller@uni-mainz.de (W.E.G.M.)

Abstract: The two marine inorganic polymers, biosilica (BS), enzymatically synthesized from ortho-silicate, and polyphosphate (polyP), a likewise enzymatically synthesized polymer consisting of 10 to >100 phosphate residues linked by high-energy phosphoanhydride bonds, have previously been shown to display a morphogenetic effect on osteoblasts. In the present study, the effect of these polymers on the differential differentiation of human multipotent stromal cells (hMSC), mesenchymal stem cells, that had been encapsulated into beads of the biocompatible plant polymer alginate, was studied. The differentiation of the hMSCs in the alginate beads was directed either to the osteogenic cell lineage by exposure to an osteogenic medium (mineralization activation cocktail; differentiation into osteoblasts) or to the chondrogenic cell lineage by incubating in chondrocyte differentiation medium (triggering chondrocyte maturation). Both biosilica and polyP, applied as Ca^{2+} salts, were found to induce an increased mineralization in osteogenic cells; these inorganic polymers display also morphogenetic potential. The effects were substantiated by gene expression studies, which revealed that biosilica and polyP strongly and significantly increase the expression of bone morphogenetic protein 2 (BMP-2) and alkaline phosphatase (ALP) in osteogenic cells, which was significantly more pronounced in osteogenic *versus* chondrogenic cells. A differential effect of the two polymers was seen on the expression of the two collagen types, I and II. While collagen Type I is highly expressed in osteogenic cells, but not in chondrogenic cells after exposure to biosilica or polyP, the upregulation of the steady-state level of collagen Type II transcripts in chondrogenic cells is comparably stronger than in osteogenic cells. It is concluded that the two polymers, biosilica and polyP, are morphogenetically active additives for the otherwise biologically inert alginate polymer. It is proposed that alginate, supplemented with polyP and/or biosilica, is a suitable biomaterial that promotes the growth and differentiation of hMSCs and might be beneficial for application in 3D tissue printing of hMSCs and for the delivery of hMSCs in fractures, surgically created during distraction osteogenesis.

Keywords: biosilica; polyphosphate; multipotent stromal cells; mesenchymal stem cells; alkaline phosphatase; 3D cell/tissue printing; distraction osteogenesis

---

1. Introduction

Bone formation is a complex process involving several cell lineages and growth factors, as well as an ordered scaffold, comprising a fibrillar organic network. The major categories of bone cells are the bone forming osteoblasts and the bone resorbing osteoclasts. In addition, osteocytes are found in mature bone, which do not divide and are derived from osteoprogenitors. Finally, lining cells cover the bone surface. During bone formation, a controlled cross-talk and coupling between the major cells takes place that maintains the balance between the anabolic osteoblasts and the catabolic osteoclasts [1]. While human osteoblasts derive from multipotent stromal cells (hMSC), previously also termed mesenchymal stem cells, osteoclasts differentiate from the monocyte/macrophage hematopoietic lineage and develop and adhere onto bone matrix.

Human MSCs (hMSCs), discovered in 1968 by Friedenstein [2], have the capacity to readily differentiate into the osteogenic, chondrogenic, adipogenic or myogenic cell lineage, depending on the activation of specific transcription factors. They can be efficiently expanded in culture; even after million-fold expansion, they retain the ability to differentiate (see [3]). Since hMSCs are very contact inhibited, they need to be expanded *ex vivo*. hMSCs are considered to be very promising candidates for bone and cartilage regeneration [4], due to their high osteogenic differentiation capacity. A promising approach to accelerate the healing of bone defects, including fractures created during bone lengthening by distraction osteogenesis (DO), might be the delivery of autologous MSCs directly to the damaged site [5]. This approach would provide a new strategy to manage the major, well-known problems caused by the common treatment of bone defects by autogenic bone grafting (e.g., [6]). At present, the most promising method for the application of MSCs appears to be the injection/delivery of the cells, embedded into platelet lysate from whole blood-derived pooled platelet concentrates and apheresis-derived platelet concentrates for the isolation and expansion of human bone marrow mesenchymal stromal cells [7]. Even though this technique sounds straightforward, it is presently difficult to deliver the cells to the desired location. Furthermore, a suitable, more solid-state matrix for embedding the MSCs during the injection process would improve the desired outcome.

Likewise important is the composition of a morphogenetically active scaffold for the three-dimensional (3D) growth of MSCs. A solution for formulating a suitable scaffold for MSCs would provide the basis for a future 3D printing/bioprinting of cells and human tissue/organs [8]. As a first step to reach this goal, the application of 3D printing in prosthetics, for the fabrication of human tissue bioprints inserted into lesions created during DO, is proposed. DO is an established and widely used technique for regenerating endogenous bone in orthopedic and maxillofacial surgery [9]. This surgical procedure aims to elicit a controlled regenerative process by the application of an active mechanical strain in order to initiate and enhance the biological response in the injured tissues to form new bone tissue. The regeneration process during DO includes four distinct stages [10]: (i) osteotomy; (ii) the latency phase, comprising the period between osteotomy and distraction (during this period, soft callus is formed); (iii) the distraction phase, during which traction is applied to transport bone; the formation of new immature woven and parallel-fibered bone islands commences; (iv) the consolidation phase, during which maturation and corticalization of the regenerating bone occurs. During the distraction phase, a dynamic microenvironment is established, characterized by an increased angiogenesis, which is paralleled and followed by an increased proliferation of spindle-shaped fibroblasts. In turn, collagen (mostly Type I) is formed alongside the angiogenic centers, allowing intramembranous, but not endochondral, ossification [11]. Besides an intense osteoblastic differentiation activity [12], occasionally, also, trans-differentiation processes of chondroblasts, as well as of fibroblasts into osteoblasts have been reported. During the distraction phase, as well as during the consolidation phase, the high expression of the genes encoding for the bone morphogenetic proteins 2 and 4 (BMP-2 and BMP-4) has been described [13]. A series of complications associated with DO has been reported that, besides intra-operative difficulties, include, in particular, intra-distraction complications, which arise during distraction, as well as post-distraction complications, which concern the late problems arising during the period of splinting, e.g., malunion or relapse [14].

Recently, we developed a morphogenetically active scaffold, based on biosilica-alginate hydrogel [8,15]. Alginates, polysaccharides, allow the fabrication of a variety of biomaterials suitable for tissue engineering, e.g., gels and fibers, and are suitable vehicles for injectable solutions as pastes [16]. If those alginates are enriched with biosilica, the fabricated hydrogel has been shown to provide a morphogenetically active scaffold for bone-related SaOS-2 cells *in vitro* [8]. Biosilica is a naturally occurring polymer used by the oldest metazoans, the sponges (phylum: Porifera), as elements for their spicule formation (reviewed in [17,18]). A likewise polymeric inorganic material is polyphosphate (polyP), which occurs in any living organisms and at high concentrations in sponges, as well (see [17]).

Based on initial studies [19,20], we discovered that biosilica, enzymatically formed from ortho-silicate by the enzyme silicatein [18], displays an inductive anabolic bone-forming effect on SaOS-2 cells. This polymer causes a significant shift of the OPG-RANKL (osteoprotegerin:receptor activator of nuclear factor-κB ligand) ratio [21], resulting in an inhibition of the differentiation pathway of pre-osteoclasts into mature osteoclasts. In addition to an increased mineralization, biosilica has been shown to increase the expression of BMP-2 in SaOS-2 cells [22]. Finally, biosilica shows osteogenic potential [21]. These data have been supported recently [23] using hMSCs.

PolyP is known to act as a storage substance of energy, a chelator for metal cations, a phosphate donor for sugars and adenylate kinase and an inducer of apoptosis; in addition, it is involved in mineralization processes of bone tissue (reviewed in [17]). Moreover, polyP acts as a modulator of gene expression, e.g., in the osteoblast-like cell lines, MC3T3-E1 and SaOS-2 cells, and in hMSCs, and causes an increased expression of the genes encoding for osteocalcin, osterix, bone sialoprotein, BMP-2 and tissue nonspecific alkaline phosphatase, all proteins that are crucial for bone formation ([15]; reviewed in [24]).

Figure 1. Multipotent differentiation of human multipotent stromal cells (hMSC). Specific transcription factors determine both the commitment and the differentiation of hMSCs towards the osteogenic, chondrogenic, adipogenic or myogenic lineage. The osteogenic and the chondrogenic lineages are involved in the restorative repair of bone and cartilage tissue (osteochondral tissue reconstitution). Biosilica and polyphosphate (polyP) display anabolic, morphogenetic effects on those two differentiation lines.

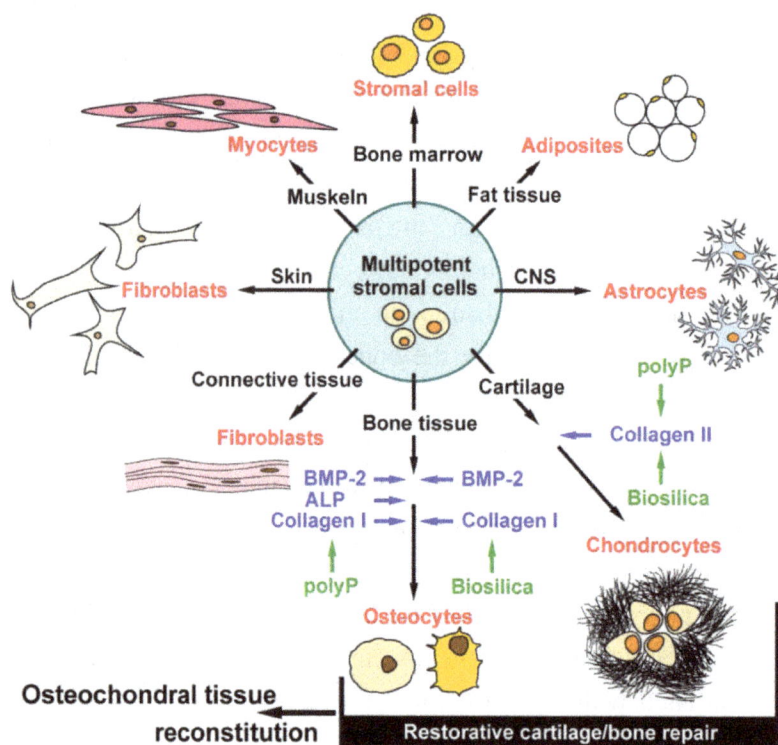

The available data indicate that both SaOS-2 cells and hMSCs, after encapsulation into alginate hydrogels, can retain their proliferation and differentiation-promoting activity if the matrix had been supplemented with biosilica and polyP. hMSCs can differentiate into several lineages (Figure 1), dependent on the inducers added to the assay system [25]. Osteogenic differentiation is triggered by incubation in medium/fetal calf serum (FCS), supplemented with dexamethasone, ascorbic acid and sodium β-glycerophosphate. Chondrogenic differentiation occurs in medium/serum, supplemented with transforming growth factor-β1, insulin, transferrin, dexamethasone and ascorbic acid. Adipogenic differentiation is promoted by medium/FCS, indomethacin, dexamethasone and 3-isobutyl-1-methylxanthine and insulin. Neurogenic differentiation is favored if the cells are incubated with β-mercaptoethanol.

The hMSCs provide a suitable cell source for osteochondral tissue reconstruction [26], required for an acceleration of the ossification processes during OD or after the transplantation of 3D tissue-like implants (Figure 1). Figure 1 also highlights recent findings that biosilica and polyP acts in an organic scaffold, like alginate, as a morphogenetically active inorganic polymer.

In the present study, we studied the differentiation of hMSCs towards the osteocyte and chondrocyte lineages. Both cell lineages are involved in bone formation and cartilage repair, two processes intimately involved in bone growth [27]. In turn, hMSCs are, due to their osteogenic and chondrogenic potential, attractive candidates for restorative cartilage/bone repair. Chondrocytes are known to express high levels of collagen Type II, while osteoblasts express collagen Type I [28]. Here, we used hMSCs to elucidate the morphogenetic potential of the two polymers of marine origin, biosilica and polyP, with respect to their differentiation capacity on the osteogenic differentiation, as well as on the chondrogenic differentiation lineage. The data show that these two polymers display a morphogenetic effect on both cell lineages.

2. Results and Discussion

2.1. Cultivation of hMSCs and Encapsulation into Alginate Beads

hMSCs were incubated in an α-MEM/FCS medium, as described in "Experimental Section". They show the characteristic plastic adherent properties (higher passage number) (Figure 2A,B). The hMSCs were subsequently induced to osteoblast- or chondrocyte-like cells and encapsulated into alginate beads of a diameter of ~1 mm. At the beginning, the cells were scattered within the alginate; after five days, the newly divided cells aggregate together and form clumps (Figure 2C,D).

Figure 2. Light microscopic images of hMSCs, cultured in a monolayer (**A** and **B**), adherent to the plastic surface. The hMSCs induced to the osteogenic or the chondrogenic lineages were embedded into alginate beads (**C** and **D**).

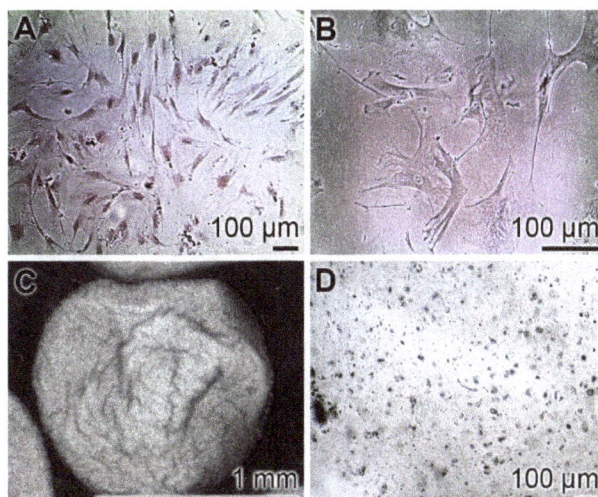

2.2. Mineralization of the Cells, Triggered to Osteogenic Differentiation

The hMSCs were encapsulated into alginate, either free of additional components "control-alginate" or containing either enzymatically synthesized biosilica "BS-alginate" or polyP (Ca^{2+} salt) "polyP-alginate". In our previous studies we used 100 to 400 μM prehydrolyzed TEOS (tetraethyl orthosilicate), together with silicatein [29] and polyP in the range of 50 to 100 μM [30] to activate cell metabolism and the gene expression of osteoblasts. In the present study, we exposed the hMSCs to 200 μM prehydrolyzed TEOS, together with 20 μg/mL of silicatein or to 50 μM polyP (Ca^{2+} salt).

The hMSCs were incubated as "control-alginate", as "BS-alginate" or as "polyP-alginate" beads for 10 days in medium/serum in the absence or in the presence of osteogenic medium. This cocktail is composed of dexamethasone, ascorbic acid and sodium β-glycerophosphate [22], in order to trigger the hMSCs to form the osteogenic differentiation lineage. During the incubation of the biosilica-containing or polyP-containing beads, those polymers (biosilica or polyP) were present also in the medium at the same concentrations.

Figure 3. Induction of mineralization of hMSCs, embedded in alginate matrix, if incubated in the presence of osteogenic medium (OM), composed of dexamethasone, ascorbic acid and sodium β-glycerophosphate. The hMSCs were transferred to alginate beads and were incubated in the absence of osteogenic medium (−OM) or the presence of osteogenic medium (+OM) for 10 days. In two series of experiments, the cultures remained either without biosilica (−BS) or without polyP (Ca^{2+} salt) (−polyP/Ca) or were exposed to those polymers (+BS; +polyP/Ca). Subsequently, the cross-linkages of alginate matrix were partially dissolved with Na-citrate and reacted with Alizarin Red S. The cells exposed to osteogenic medium and either biosilica or polyP (Ca^{2+} salt) (+OM +BS; +OM +polyP/Ca) showed the strongest staining intensity to the dye. All images are in the same magnification; the scale is shown in (A).

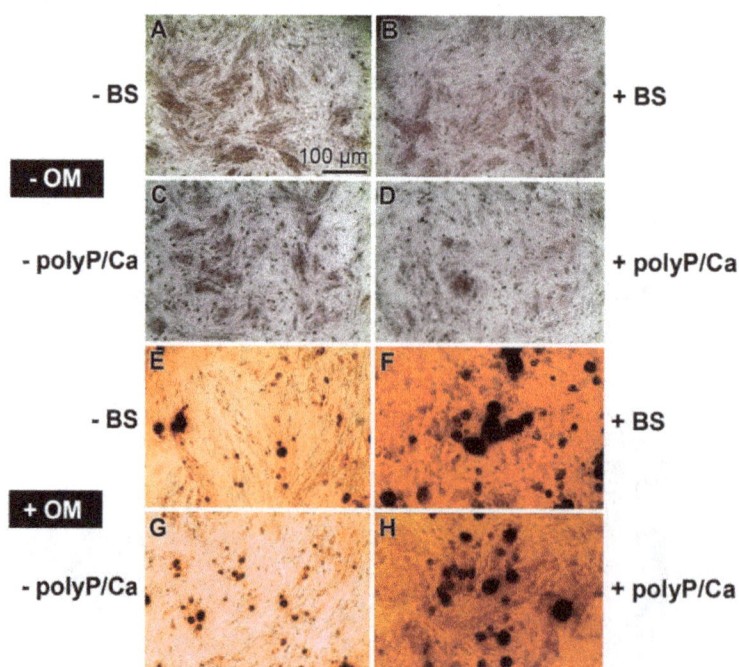

After a 10-day's incubation period, the alginate matrix was partially de-cross-linked with Na-citrate for 15 min. Then, the cells were transferred to a microscope slide and stained with Alizarin Red S. The results show that the cells, not treated with osteogenic medium, are not stained, irrespective of whether they were exposed to biosilica or to polyP or remained without these polymers (Figure 3A–D). However, the intensity of the red staining of the cells increased strongly if the cells were incubated with osteogenic medium (Figure 3E–H). The level of intensity further increased if the hMSCs were incubated, in addition to osteogenic medium (Figure 3E,G), either with biosilica or with polyP (Ca^{2+} salt) (Figure 3F,H).

2.3. Osteogenic versus *Chondrogenic Differentiation: Effect of Biosilica and Polyp on* BMP-2 *Expression*

Earlier, we reported that both biosilica and polyP cause *BMP-2* gene induction in SaOS-2 cells, if they grow in liquid medium [22]. In the present study, using alginate-encapsulated hMSCs, we exposed the beads either to osteogenic medium, to induce osteogenic differentiation, or to chondrocyte differentiation medium, to direct hMSCs to differentiation in the chondrogenic direction. The two sets of bead cultures were kept in the absence ("control-alginate") or the presence of either biosilica ("BS-alginate") or polyP (Ca^{2+} salt) ("polyP-alginate"). Samples were taken after one day, five days or 10 days. The alginate matrix around the cells was solubilized, and the released cells were subjected to real-time RT (reverse transcription)-PCR (qRT-PCR) to quantitatively determine the expression level of *BMP-2*. The expression was correlated to the expression of the house-keeping gene, *GAPDH*.

Figure 4. Levels of *BMP-2* transcripts in hMSCs after (**A**) induction with osteogenic medium to the osteogenic lineage or (**B**) triggering to form the chondrogenic lineage with chondrocyte differentiation medium (CDM). The cells were embedded into alginate beads and exposed to biosilica (BS; squared bars) or polyP (Ca^{2+} salt; closed bars) (polyP), as described in "Experimental Section"; the controls (open bars) did not receive those polymers (BS or polyP). Samples of beads were collected at Day 1, Day 5 and Day 10, after starting the experiments. The cells were released from the alginate matrix and, after isolation of the RNA, subjected to real-time RT (reverse transcription)-PCR (qRT-PCR). The expression level of *BMP-2* was normalized to the expression of *GAPDH*. Data are expressed as mean values ± SD for four independent experiments. Differences between the groups were evaluated using the unpaired *t*-test. * $p < 0.05$.

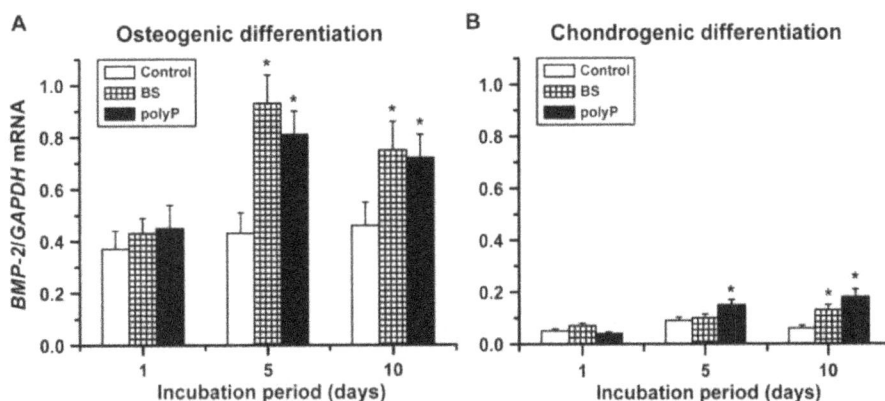

The experiments revealed that in the osteoblasts lineage, a high steady-state level of *BMP-2* transcription occurred already after five days if the cells in the beads were exposed either to biosilica or to polyP (Ca^{2+} salt). This increase is about two-fold, if compared to control cultures. After an extended incubation for 10 days, the transcription level drops to 1.5-fold, with respect to the controls (Figure 4A). In contrast, the expression level of *BMP-2* in the chondrogenic lineage is low, and (measured in the controls) it is approximately 15% of the level seen in the osteogenic lineage. If the cells/beads are exposed for five days or 10 days to biosilica or polyP (Ca^{2+} salt), the level of *BMP-2* in the chondrogenic lineage increases only up to about 25% of the levels found in the osteogenic lineage (Figure 4B).

2.4. Osteogenic versus Chondrogenic Differentiation: Expression of ALP

As expected, the basal level of the *ALP* expression is already about 1.4-fold higher in the osteogenic cells (Figure 5A), compared to the transcription level in the chondrogenic cells (Figure 5B). This level is significantly increased to two-fold after the five-day incubation period in the presence of biosilica or in the presence of polyP (Ca^{2+} salt) to three-fold. After an incubation period of 10 days, this increase is even enhanced; the level in the presence of polyP (Ca^{2+} salt) is 5.9-fold compared to the controls.

Furthermore, chondrocytes contain the ALP that cleaves pyrophosphate [31], even though at a lower level. The overall *ALP* transcription level in chondrogenic cells is up to about 30% lower compared to the osteogenic cells (Figure 5B). The inducing activity of both biosilica and polyP (Ca^{2+} salt) in chondrogenic cells is significant. However, compared to the osteogenic cells, the level of induction is about three-fold lower in chondrogenic cells.

Figure 5. The induction *alkaline phosphatase* (*ALP*) gene in alginate-encapsulated hMSCs induced to either (**A**) the osteogenic lineage or (**B**) the chondrogenic lineage. The cells in those beads were incubated in the absence (controls; open bars), or the presence of either biosilica (BS; squared bars) or polyP (Ca^{2+} salt) (polyP; closed bars). * $p < 0.05$.

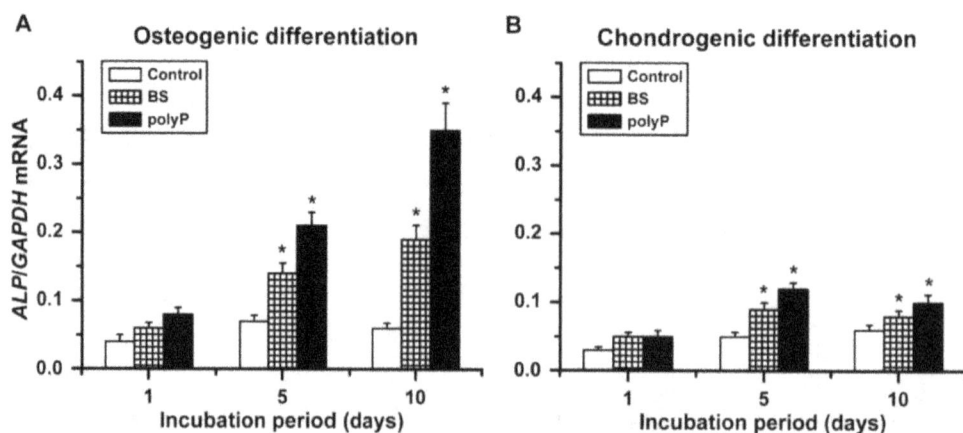

2.5. Osteogenic versus Chondrogenic Differentiation: Expression of Collagen Type I

Osteoblasts can be differentiated in their expression levels of the collagen isoforms. While during osteogenic differentiation, the *collagen type I* gene undergoes a strong and increased expression [15,32], chondrocytes express primarily *collagen type II* [33].

The steady-state expression of the *collagen type I* gene is higher in the osteogenic cells (Figure 6A), compared to the chondrogenic cells (Figure 6B). If the osteogenic cells are exposed to both biosilica and polyP (Ca^{2+} salt), an increased steady-state mRNA level is seen at Day 5 and Day 10; this increase is more pronounced in cultures supplemented with biosilica. No significant alteration of the *collagen type I* transcription level was measured in response to biosilica or polyP (Ca^{2+} salt) in chondrogenic cells (Figure 6B).

Figure 6. Effect of biosilica or polyP (Ca^{2+} salt) on the expression level of *collagen type I*. The osteogenic cells (**A**) or the chondrogenic cells (**B**) were incubated with biosilica, or with polyP (Ca^{2+} salt) or remained untreated (controls). * $p < 0.05$.

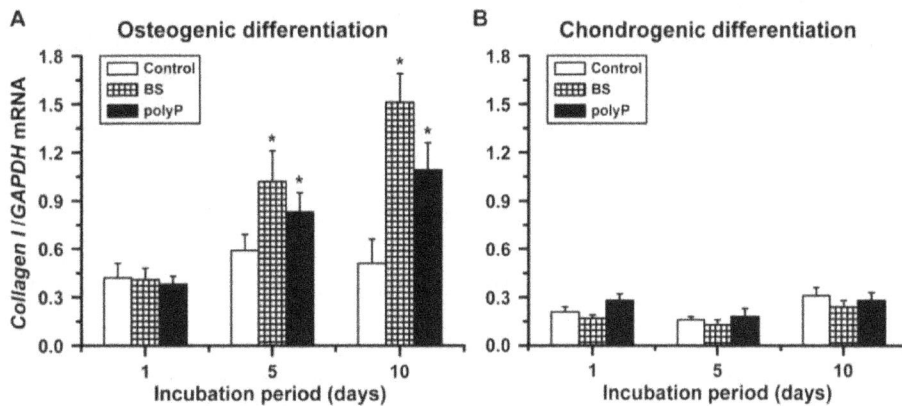

2.6. Osteogenic versus *Chondrogenic Differentiation: Expression of the* Collagen Type II

The overall gene expression level of *collagen type II* in osteogenic cells (Figure 7A) is about half the one measured in the chondrogenic cells (Figure 7B). The inductive effect of biosilica and polyP (Ca^{2+} salt) is low in osteogenic cells, compared to the one measured for chondrogenic cells. Biosilica and polyP (Ca^{2+} salt) cause, at Day 5, a significant *collagen type II* gene induction in osteogenic cells. This increase of the steady-state expression of *collagen type II* is low in osteogenic cells compared to the expression seen for chondrogenic cells.

Figure 7. Gene expression of *collagen type II* in osteogenic cells (**A**) *versus* chondrogenic cells (**B**). As marked, the induced hMSCs were incubated with either biosilica, or polyP (Ca^{2+} salt) or remained without these polymers (controls). * $p < 0.05$.

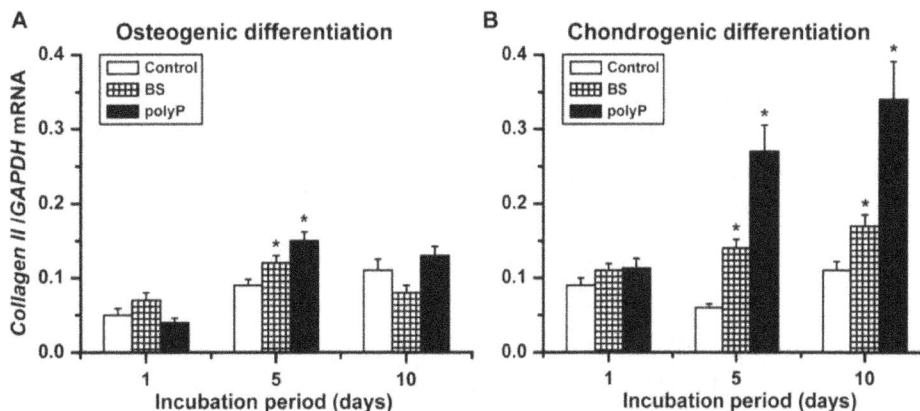

2.7. Discussion

The major challenge to realize an effective 3D printing of custom-built tissue implants is the development of a suitable morphogenetically active scaffold that allows the cells to be embedded to communicate with each other and to direct the individual cells to undergo differentiation into functional "terminally"-differentiated cells. Those cross-talking cells in a fluid/solid matrix will not only fill the space within the damaged tissue, but will also facilitate the replacement of the implanted material by physiologically developing cell assemblies, giving rise to spatially organized multicomponent tissues and structures. Alginate has been found to be a promising matrix for cells, potentially to be used for 3D printing [34]; this plant-derived polymer is, however, morphogenetically inactive. However, if alginate is enriched with organic cytokines/morphogens, e.g., BMP-2 [35], this inert matrix acquires the potency to facilitate the differentiation of stem cells *in vitro*. A further approach, which is likewise physiological, but by far less expensive, is to substitute the organic growth factors in the alginate with inorganic polymers, e.g., biosilica or polyP. Both inorganic polymers are physiologically formed in metazoans [24,36].

Figure 8. Distraction osteogenesis scheme. (**A**) The healthy part of the bone is broken into two segments with an external instrument. A distraction gap is formed; (**B**) distraction osteogenesis phases: osteotomy/latency, distraction, consolidation. During the distraction phase, hMSCs, embedded into alginate, are proposed to be injected into the fracture zone. These supplements are hoped to accelerate the velocity of bone formation under shortening the consolidation period. This alginate matrix is proposed to be supplemented with the morphogenetically active inorganic polymers, biosilica and polyP.

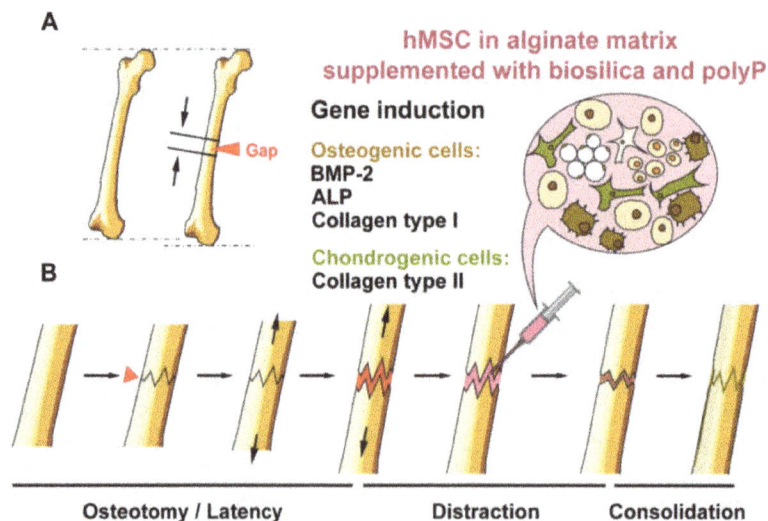

Besides 3D tissue printing, a morphogenetically active matrix is also needed to accelerate the regeneration step during DO. This technique, DO, requires a long time, during which the external fixator is attached at the limb during the consolidation phase. To ameliorate the medical, psychological, social and financial problems for the patient, successful attempts have been made to administer organic cytokines/morphogens, e.g., BMP-2, to accelerate bone formation [37]. During the first phase of DO/osteotomy, the latency period, distraction starts under the formation of soft callus and initiation of bone repair processes [14] (Figure 8). During the distraction phase and the subsequent consolidation period, the bone-lacking gap is filled with bone cells until the bone is solidified and the fractures are healed.

The present study shows that both biosilica and polyP provide hMSC-containing alginate hydrogels with a morphogenetic/osteogenic potential, resulting in an increased growth and differentiation potential of the osteoblasts and their precursor cells. In detail, the genes encoding BMP-2, ALP and collagen Type I are induced, three functional and structural prerequisites for a functional activity of the osteoblasts. Furthermore, BMP-2 is known to induce ectopic bone formation and increases bone repair in several animal models. BMP-2 and the related morphogens, BMP-4 and BMP-7, promote cell differentiation into osteoblasts (see [38]). ALP, for a long time already implicated in biomineralization, is a feature of the osteoblast phenotype [39]. Moreover, experimental evidence is available that ALP, which is associated with cell membranes and matrix vesicles, is involved in bio-seed formation during hydroxyapatite formation (see [40]). *Collagen type I* gene expression precedes ALP formation in osteoblasts and in a back-circuit, promotes osteoblast differentiation from precursor cells [41].

Biosilica and polyP also act in a stimulatory manner on the differentiation of hMSCs to chondrogenic cells. Both polymers have been shown to induce in hMSCs, embedded into alginate, the gene encoding for collagen Type II. This fibrillar structural protein, again in a back-circle, promotes chondrogenic differentiation from hMSCs [42].

3. Experimental Section

3.1. Isolation and Cultivation of Human MSCs

The hMSCs were isolated using previously described methods [25,43]. The human cells were obtained, after approval from the ethics committee, from bone marrow aspirations after informed consent of the donors. The deep-frozen, preserved hMSCs were thawed, and 1×10^6 cells were suspended in one 75-cm^2 flask (Cat. no. 658175; Greiner, Frickenhausen, Germany) and cultivated in α-MEM (Cat. no. F0915; Biochrom, Berlin, Germany), supplemented with 20% fetal calf serum (FCS; Gibco Invitrogen, Carlsbad, CA, USA), as well as with 0.5 mg/mL of gentamycin, 100 units penicillin and 100 μg/mL of streptomycin, as well as 1 mM pyruvate (Sigma, Taufkirchen, Germany). The characteristics of the hMSCs with respect to their osteoblast, as well as adipogenic and chondrogenic differentiation potentials, have been given in detail [25,43]. The incubation was performed in a humidified incubator at 37 °C and 5% CO_2. After 2 days, the non-adherent cells were discarded, and the adherent cells continued to be incubated with α-MEM/FCS. Then, the culture medium was renewed every 3 days. At ~80% confluence, the cells were suspended using a 0.25% trypsin/ 0.02% EDTA solution (Sigma) and plated at a concentration of ~5000 cells/cm^2. The cultures were split at a ratio of 1:3 every 5 to 6 days after the first passage. Those cells were used for the experiments, mainly for encapsulation into alginate.

The assays for staining the cells were performed in 24-well plates (Cat. no. 662160; Greiner), while the experiments with cells subjected to qRT-PCR were performed either in 48-well plates (Cat. no. 677102; Greiner) or in 25-cm^2 flasks (Cat. no. 690175; Greiner). At the time of harvesting the cells, the cultures were approximately 80% confluent.

3.2. Preparation of Alginate/Silica and Alginate/PolyP (Ca²⁺ Salt) Composite Hydrogel Beads

Adherent cells from ~80% confluent cultures were released from the plates and suspended in 1.2% (w/v) alginate (Na-alginate, Cat. no. CAA W20,150-2, low viscosity, dissolved in PBS, pH 7.4; Sigma) and processed as described [34]. In brief, the cell-hydrogel suspensions were passed through a needle (dimension: 0.45 × 6 × 23 mm) attached to a 1 mL syringe and dropped into a 1.5% (w/v) $CaCl_2$ solution. After 5 min, the approximately 1 mm-large beads were washed three times in saline and then twice in α-MEM medium. The beads (approximately 60 beads/well) were placed into 24-well plates (Nunc, Langenselbold, Germany) and incubated in 3 mL of α-MEM/FCS in the presence of the respective induction medium. Those cultures were termed "control-alginate".

After termination of the experiments, the cells were released from the beads by washing twice with PBS and transferring them into 2 mL of 55 mM Na-citrate; after 30 min at 37 °C, the cells were collected by centrifugation (900× g, 5 min; [34,44]) and used for quantitative real-time RT-PCR determinations. For staining the cells with Alizarin Red S, the period for de-cross-linking of alginate with Na-citrate was shortened to 15 min. Then, the cells were stained.

Where indicated, the alginate beads were supplemented with biosilica or with polyP. Biosilica was prepared as described [29]. Prehydrolyzed TEOS (tetraethyl orthosilicate; Sigma) in a concentration of 200 μM was added to 20 μg/mL of recombinant silicatein; these components were added to the alginate matrix. After each medium change, prehydrolyzed TEOS/recombinant silicatein was added again to the medium. The biosilica-containing beads were termed "BS-alginate". In a control experiment, silicatein was replaced in the assays by 20 μg/mL of bovine serum albumin (BSA); those beads displayed no effect on the expression of the genes selected in the present study (data not shown). The beads were incubated for up to 10 days.

Na-polyP (average chain of approximately 40 phosphate units) was obtained from Chemische Fabrik Budenheim (Budenheim, Germany). To compensate for any effect, caused by a potential chelating activity of polyP to Ca^{2+}, the polymer was mixed together with $CaCl_2$ in a stoichiometric ratio of 2:1 (polyP:$CaCl_2$), as described [30]; the salt, designated as "polyP (Ca^{2+} salt)", was added to the beads and the medium at a concentration of 50 μM. Incubation was again for up to 10 days. The beads containing polyP were termed "polyP-alginate".

3.3. Differentiation Assays in Vitro

Osteogenic differentiation of hMSCs, encapsulated into hydrogel beads, was performed by exposure of the cells to the osteogenic medium, as described [22]. Osteogenic medium contained in the α-MEM/FCS, dexamethasone (Sigma), ascorbic acid (Sigma) and sodium β-glycerophosphate (Sigma). As end-point marker for osteoblasts, the extent of mineralization based on Alizarin Red S staining, as well as the gene expression levels of selected marker genes was determined.

Chondrogenic differentiation of the stem cells was induced in α-MEM/FCS supplemented with chondrocyte differentiation medium, as described [3,27]. It consists of premix tissue culture supplement (Becton Dickinson, Heidelberg, Germany), dexamethasone, ascorbate-2-phosphate, pyruvate and transforming growth factor-β1 (Sigma), human insulin (Sigma) and transferrin (Sigma).

The described experiments were performed after having transferred the hMSCs to the respective activation medium.

The cells/beads were inspected with light microscopically using either a Keyence BZ-8000 epifluorescence microscope or a Keyence VK-8710K, color 3D laser microscope (Neu-Isenburg, Germany).

3.4. Mineralization Assay with Alizarin Red S

Mineralization by differentiated hMSCs was qualitatively assessed by staining the cell cultures on the coverslips with 10% Alizarin Red S, after fixation with ethanol [45].

3.5. Quantitative Real-Time RT-PCR (qRT-PCR) Analysis

The technique of quantitative real-time RT (reverse transcription)-PCR (qRT-PCR) was applied to determine the levels of transcription of the following genes: *BMP-2* (*bone morphogenetic protein-2*; NM_001200.2) Fwd: 5′-ACCCTTTGTACGTGGACTTC-3′ (nt_{1681} to nt_{1700}); and Rev: 5′-GTGGAG TTCAGATGATCAGC-3′ (nt_{1785} to nt_{1804}; 124 bp); *ALP* (*alkaline phosphatase*; NM_000478.4) Fwd: 5′-TGCAGTACGAGCTGAACAGGAACA-3′ (nt_{1141} to nt_{1164}); and Rev: 5′-TCCACCAAATGTG AAGACGTGGGA-3′ (nt_{1418} to nt_{1395}; 278 bp); *COLI* (*collagen type I*; NM_000088) Fwd: 5′-TATGG GACCCCAAGGACCAAAAGG-3′ (nt_{1122} to nt_{1145}); and Rev: 5′-TTTTCCATCTGACCCAGGG GAACC-3′ (nt_{1234} to nt_{1257}) (136 bp); *COLII* (*collagen type II*, alpha 1 (COL2A1), transcript variant 1, mRNA: NM_001844) Fwd: 5′-TCCATTCATCCCACCCTCTCAC-3′ (nt_{4755} to nt_{4776}); and Rev: 5′-TTTCCTGCCTCTGCCTTGACC-3′ (nt_{4902} to nt_{4882}; 148 bp). As a reference gene, GAPDH (glyceraldehyde 3-phosphate dehydrogenase; NM_002046.3) (Fwd: 5′-CCGTCTAGAAAAACC TGCC-3′ (nt_{845} to nt_{863}); and Rev: 5′-GCCAAATTCGTTGTCATACC-3′ (nt_{1059} to nt_{1078}; 215 bp)) was used.

The cells were released from the hydrogel samples and collected by centrifugation. After RNA extraction using the TRIzol reagent (Invitrogen GmbH, Darmstadt, Germany), the samples were subjected to qRT-PCR. For that, 2 μL of the appropriate dilution were employed as a template in the 30 μL qRT-PCR assays. All reactions were run with an initial denaturation at 95 °C for 3 min, followed by 40 cycles, each with 95 °C for 20 s, 58 °C for 20 s, 72 °C for 20 s and 80 °C for 20 s. Fluorescence data were collected at the 80 °C step. The runs were performed in an iCycler (Bio-Rad, Hercules, CA, USA). The mean C_t values and efficiencies were calculated by applying the iCycler software (Bio-Rad); the estimated PCR efficiencies were in the range of 93%–103%. Expression levels were correlated to the GAPDH reference gene to determine relative expression, as described [21].

3.6. Further Analyses

The results were statistically evaluated using the paired Student's *t*-test [46]. DNA content was determined by application of the PicoGreen method, as described [45], using calf thymus DNA as a standard.

4. Conclusion

In conclusion, for the two inorganic polymers, biosilica and polyP, which are abundantly found in marine organisms, our findings extend our earlier results obtained with (almost) terminally differentiated osteoblasts (reviewed in [17]), establishing that, both inorganic polymers, biosilica and polyP, are potent morphogenetically active additives of the alginate matrix.

Acknowledgments

We thank Sandra Ritz (Max-Planck-Institute for Polymer Research, Mainz, Germany) for supplying us with the hMSCs, as well as for introducing us to the technique of stem cell cultivation. W.E.G.M. is a holder of an ERC Advanced Investigator Grant (no. 268476 BIOSILICA), as well as of an ERC proof-of-concept grant (no. 324564; Silica-based nanobiomedical approaches for treatment of bone diseases). This work was supported by grants from the European Commission (no. 311848 "SPECIAL", and large-scale integrating project no. FP7-KBBE-2010-4-266033 "BlueGenics", as well as European-Chinese Research Staff Exchange Cluster PIRSES-GA-2009-246987 "MarBioTec* EU-CN*"), the International Human Frontier Science Program, the German Bundesministerium für Bildung und Forschung, International Bureau (no. CHN 09/1AP, German-Chinese Joint Lab on Bio-Nano-Composites), and the BiomaTiCS research initiative of the University Medical Center, Mainz.

References

1. Parra-Torres, A.Y.; Valdés-Flores, M.; Orozco, L.; Velázquez-Cruz, R. Molecular Aspects of Bone Remodeling. In *Topics in Osteoporosis*; InTech: Rijeka, Croatia, 2013; pp. 1–27.

2. Friedenstein, A.J.; Petrakova, K.V.; Kurolesova, A.I.; Frolova, G.P. Heterotopic of bone marrow. Analysis of precursor cells for osteogenic and hematopoietic tissues. *Transplantation* **1968**, *6*, 230–247.

3. Pittenger, M.F.; Mackay, A.M.; Beck, S.C.; Jaiswal, R.K.; Douglas, R.; Mosca, J.D.; Moorman, M.A.; Simonetti, D.W.; Craig, S.; Marshak, D.R. Multilineage potential of adult human mesenchymal stem cells. *Science* **1999**, *284*, 143–147.

4. Beyth, S.; Schröder, J.; Liebergall, M. Stem cells in bone diseases: Current clinical practice. *Br. Med. Bull.* **2011**, *99*, 199–210.

5. Sunay, O.; Can, G.; Cakir, Z.; Denek, Z.; Kozanoglu, I.; Erbil, G.; Yilmaz, M.; Baran, Y. Autologous rabbit adipose tissue-derived mesenchymal stromal cells for the treatment of bone injuries with distraction osteogenesis. *Cytotherapy* **2013**, *15*, 690–702.

6. Panetta, N.J.; Gupta, D.M.; Slater, B.J.; Kwan, M.D.; Liu, K.J.; Longaker, M.T. Tissue engineering in cleft palate and other congenital malformations. *Pediatr. Res.* **2008**, *63*, 545–551.

7. Gessmann, J.; Köller, M.; Godry, H.; Schildhauer, T.A.; Seybold, D. Regenerate augmentation with bone marrow concentrate after traumatic bone loss. *Orthop. Rev.* **2012**, *4*, e14.

8. Müller, W.E.G.; Schröder, H.C.; Feng, Q.L.; Schloßmacher, U.; Link, T.; Wang, X.H. Development of a morphogenetically active scaffold for three-dimensional growth of bone cells: Biosilica/Alginate hydrogel for SaOS-2 cell cultivation. *J. Tissue Engin. Regener. Med.* **2013**, doi:10.1002/term.1745.

9. Ilizarov, G.A. Clinical application of the tension–stress effect for limblengthening. *Clin. Orthop. Relat. Res.* **1990**, *250*, 8–26.

10. Cope, J.B.; Samchukov, M.L.; Cherkashin, A.M. Mandibular distraction osteogenesis: A historic perspective and future directions. *Am. J. Orthod. Dentofacial Orthop.* **1999**, *115*, 448–460.

11. Jazrawi, L.M.; Majeska, R.J.; Klein, M.L.; Kagel, E.; Stromberg, L.; Einhorn, T.A. Bone and cartilage formation in an experimental model of distraction osteogenesis. *J. Orthop. Trauma* **1998**, *12*, 111–116.

12. Sato, M.; Yasui, N.; Nakase, T.; Kawahata, H.; Sugimoto, M.; Hirota, S.; Kitamura, Y.; Nomura, S.; Ochi, T. Expression of bone matrix proteins mRNA during distraction osteogenesis. *J. Bone Miner. Res.* **1998**, *13*, 1221–1231.

13. Matsubara, H.; Hogan, D.E.; Morgan, E.F.; Mortlock, D.P.; Einhorn, T.A.; Gerstenfeld, L.C. Vascular tissues are a primary source of BMP2 expression during bone formation induced by distraction osteogenesis. *Bone* **2012**, *51*, 168–180.

14. Hegab, A.F.; Shuman, M.A. Distraction osteogenesis of the maxillofacial skeleton: biomechanics and clinical implications. *Sci. Rep.* **2012**, *1*, 1–12.

15. Müller, W.E.G.; Wang, X.H.; Grebenjuk, V.; Diehl-Seifert, B.; Steffen, R.; Schloßmacher, U.; Trautwein, A.; Neumann, S.; Schröder, H.C. Silica as a morphogenetically active inorganic polymer: effect on the BMP-2-dependent and RUNX2-independent pathway in osteoblast-like SaOS-2 cells. *Biomater. Sci.* **2013**, *1*, 669–678.

16. Andersen, T.; Strand, B.L.; Formo, K.; Alsberg, E.; Christensen, B.E. Alginates as biomaterials in tissue engineering. *Carbohydr. Chem.* **2012**, *37*, 227–258.

17. Wang, X.H.; Schröder, H.C.; Wiens, M.; Ushijima, H.; Müller, W.E.G. Bio-silica and bio-polyphosphate: Applications in biomedicine (bone formation). *Curr. Opin. Biotechnol.* **2012**, *23*, 570–578.

18. Müller, W.E.G.; Schröder, H.C.; Burghard, Z.; Pisignano, D.; Wang, X.H. Silicateins—A novel paradigm in bioinorganic chemistry: Enzymatic synthesis of inorganic polymeric silica. *Chem. Eur. J.* **2013**, *19*, 5790–5804.

19. Carlisle, E.M. Silicon: An essential element for the chick. *Science* **1972**, *178*, 619–621.

20. Schwarz, K.; Milne, D.B. Growth-promoting effects of silicon in rats. *Nature* **1972**, *239*, 333–334.

21. Wiens, M.; Wang, X.H.; Schröder, H.C.; Kolb, U.; Schloßmacher, U.; Ushijima, H.; Müller, W.E.G. The role of biosilica in the osteoprotegerin/RANKL ratio in human osteoblast-like cells. *Biomaterials* **2010**, *31*, 7716–7725.

22. Wiens, M.; Wang, X.H.; Schloßmacher, U.; Lieberwirth, I.; Glasser, G.; Ushijima, H.; Schröder, H.C.; Müller, W.E.G. Osteogenic potential of bio-silica on human osteoblast-like (SaOS-2) cells. *Calcif. Tissue Int.* **2010**, *87*, 513–524.

23. Han, P.; Wu, C.; Xiao, Y. The effect of silicate ions on proliferation, osteogenic differentiation and cell signalling pathways (WNT and SHH) of bone marrow stromal cells. *Biomater. Sci.* **2013**, *1*, 379–392.

24. Müller, W.E.G.; Albert, O.; Schröder, H.C.; Wang, X.H. Bio-inorganic nanomaterials for biomedical applications (Bio-silica and polyphosphate). In *Handbook of Nanomaterials Properties*; Bhushan, B., Luo, D., Schricker, S., Sigmund, W., Zauscher, S., Eds.; Springer-Press: Berlin, Germany, 2014; in press.

25. Pittenger, M.F.; Martin, B.J. Mesenchymal stem cells and their potential as cardiac therapeutics. *Circ. Res.* **2004**, *95*, 9–20.

26. Caplan, A.I. Mesenchymal stem cells. *J. Orthop. Res.* **1991**, *9*, 641–50.

27. Solchaga, L.A.; Penick, K.J.; Welter, J.F. Chondrogenic differentiation of bone marrow-derived mesenchymal stem cells: tips and tricks. *Methods Mol. Biol.* **2011**, *698*, 253–278.

28. Deshmukh, K.; Sawyer, B.D. Synthesis of collagen by chondrocytes in suspension culture: modulation by calcium, 3':5'-cyclic AMP, and prostaglandins. *Proc. Natl. Acad. Sci. USA* **1977**, *74*, 3864–3868.

29. Schlossmacher, U.; Wiens, M.; Schröder, H.C.; Wang, X.H.; Jochum, K.P.; Müller, W.E.G. Silintaphin-1: Interaction with silicatein during structureguiding biosilica formation. *FEBS J.* **2011**, *278*, 1145–1155.

30. Müller, W.E.G.; Wang, X.H.; Diehl-Seifert, B.; Kropf, K.; Schloßmacher, U.; Lieberwirth, I.; Glasser, G.; Wiens, M.; Schröder, H.C. Inorganic polymeric phosphate/polyphosphate as an inducer of alkaline phosphatase and a modulator of intracellular Ca^{2+} level in osteoblasts (SaOS-2 cells) *in vitro*. *Acta Biomater.* **2011**, *7*, 2661–2671.

31. Xu, Y.; Pritzker, K.P.; Cruz, T.F. Characterization of chondrocyte alkaline phosphatase as a potential mediator in the dissolution of calcium pyrophosphate dihydrate crystals. *J. Rheumatol.* **1994**, *21*, 912–919.

32. Reffitt, D.M.; Ogston, N.; Jugdaohsingh, R.; Cheung, H.F.; Evans, B.A.; Thompson, R.P.; Powell, J.J.; Hampson, G.N. Orthosilicic acid stimulates collagen type 1 synthesis and osteoblastic differentiation in human osteoblast-like cells *in vitro*. *Bone* **2003**, *32*, 127–135.

33. Ma, H.L.; Hung, S.C.; Lin, S.Y.; Chen, Y.L.; Lo, W.H. Chondrogenesis of human mesenchymal stem cells encapsulated in alginate beads. *J. Biomed. Mater. Res. A* **2003**, *64*, 273–281.

34. Schloßmacher, U.; Schröder, H.C.; Wang, X.H.; Feng, Q.; Diehl-Seifert, B.; Neumann, S.; Trautwein, A.; Müller, W.E.G. Alginate/silica composite hydrogel as a potential morphogenetically active scaffold for three-dimensional tissue engineering. *RSC Adv.* **2013**, *3*, 11185–11194.

35. Lim, H.J.; Ghim, H.D.; Choi, J.H.; Chung, H.Y.; Lim, J.O. Controlled release of BMP-2 from alginate nanohydrogels enhanced osteogenic differentiation of human bone marrow stromal cells. *Macromol. Res.* **2010**, *18*, 787–792.

36. Wang, X.H.; Schröder, H.C.; Diehl-Seifert, B.; Kropf, K.; Schloßmacher, U.; Wiens, M.; Müller, W.E.G. Dual effect of inorganic polymeric phosphate/polyphosphate on osteoblasts and osteoclasts *in vitro*. *J. Tissue Eng. Regen. Med.* **2013**, *7*, 767–776.

37. Mandu-Hrit, M.; Haque, T.; Lauzier, D.; Kotsiopriftis, M.; Rauch, F.; Tabrizian, M.; Henderson, J.E.; Hamdy, R.C. Early injection of OP-1 during distraction osteogenesis accelerates new bone formation in rabbits. *Growth Factors* **2006**, *24*, 172–183.

38. Osyczka, A.M.; Diefenderfer, D.L.; Bhargave, G.; Leboy, P.S. Different effects of BMP-2 on marrow stromal cells from human and rat bone. *Cells Tissues Organs* **2004**, *176*, 109–119.

39. Fedarko, N.S.; Bianco, P.; Vetter, U.; Robey, P.G. Human bone cell enzyme expression and cellular heterogeneity: Correlation of alkaline phosphatase enzyme activity with cell cycle. *J. Cell Physiol* **1990** *144*, 115 121

40. Golub, E.E.; Boesze-Battaglia, K. The role of alkaline phosphatase in mineralization. *Curr. Opin. Orthop.* **2007**, *18*, 444–448.

41. Shi, S.; Kirk, M.; Kahn, A.J. The role of type I collagen in the regulation of the osteoblast phenotype. *J. Bone Miner. Res.* **1996**, *11*, 1139–1145.

42. Chen, C.W.; Tsai, Y.H.; Deng, W.P.; Shih, S.N.; Fang, C.L.; Burch, J.G.; Chen, W.H.; Lai, W.F. Type I and II collagen regulation of chondrogenic differentiation by mesenchymal progenitor cells. *J. Orthop. Res.* **2005**, *23*, 446–453.

43. Lorenz, M.R.; Holzapfel, V.; Musyanovych, A.; Nothelfer, K.; Walther, P.; Frank, H.; Landfester, K.; Schrezenmeier, H.; Mailänder, V. Uptake of functionalized, fluorescent-labeled polymeric particles in different cell lines and stem cells. *Biomaterials* **2006**, *27*, 2820–2828.

44. Shoichet, M.S.; Li, R.H.; White, M.L.; Winn, S.R. Stability of hydrogels used in cell encapsulation: An *in vitro* comparison of alginate and agarose. *Biotechnol. Bioeng.* **1996**, *50*, 374–381.

45. Schröder, H.C.; Borejko, A.; Krasko, A.; Reiber, A.; Schwertner, H.; Müller, W.E.G. Mineralization of SaOS-2 cells on enzymatically (Silicatein) modified bioactive osteoblast-stimulating surfaces. *J. Biomed. Mat. Res. B* **2005**, *75B*, 387–392.

46. Sachs, L. *Angewandte Statistik*; Springer: Berlin, Germany, 1984; p. 242.

8

Development of Pedigree Classification using Microsatellite and Mitochondrial Markers for Giant Grouper Broodstock (*Epinephelus lanceolatus*) Management in Taiwan

Hsiao-Che Kuo [1,2,3,†], **Hao-Hsuan Hsu** [1,2,3,†], **Chee Shin Chua** [1,3], **Ting-Yu Wang** [1,2], **Young-Mao Chen** [1,2,3] and **Tzong-Yueh Chen** [1,2,3,4,5,]*

[1] Laboratory of Molecular Genetics, Institute of Biotechnology, College of Bioscience and Biotechnology, National Cheng Kung University, Tainan 70101, Taiwan;
E-Mails: shoujer@gmail.com (H.-C.K.); samuel0801@msn.com (H.-H.H.); cheeshinc@gmail.com (C.S.C.); nnigdd@gmail.com (T.-Y.W.); ymc868@yahoo.com.tw (Y.-M.C.)

[2] Translational Center for Marine Biotechnology, National Cheng Kung University, Tainan 70101, Taiwan

[3] Agriculture Biotechnology Research Center, National Cheng Kung University, Tainan 70101, Taiwan

[4] University Center of Bioscience and Biotechnology, National Cheng Kung University, Tainan 70101, Taiwan

[5] Research Center of Ocean Environment and Technology, National Cheng Kung University, Tainan 70101, Taiwan

[†] These authors contributed equally to this work.

[*] Author to whom correspondence should be addressed; E-Mail: ibcty@mail.ncku.edu.tw

Abstract: Most giant groupers in the market are derived from inbred stock. Inbreeding can cause trait depression, compromising the animals' fitness and disease resistance, obligating farmers to apply increased amounts of drugs. In order to solve this problem, a pedigree classification method is needed. Here, microsatellite and mitochondrial DNA were used as genetic markers to analyze the genetic relationships among giant grouper broodstocks. The 776-bp fragment of high polymorphic mitochondrial D-loop sequence was selected for measuring sibling relatedness. In a sample of 118 giant groupers, 42 haplotypes were categorized, with nucleotide diversity (π) of 0.00773 and haplotype diversity (HD) of 0.983. Furthermore, microsatellites were used for investigation of parentage. Six out of

33 microsatellite loci were selected as markers based on having a high number of alleles and compliance with Hardy-Weinberg equilibrium. Microsatellite profiles based on these loci provide high variability with low combined non-exclusion probability, permitting practical use in aquaculture. The method described here could be used to improve grouper broodstock management and lower the chances of inbreeding. This approach is expected to lead to production of higher quality groupers with higher disease resistance, thereby reducing the need for drug application.

Keywords: giant grouper; inbreeding; mitochondria; microsatellite

1. Introduction

The numbers of domesticated and farmed fish are increasing to facilitate feeding of the burgeoning human population [1]. More than 50 different species of grouper inhabit the tropical waters around Taiwan and some have been farmed since 1979. Among them, giant grouper is the most valuable grouper species in Taiwan. Although key aquaculture techniques on an industrial scale have been well established, outbreaks of various diseases remain a major unsolved problem [2–4]. However, most grouper broodstocks are second generation from wild-caught parent fish [5]. The larvae (third generation) cultivated in the fish farm for sale are derived from those broodstocks, and some of these larvae may be kept and used as broodstocks. The mixture of second- and third-generation broodstocks in the same pond means that the larvae (fourth generation) will be derived by inbreeding. Inbreeding is expected to lead to the appearance of defective recessive alleles that will reduce the trait quality and survival rate, resulting in growth depression and sensitive to environmental stress [6–9]. To solve this problem, a systematic broodstock management platform that can track the family tree by using genetic markers must be established [10]. An ideal genetic marker would be one that incorporates aspects of variability, heritability, stability, and accessibility during identification [11]. Microsatellite DNA is commonly used as a genetic identification tool due to its high polymorphism, co-dominant features and neutral mutation [12,13]. Microsatellite markers can reveal the genetic inheritance of an individual within a population [14,15], given that there is a low probability of different and non-related individuals exhibiting the same microsatellite pattern [16]. Besides, the use of the mitochondrial D-loop has been proposed as a genetic marker as well; this segment of the mitochondrion, a maternally inherited genetic material, exhibits high variability and can be used to identify sibling relationship within a population [17].

Microsatellite and mitochondria markers have been used previously for tracking the genetic history of groupers [18–22]. Although there are many successful applications of both microsatellites and mitochondrial D-loop as genetic markers for identification of individual groupers, the management of giant grouper broodstocks in Taiwan still lacks a practical methodology. In this study, we used both microsatellite markers and mitochondrial D-loop sequences to develop a system for identification of parentage and sibling relationships. Establishment of these genetic markers is expected to ensure trait quality of broodstocks while not only preventing inbreeding depression like physical and health defects but also reducing drug and therapeutant use in aquaculture.

2. Results and Discussion

2.1. Analysis of Mitochondrial D-Loop Region

Within the 776-nt D-loop fragment examined here, 56 nucleotides exhibited variations, including insertion and deletion, among the 118 giant grouper broodstocks (Table 1). Among those 56 nucleotide variations, 42 haplotypes can be identified (a single mutation site correspond to a single genotype). The nucleotide diversity (π) was 0.00773, and the haplotype diversity (HD) was 0.983 (Table 1). The genetic distance of haplotypes ranged between 0.215 and 0.0013, with an average of 0.0079 (Table 1). From 118 samples, the variation of nucleotides ranged between 16.684 and 1.0088 (average = 6.1304). These values indicate that the D-loop region used in this study is reliable for biogeographic analysis and sibling relationship determination.

Table 1. Mitochondrial D-loop-based genetic diversity of giant groupers in Taiwan aquaculture industry.

Parameter to the Genetic Diversity	
Farms	3
Samples	118
Haplotypes	42
Length (base pair)	776
Nucleotide variations	56
Haplotype diversity(Hd)	0.983
Nucleotide diversity(π)	0.00773
Maximum distance between each two haplotypes	0.0215
Minimum distance between each two haplotypes	0.0013
Mean distance between each two haplotypes	0.0079

In other developed systems, either of two distinct genetic markers (the mitochondrial D-loop and multiallelic microsatellites) has been used to assess variation within populations and to identify individuals. Low variation in D-loop nucleotide sequences of this fish species from our investigation suggested that the broodstock inbreeding level in Taiwan is high.

2.2. Analysis of Giant Grouper Microsatellite Loci

Examination of the results for microsatellite loci in giant grouper (Table 2) revealed that there were only 21 amplifiable sequences. Those loci were analyzed for allele number, allele distribution, expected heterozygosity (EH), and observed heterozygosity (OH) (Table 2). Among those loci, six were shortlisted based on their high heterozygosity and variability, properties that were expected to permit reliable parentage relationship determinations with high distinguishability and low error probability. These six shortlisted loci included Efu19 (EH, 0.746063; OH, 0.8125), ELMS009 (EH, 0.77842; OH, 0.875), ELMS015 (EH, 0.83157; OH, 0.640625), RH_CA_2 (EH, 0.822466; OJ, 0.890625) and RH_GATA_3 (EH, 0.845349; OH, 0.828125) (Table 2) are all carry a high number of alleles and comply with Hardy-Weinberg Equilibrium.

Table 2. Primers used for characterization of microsatellite loci in giant grouper.

Locus *	Accession Number	Repeat Motif	Primer Sequences (5'→3')	Allele Size	Allele Number	Expected Heterozygosity (EH)	Observed Heterozygosity (OH)
Efu02 [a]	EU016533	(CA)n	F: CTGTCTCAGCTGATTTATGG R: TTTACAGTCTCGTGGTTTCA	345–371	10	0.667815	0.484375
Efu06 [a]	EU016535	(GACA)n	F: CATTGTCATTGTTGCTGTTTCTGTC R: CCCTTTGGCCAATTGATGTGAT	308	1	0	0
Efu08 [a]	EU016537	(CA)n	F: TGGAGAAGCCTGTAGATTATTGTG R: AAGCAGGAGGAGAGTTGAAGGAGT	292–330	9	0.783465	0.796875
Efu18 [a]	EU016543	(CA)n	F: ACTGGCTCCCTTCTGTTCC R: ATTGCCACCATCGCTACC	370–384	2	0.361713	0.4375
Efu19 [a]	EU016544	(CA)n	F: GGGCGGTAACCTCTCCAG R: AGCACGCAACACCTTCTTCTCA	93–115	7	0.746063	0.8125
Efu41 [a]	EU016545	(CA)n	F: CAGCACGCAGTTTAATTTACCAG R: CAGGACCCGAGCTTCAGAA	243–249	2	0.353223	0.390625
ELMS009 [b]	EF607131	(CA)n	F: TTCCACAGCAATTAGCAGCA R: TTTCCTCCCACAGTCCAAAG	260–278	8	0.77842	0.875
ELMS015 [b]	EF607136	(TG)n	F: AAGCTGAGCGAATTTTTCA R: GCTCCTCGTGTTTCCGATTA	335–369	12	0.83157	0.640625
Epaw3 [c]	EU684479	(GT)n	F: GTCGTGTCTGTGACCATGAG R: TAAGGAGGGGGCTAAATGAT	72–76	2	0.503445	0
Epaw6 [c]	EU684482	(GT)n	F: ATGGTGTGGGAAAAGAGAGT R: TTGTTTCAGGACAAGTGAGC	146–233	5	0.619218	0.453125
Epaw19 [c]	EU684495	(GT)n	F: AGGTGGCTTGTGTGTGTATT R: GCTTCCTTGACTGCTATGAC	243–247	3	0.215428	0.234375
Epaw34 [c]	EU684510	(TG)n	F: ACAGCACCTCTACCATGAAC R: CGTCCCGTATATATCTCTG	224–248	3	0.212968	0.078125

Table 2. *Cont.*

			Primer sequence				
CA-2 [d]	AF539606	(CA)n	F: GACTTGATTCAGCAGCAAAATAAAGATG R: AGAGACGGTGCCAGTAAATGAA	150–262	6	0.350271	0.0625
CA-3 [d]	AF539605	(CA)n	F: ATGTGACACGTTGACAGGCAAGT R: GACCTTGATATTTCATTGCTTG	300	1	0	0
CA-6 [d]	AF539608	(CA)n	F: GTGTTGCTGGGGTTACTAATGAAG R: TTAGACACATTGTCACGATGGTCC	266–290	4	0.495325	0.5
RH_CA_1 [e]	DQ223785	(CA)n	F: CGAGATAAGCCCTGGTGAAA R: AGTCCCGATGTGGTAACGAG	376–388	3	0.452879	0.46875
RH_GATA_2 [e]	DQ223791	(GATA)n	F: CTCGACAGTGGACAAGGTCA R: AAGGGCATGATGGGAAATG	132	1	0	0
RH_CA_2 [e]	DQ223785	(CA)n	F: CTCGTTACCACATCGGGACT R: AACACTGGCTGGTTTGCACT	135–175	13	0.822466	0.890625
RH_GATA_3 [e]	DQ223790	(GATA)n	F: GGGCAATTTGGTTCTTCACA R: TGTCAATGCCACAGGATACA	225–273	10	0.845349	0.828125
RH_CA_7 [e]	DQ223786	(CA)n	F: CAGAAACATCTCCCCCAAAA R: CTGGCAGAGCAATTAGAGGC	259–335	12	0.639887	0.34375
RH_CA_8 [e]	DQ223787	(CA)n	F: AGTTGCCCAGGTTACACGAG R: TTGGGTCCTGGCATTTAGAG	219–227	4	0.554995	0.515625

* The loci listed here represent 21 microsatellite loci selected from previous studies of various species: (i) Tiger grouper: Efu02, Efu06, Efu08, Efu18, Efu19, Efu41; (ii) Giant grouper: ELMS009, ELMS015; (iii) Banded grouper: Epaw3, Epaw6, Epaw19, Epaw34; (iv) Hawaiian grouper: CA-2, CA-3, CA-6; (v) Red-spotted grouper: RH_CA_1, RH_CA_2, RH_CA_7, RH_CA_8, RH_GATA_2, RH_GATA_3; The references used for microsatellite loci primer design are [a] Lo and Yue, 2008 [23]; [b] Zeng *et al.*, 2008 [24]; [c] Zhao *et al.*, 2009 [25]; [d] Rivera *et al.*, 2003 [21]; [e] Ramirez *et al.*, 2006 [26], respectively.

By applying these six loci to a subset of 64 fish, the first parent non-exclusion probability is 0.01138738 (or 98.86% accurate) which we can identify the parentage relationship without knowing the genotypes of both parent fish. The non-exclusion probability (second parent) will lower (0.00089627 or 99.91% accurate) when one of the parent fish genotype is known (Table 3). It is important to obtain low combined non-exclusion probability (first and second parent) which means the combination of all six loci is predicted to determine parentage with low chance of misjudgment.

Knowing the origin of the broodstock can help to preclude collection of genetically closely related fish. We collected the samples from farms known to constantly replenish the broodstock from wild-caught fish or by purchasing small numbers of giant groupers from each of multiple different fish farms. This approach is recommended, since it can increase genetic variation within the fish population.

Table 3. Paternity exclusion probabilities based on six selected microsatellite loci.

Paternity Exclusion Probabilities	
Combined non-exclusion probability (first parent)	0.01138738
Combined non-exclusion probability (second parent)	0.00089627

The observed large numbers of haplotypes (from the mitochondrial D-loop sequence) and high allele numbers (among the six multiallelic microsatellite loci) indicate that our samples were derived from different genetic pools. Nevertheless, the present study analyzed only a small number of giant groupers; the sample sizes of future studies will affect the determination of the number of D-loop haplotypes, number of microsatellite loci alleles, and their combined non-exclusive probabilities. Analysis of mitochondrial D-loop alone may have limitations. Offspring from a single female are expected to share a single mitochondrial haplotype [27]. However, in practical giant grouper breeding, the occurrence of transsexuality is expected to confuse the derivation of mitochondrial haplotypes. Such transsexual fish represents a problem, since these events may cause misidentification of the parent. However, the inclusion of the six selected microsatellite loci is expected to resolve this shortcoming. To determine sibling relationships between any two individuals, microsatellite loci of one parent must be known. However, there is no microsatellite loci database for grouper broodstock.

3. Experimental Section

3.1. Fish Samples

The fin tissue samples from 118 individual broodstock animals were collected in three different locations (Linbian (44 fish), Jiadong (64 fish), and Fangliao (10 fish); Figure 1). Each fish in the same fishing pond was implanted and tagged with radio frequency identification (RFID) chips.

3.2. DNA Extraction

A piece of fin tissue (100 mg) sampled from each grouper was subjected to the following DNA extraction procedure. The fin tissue was homogenized in 1 mL of extraction buffer (10 mM Tris-HCl, pH = 8.0, 2 mM EDTA, 10 mM NaCl, 10 M DTT, 1% SDS) with 100 µg mL^{-1} of proteinase K, then incubated at 55 °C in a water bath until the tissue was completely dissolved. After lysis was completed, 20 µL of RNaes A (10 µg/mL) was added to the tube, and then incubated at 37 °C for 1 h.

DNA extraction was performed using the phenol-chloroform (phenol:chloroform:isoamyl alcohol = 25:24:1) method. Following precipitation in 99% ethanol, air dry and resuspension in TE buffer (10 mM Tris-HCl, pH = 8.0, 2 mM EDTA).

Figure 1. Giant grouper samples collected from farms in Pingtung region, Taiwan. Map (Modified from a map [28] under a Creative Commons Attribution-ShareAlike 1.0 License) inset shows locations of three different farms used for sampling: Linbian ($n = 44$), Jiadong ($n = 64$), and Fangliao ($n = 10$), for a total of 118 fish.

3.3. Establishment of Haplotype Database

Full-length (16,642 bp) giant grouper mitochondrial DNA, as sequenced by our laboratory (accession number: KJ451389), was used for primer design. The forward (ATCATCGGCCA AATCGCATC) and reverse (GAACTGTAGGGCATTCTCAC) primers were designed to amplify a 1259-bp fragment. We then focused on the 776-bp D-loop sequence. Genetic analysis, including nucleotide composition, mutation rate, nucleotide diversity (π) [29], haplotype diversity (Hd) [30] and genetic distance (Kimura's two-parameter model), were performed using MEGA 5.1 [31] software. DAMBE 5.3.10 [24] software was used to analyze the transition/transversion ratio of D-loop mutations.

3.4. PCR and Analysis of Microsatellite Loci

The selection of the potential microsatellite loci as genetic markers in our species was based on previous studies. Primer design focused on several loci described in previous studies for various species (Table 2), including *Epinephelus lanceolatus* [26], *E. guttatus* [23], *E. fuscoguttatus* [25], *E. coioides* [20], *E. quernus* [21] and *E. awoara* [32]. The PCR reaction consisted of an initial round at

94 °C for 5 min, followed by 30 cycles of 94 °C for 1 min, 55 °C for 1 min, and 72 °C for 1 min, with a final extension at 72 °C for 10 min. In order to get a better resolution, ten microliters of the PCR product then was analyzed by electrophoresis on a 12% polyacrylamide gel at 100 volts for 75 min and visualized by ethidium bromide. Following clean-up, the PCR products were sequenced using an ABI PRISM 3730 DNA analyzer (Life Technologies Co., Carlsbad, CA, USA).

The number of alleles for each locus, distribution of alleles, observed heterozygosity (OH), expected heterozygosity (EH), and agreement with Hardy-Weinberg equilibrium was measured using MSA 4.05 [33] software. Cervus 3.0.3 [34] software was used to calculate the combination of non-exclusion probability of the loci as parentage identification. The primers used in this study are listed in Table 2.

4. Conclusions

So far, there are about 300 male giant grouper broodstocks in Taiwan. Collection the genetic information from microsatellite to develop markers for each broodstock will be crucial for addressing the problem of inbreeding. For the techniques described in the present work, the mitochondrial D-loop should be used first to determine the potential sibling relationships among fish; biogeographic analysis then can be used to identify the individual broodstock to reduce inbreeding. Furthermore, the marker can be used to preclude the offspring from parent fish which shared the same microsatellite pattern but different in mitochondrial DNA sequence. In fact, the numbers of male broodstock typically are much lower than those of female broodstock, so maintenance of a separate male broodstock (along with genetic screening) may constitute an effective way to manage the genetic variation of farmed giant grouper.

Acknowledgments

We thank Chien-Hsien Kuo for his assistance with data analysis. We also thank the fish farmers for providing samples for this study. This study was supported by a grant from the Council of Agriculture, Taiwan (100AS-1.1.2-FA-F1, 101AS-11.3.1-FA-F5, 102AS-11.3.3-FA-F3) and National Science Council, Taiwan (NSC 101-2321-B-006-015, NSC 102-2321-B-006-016, NSC 103-2321-B-006-013).

Author Contributions

Conceived and designed the experiments: TYC. Performed the experiments: HCK, HHH, CSC, YMC, TYW. Analyzed the data: HCK, HHH, YMC, TYW, TYC. Wrote the paper: HCK, HHH, TYW, TYC. Assisted in drafting of text and figures of manuscript: HCK, HHH. Revised manuscript critically for important intellectual content: HCK, HHH, TYC.

References

1. Duarte, C.M.; Marbá, N.; Holmer, M. Rapid domestication of marine species. *Science* **2007**, *316*, 382–383.

2. Kuo, H.C.; Wang, T.Y.; Chen, P.P.; Chen, Y.M.; Chuang, H.C.; Chen, T.Y. Real-time quantitative PCR assay for monitoring of nervous necrosis virus infection in grouper aquaculture. *J. Clin. Microbiol.* **2011**, *49*, 1090–1096.

3. Kuo, H.C.; Wang, T.Y.; Hsu, H.H.; Lee, S.H.; Chen, Y.M.; Tsai, T.J. Ou, M.C.; Ku, H.T.; Lee, G.B.; Chen, T.Y. An automated microfluidic chip system for detection of piscine nodavirus and characterization of its potential carrier in grouper farms. *PLoS One* **2012**, *7*, e42203.

4. Kuo, H.C.; Wang, T.Y.; Hsu, H.H.; Chen, P.P.; Lee, S.H.; Chen, Y.M.; Tsai, T.J.; Wang, C.K.; Ku, H.T.; Lee, G.B.; *et al.* Nervous necrosis virus replicates following the embryo development and dual infection with iridovirus at juvenile stage in grouper. *PLoS One* **2012**, *7*, e36183.

5. Pierre, S.; Gaillard, S.; Prevot-D'Alvise, N.; Aubert, J.; Rostaing-Capaillon, O.; Leung-Tack, D.; Grillasca, J.P. Grouper aquaculture: Asian success and Mediterranean trials. *Aquat. Conserv. Mar. Freshw. Ecosyst.* **2008**, *18*, 297–308.

6. Gjerde, B.; Gunnes, K.; Gjedrem, T. Effect of inbreeding on survival and growth in rainbow trout. *Aquaculture* **1983**, *34*, 327–332.

7. Pante, M.J.R.; Gjerde, B.; McMillan, I. Effect of inbreeding on body weight at harvest in rainbow trout, *Oncorhynchus mykiss. Aquaculture* **2001**, *192*, 201–211.

8. Chiang, T.Y.; Lee, T.W.; Hsu, K.C.; Kuo, C.H.; Lin, D.Y.; Lin, H.D. Population structure in the endangered cyprinid fish *Pararasbora moltrechti* in Taiwan, based on mitochondrial and microsatellite DNAs. *Zool. Sci.* **2011**, *28*, 642–651.

9. Jackson, T.R.; Martin-Robichaud, D.J.; Reith, M.E. Application of DNA markers to the management of Atlantic halibut (*Hippoglossus hippoglossus*) broodstock. *Aquaculture* **2003**, *220*, 245–259.

10. Jonsson, B.; Jonsson, N. Cultured Atlantic salmon in nature: A review of their ecology and interaction with wild fish. *ICES J. Mar. Sci.* **2006**, *63*, 1162–1181.

11. Sunnucks, P. Efficient genetic markers for population biology. *Trends Ecol. Evol.* **2000**, *15*, 199–203.

12. Sekino, M.; Takagi, N.; Hara, M.; Takahashi, H. Analysis of microsatellite DNA polymorphisms in rockfish *Sebastes thompsoni* and application to population genetics studies. *Mar. Biotechnol.* **2001**, *3*, 45–52.

13. Powell, W.; Morgante, M.; Andre, C.; McNicol, J.W.; Machray, G.C.; Doyle, J.J.; Tingey, S.V.; Rafalski, J.A. Hypervariable microsatellites provide a general source of polymorphic DNA markers for the chloroplast genome. *Curr. Biol.* **1995**, *5*, 1023–1029.

14. Beacham, T.D.; Pollard, S.; Le, K.D. Microsatellite DNA population structure and stock identification of steelhead trout (*Oncorhynchus mykiss*) in the Nass and Skeena Rivers in northern British Columbia. *Mar. Biotechnol.* **2000**, *2*, 587–600.

15. Olsen, J.B.; Bentzen, P.; Banks, M.A.; Shaklee, J.B.; Young, S. Microsatellites reveal population identity of individual pink salmon to allow supportive breeding of a population at risk of extinction. *Trans. Am. Fish. Soc.* **2000**, *129*, 232–242.

16. Bierne, N.; Beuzart, I.; Vonau, V.; Bonhomme, F.; Bédier, E.A. Microsatellite-associated heterosis in hatchery-propagated stocks of the shrimp *Penaeus stylirostris. Aquaculture* **2000**, *184* 203–219.

17. Sang, T.K.; Chang, H.Y.; Chen, C.T.; Hui, C.F. Population structure of the Japanese eel, *Anguilla japonica*. *Mol. Biol. Evol.* **1994**, *11*, 250–260.

18. Zhu, Z.Y.; Lo, L.C.; Lin, G.; Xu, Y.X.; Yue, G.H. Isolation and characterization of polymorphic microsatellites from red coral grouper (*Plectropomus maculatus*). *Mol. Ecol. Notes* **2005**, *5*, 579–581.

19. An, H.S.; Kim, J.W.; Lee, J.W.; Kim, S.K.; Lee, B.I.; Kim, D.J.; Kim, Y.C. Development and characterization of microsatellite markers for an endangered species, *Epinephelus bruneus*, to establish a conservation program. *Anim. Cell Syst.* **2012**, *16*, 50–56.

20. Wang, L.; Meng, Z.; Liu, X.; Zhang, Y.; Lin, H. Genetic diversity and differentiation of the orange-spotted grouper (*Epinephelus coioides*) between and within cultured stocks and wild populations inferred from microsatellite DNA analysis. *Int. J. Mol. Sci.* **2011**, *12*, 4378–4394.

21. Rivera, M.A.; Graham, G.C.; Roderick, G.K. Isolation and characterization of nine microsatellite loci from the Hawaiian grouper *Epinephelus quernus* (*Serranidae*) for population genetic analyses. *Mar. Biotechnol.* **2003**, *5*, 126–129.

22. Han, J.; Lv, F.; Cai, H. Detection of species-specific long VNTRs in mitochondrial control region and their application to identifying sympatric Hong Kong grouper (*Epinephelus akaara*) and yellow grouper (*Epinephelus awoara*). *Mol. Ecol. Resour.* **2011**, *11*, 215–218.

23. Ramírez, M.A.; Patricia-Acevedo, J.; Planas, S.; Carlin, J.L.; Funk, S.M.; McMillan, W.O. New microsatellite resources for groupers (*Serranidae*). *Mol. Ecol. Notes* **2006**, *6*, 813–817.

24. Xia, X.; Xie, Z. DAMBE: Software package for data analysis in molecular biology and evolution. *J. Heredity* **2001**, *92*, 371–373.

25. Lo, L.C.; Yue, G.H. Microsatellites for broodstock management of the Tiger grouper, *Epinephelus fuscoguttatus*. *Anim. Genet.* **2008**, *39*, 90–91.

26. Zeng, H.S.; Ding, S.X.; Wang, J.; Su, Y.Q. Characterization of eight polymorphic microsatellite loci for the giant grouper (*Epinephelus lanceolatus* Bloch). *Mol. Ecol. Resour.* **2008**, *8*, 805–807.

27. Lin, L.Y.; Cheng, I.P.; Tzeng, C.S.; Huang, P.C. Maternal transmission of mitochondrial DNA in ducks. *Biochem. Biophys. Res. Comm.* **1990**, *168*, 188–193.

28. Mapsof.net. Available online: http://mapsof.net (accessed on 15 February 2013).

29. Nei, M.; Gojobori, T. Simple methods for estimating the numbers of synonymous and nonsynonymous nucleotide substitutions. *Mol. Biol. Evol.* **1986**, *3*, 418–426.

30. Nei, M.; Tajima, F. DNA polymorphism detectable by restriction endonucleases. *Genetics* **1981**, *97*, 145–163.

31. Tamura, K.; Dudley, J.; Nei, M.; Kumar, S. MEGA4: Molecular Evolutionary Genetics Analysis (MEGA) software version 4.0. *Mol. Biol. Evol.* **2007**, *24*, 1596–1599.

32. Zhao, L.; Shao, C.; Liao, X.; Ma, H.; Zhu, X.; Chen, S. Twelve novel polymorphic microsatellite loci for the Yellow grouper (*Epinephelus awoara*) and cross-species amplifications. *Conserv. Genet.* **2009**, *10*, 743–745.

33. Dieringer, D.; Schlötterer, C. MICROSATELLITE ANALYSER (MSA): A platform independent analysis tool for large microsatellite data sets. *Mol. Ecol. Notes* **2003**, *3*, 167–169.

34. Kalinowski, S.T.; Taper, M.L.; Marshall, T.C. Revising how the computer program CERVUS accommodates genotyping error increases success in paternity assignment. *Mol. Ecol.* **2007**, *16*, 1099–1106.

Marine Invertebrate Xenobiotic-Activated Nuclear Receptors: Their Application as Sensor Elements in High-Throughput Bioassays for Marine Bioactive Compounds

Ingrid Richter [1,2,*] **and Andrew E. Fidler** [1,3,4]

[1] Environmental Technology Group, Cawthron Institute, Private Bag 2, Nelson 7012, New Zealand;
E-Mail: andrew.fidler@cawthron.org.nz

[2] Environmental Technology Group, Victoria University of Wellington, P.O. Box 600,
Wellington 6140, New Zealand

[3] Maurice Wilkins Centre for Molecular Biodiscovery, University of Auckland,
Auckland 1142, New Zealand

[4] Institute of Marine Science, University of Auckland, Auckland 1142, New Zealand

* Author to whom correspondence should be addressed; E-Mail: ingrid.richter@cawthron.org.nz

External Editor: Paul Long

Abstract: Developing high-throughput assays to screen marine extracts for bioactive compounds presents both conceptual and technical challenges. One major challenge is to develop assays that have well-grounded ecological and evolutionary rationales. In this review we propose that a specific group of ligand-activated transcription factors are particularly well-suited to act as sensors in such bioassays. More specifically, xenobiotic-activated nuclear receptors (XANRs) regulate transcription of genes involved in xenobiotic detoxification. XANR ligand-binding domains (LBDs) may adaptively evolve to bind those bioactive, and potentially toxic, compounds to which organisms are normally exposed to through their specific diets. A brief overview of the function and taxonomic distribution of both vertebrate and invertebrate XANRs is first provided. Proof-of-concept experiments are then described which confirm that a filter-feeding marine invertebrate XANR LBD is activated by marine bioactive compounds. We speculate that increasing access to marine invertebrate genome sequence data, in combination with the expression of functional recombinant marine invertebrate XANR LBDs, will facilitate the generation of high-throughput bioassays/biosensors of widely differing specificities, but all based on activation of XANR LBDs. Such assays may find application in screening marine extracts for bioactive compounds that could act as drug lead compounds.

Keywords: xenobiotic; nuclear receptor; bioassay; marine; bioactive; invertebrate

Abbreviations

AD, activation domain; AhR, aryl hydrocarbon receptor; BaP, benzo [α] pyrene; *Bs*VDR/PXRα, *Botryllus schlosseri* VDR/PXR orthologue α; CAR, constitutive androstane receptor; *Ci*VDR/PXRα, *Ciona intestinalis* VDR/PXR orthologue α; *Ci*VDR/PXRβ, *Ciona intestinalis* VDR/PXR orthologue β; CPRG, chlorophenol red-β-D-galactopyranoside; CYP, cytochrome P450 gene; DBD, DNA-binding domain; DDT, 1,1,1-trichloro-2,2-di(4-chlorophenyl)ethane; ER, estrogen receptor; GAL4, yeast DNA-binding transcription factor; GST, glutathione S-transferases; HR96, nuclear hormone receptor 96; LBD, ligand-binding domain; MRP, multidrug resistance-associated protein; NR, nuclear receptor; NR1I, nuclear receptor sub-family 1, class I; NR1J, nuclear receptor sub-family 1, class J; NR1H, nuclear receptor sub-family 1, class H; OA, okadaic acid; PAHs, polycyclic aromatic hydrocarbons; PTX-2, pectenotoxin-2; PCN, pregnenolone 16α-carbonitrile; PXR, pregnane X receptor; RXR, retinoid X receptor; SRC-1, steroid co-activator 1; TCPOBOP, 1,4-Bis[2-(3,5-dichloropyridyloxy)]benzene; VDR, vitamin D receptor; VP16, viral protein 16; XANR, xenobiotic-activated nuclear receptor; XR, xenobiotic receptor.

1. Introduction

A major challenge facing researchers investigating marine natural products, with a view to identify potential drug lead compounds, is the selection and/or development of suitable bioassays. Typically the bioassays selected reflect the researcher's long-term, applied science goals but often they have little ecological or evolutionary rationale. The goal of this review is to promote the idea that chemical detection mechanisms, which adaptively evolve to allow marine animals to detect dietary bioactive chemicals, can be used in bioassays for marine bioactive chemicals. More specifically, we propose that nuclear receptor (NR) proteins may provide "sensor elements" that can be utilized in bioassays. Briefly, in the envisaged bioassays the "sensor element" would be a NR ligand-binding domain (LBD), which binds a bioactive dietary chemical and the resulting conformational change is then transduced into an output signal.

Having stated the overall goal of this review, its limits should also be made explicit. First, the term "xenobiotic receptor" (XR) will be used in this review to denote members of the zinc-finger NR transcription factor super-family that are activated by xenobiotic chemicals (xenobiotic, from the Greek *xenos*: foreigner; *bios*: life). Under this definition we have excluded genuine xenobiotic receptors such as the aryl hydrocarbon receptor (AhR), which are activated by xenobiotics but do not belong to the NR super-family because they bind DNA by a different mechanism [1–3]. In principle, such non-NR xenobiotic receptors could also be utilized in similar high-throughput screens to those proposed here but this is beyond the scope of this review. To remove any ambiguities in terminology, the term "xenobiotic-activated nuclear receptor" (abbreviated XANR) will hereafter be used for the group of xenobiotic receptors considered in this review [4–6].

Marine invertebrate XANRs will be the primary focus of this review as such organisms occupy diverse ecological niches and are characterized by great taxonomic diversity. Significantly, this diversity includes both filter-feeders and intense surface-grazers, two foraging behaviors that expose animals to dietary xenobiotics at high concentrations (Figure 1). For example, it is well-established that filter-feeding marine invertebrates clear large volumes of phytoplankton from the water column and consequently bioaccumulate high concentrations of microalgal biotoxins [7,8]. It is to be expected that selective pressures associated with any bioactive compounds, particularly toxic ones, ingested by marine invertebrates may drive adaptive evolution of XANR LBDs that bind these ingested compounds. This idea is explored further in this review.

Figure 1. Examples of filter-feeding and surface-grazing marine invertebrates. Marine invertebrates that filter their diet from seawater are found in a range of taxonomic groups including the phyla *Chordata* and *Mollusca* (**A–D**). In contrast, grazing gastropods (Phylum *Mollusca*) scour their food from surfaces (**E,F**). Whatever their ecological niche, marine invertebrates require detoxification mechanisms in order to detect, and effectively metabolize, any potentially deleterious xenobiotics encountered in their diet (**A,C,E**). Schematic diagrams of the digestive tracts (dark grey) of filter-feeding tunicates (**A**), filter-feeding bivalve molluscs (**C**), and grazing gastropod molluscs (**E**) are shown and the direction of food movement is indicated by blue arrows. Examples of a filter-feeding tunicate (**B**), *Ciona intestinalis*, Phylum *Chordata*; a filter-feeding bivalve mollusc (**D**), *Mytilus edulis*, Phylum *Mollusca*; and a grazing gastropod mollusc (**F**), *Amphibola crenata*, Phylum *Mollusca*.

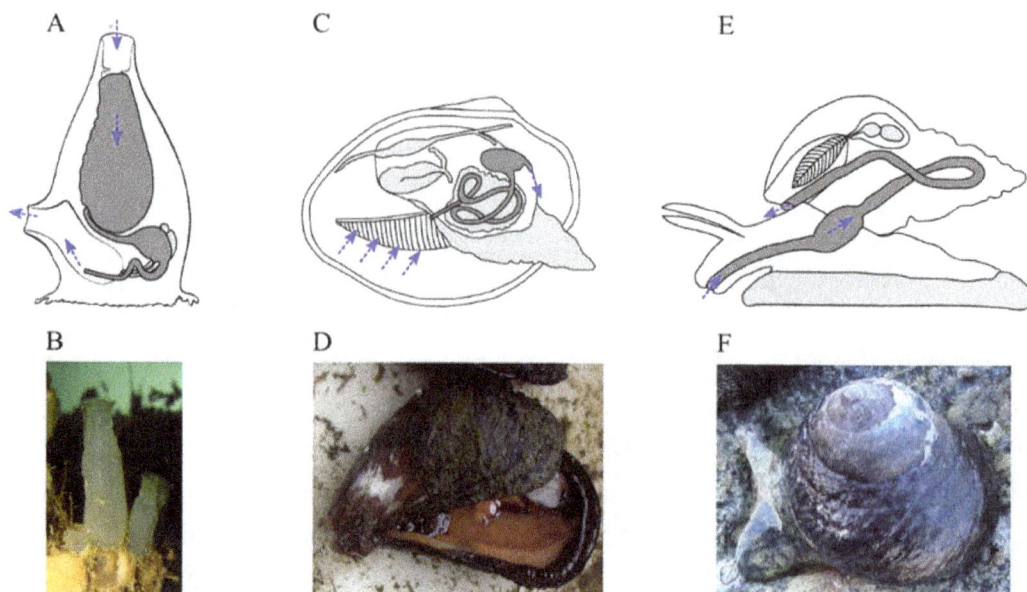

2. Bioactive Chemicals are Naturally Present in Animal Diets

Chemicals in animal diets are often viewed as simply energy sources (e.g., carbohydrates, lipids), building blocks (e.g., proteins), or biochemical pathway intermediates (e.g., vitamins). However, it is apparent that some dietary chemicals have "physiological" roles in the sense that they modulate animal biochemistry and physiology. The biological effects of such dietary chemicals range from influences on reproductive physiology and developmental transitions through to acute poisoning [9–14]. Animal taxa exposed to bioactive dietary xenobiotics evolve both behavioral and physiological traits to minimize any associated deleterious effects [15]. Many animals simply avoid eating plants/prey likely to contain toxins with such avoidance behaviors being both instinctual and learned [16,17]. For example, vertebrates tend to avoid bitter-tasting plants, as many poisonous phytochemicals (e.g., alkaloids) taste bitter [18–20]. Interestingly, there is evidence that natural selection associated with bitter taste perception may have influenced the evolution of bitter taste receptor gene repertoire sizes in vertebrate genomes [20–22]. In the marine environment, avoidance of toxic/unpalatable prey by coral reef fish is well-documented [23–25], while bivalve molluscs can limit their exposure to toxic compounds using behavioral responses, such as shell closure and restriction of filtration rate [8,26–28]. Despite avoidance behaviors, the diet of many animals will inevitably contain bioactive, and potentially toxic, chemicals that need to be metabolized and eliminated from their bodies [29–33].

3. Detoxification Pathways and Their Transcription-Level Regulation

Metazoan organisms have specialized biochemical pathways that metabolize and eliminate potentially toxic chemicals, whether endogenously synthesized or exogenously acquired. The complexities of detoxification biochemistry are beyond the scope of this review and are outlined in a number of recent reviews [34–38]. Nonetheless, a brief overview of metazoan detoxification pathways, and the transcriptional control of associated genes, is required to understand the functions of XANRs [39,40].

Detoxification pathways have been classified into three functional stages: oxidation/reduction (Phase I), conjugation (Phase II), and elimination (Phase III) [41,42]. It is to be expected that the genes encoding the functional elements of all three phases may be under both conservative and, at times, adaptive evolutionary pressures reflecting variation in the structures and mode of action of different xenobiotics/toxins associated with different animal diets [33,43]. The cat family (*Felidae*) provides an example of the apparent consequences of relaxation of conservative selective pressures associated with reduced ingestion of phytochemicals. Both domesticated and wild members of the *Felidae* are extremely susceptible to poisoning by phenolic compounds. This sensitivity is associated with a mutation in the feline orthologue of the gene encoding the enzyme UDP-glucuronosyltransferase (UGT) 1A6, a Phase II phenolic compound detoxification enzyme, leading to pseudogenization [44,45]. It is speculated that such UGT1A6 inactivation mutations are tolerated in the *Felidae*, and other hypercarnivores, because of relaxed selective pressures associated with their diet [46], as hypercarnivores are rarely exposed to plant-derived phenolic compounds [45].

In the context of this review, it is the *control of transcription* of detoxification related genes that is of particular interest. Expression of many detoxification pathway genes can be induced by the xenobiotic(s) that the pathway ultimately metabolizes [38,42,47]. Such xenobiotic-mediated induction of gene expression is of medical interest because of its implications for the efficacy, persistence, and side-effects of therapeutic drugs [34,48,49]. Xenobiotic-mediated induction of detoxification gene expression is best characterized for Phase I cytochrome P450 (CYP) enzymes, particularly members of CYP sub-families CYP1-4 that are associated with xenobiotic metabolism [34,48,50,51]. For example, levels of the human CYP enzyme CYP3A4, an enzyme responsible for oxidizing >50% of medicinal drugs, are induced by a range of therapeutic compounds, such as rifampicin, tamoxifen, and hyperforin [49,52], while human CYP1A2 enzymatic activity is induced by polycyclic aromatic hydrocarbons (PAHs) [34,53]. Interestingly, there exists striking inter-taxa variation in inductive responses to some xenobiotics [54]. For example, the steroidal drugs pregnenolone 16α-carbonitrile (PCN) and dexamethasone are highly efficacious CYP3A enzyme inducers in rodents but not so in humans [55,56]. In contrast, rifampicin is a strong inducer of human and dog CYP3A but not of rodent CYP3A [57,58]. While none of these examples are likely to be of ecological/evolutionary significance, such conspicuous inter-taxa variation in the response to xenobiotics suggests the possibility of adaptive evolution in the genetic elements that control expression of detoxification genes.

Xenobiotic induction of cytochrome P450 enzyme levels has also been reported in several invertebrate phyla, most prominently in the *Arthropoda* in the context of pesticide resistance [59–61]. In insects, the CYP enzymes induced by xenobiotics mainly belong to the sub-families CYP4, CYP6, and CYP9 [36]. In both *dipteran* and *lepidopteran* insect taxa, the barbiturate phenobarbital induces CYP enzymatic activity in association with transcription level induction of the CYP4, CYP6, and CYP9 genes [62–65]. In the honey bee (*Apis mellifera*), aflatoxin and propolis, ecologically relevant natural xenobiotics, induce CYP gene expression [66,67], while in the "model" arthropod *Drosophila melanogaster* many CYP genes are induced by caffeine and phenobarbital [68–70]. In *Drosophila mettleri* the CYP4D10 gene is induced by primary host plant alkaloids but not by similar alkaloids from a rarely utilized host plant. Clearly this finding suggests adaptive evolution of the associated gene induction mechanism(s) of *Drosophila mettleri* [71]. The soil nematode *Caenorhabditis elegans* (phylum *Nematoda*) displays up-regulated CYP gene expression in response to exposure to multiple xenobiotics [72–74]. For example, beta-naphthoflavone, polychlorinated biphenyl, PCB52, and lansoprazol are all strong inducers of almost all *C. elegans* CYP35 isoforms [72]. Rifampicin, one of the strongest inducers of the human CYP3A4 gene, is also a strong inducer of the *C. elegans* CYP13A7 gene [75]. Amongst marine invertebrates the polychaete *Perinereis nuntia* (phylum *Annelida*) shows increased levels of some CYP gene transcripts after exposure to benzo[α]pyrene (BaP) and PAHs [76]. In an ecologically more relevant context, the marine gastropod *Cyphoma gibbosum* (phylum *Mollusca*) is suggested to have adapted to feeding exclusively on highly toxic gorgonian corals by differential regulation of transcripts for two CYP enzymes, CYP4BK and CYP4BL [77].

Whilst most xenobiotic-mediated gene induction research has focused on Phase I CYP genes, Phase II glutathione S-transferases (GST) and Phase III multidrug resistance-associated proteins (MRPs) have also been reported to be inducible by some xenobiotics [35,78]. For example, expression of mouse GSTA1, MRP2, and MRP3 genes is induced by both pregnenolone 16α-carbonitrile (PCN) and 1,4-Bis[2-(3,5-dichloropyridyloxy)]benzene (TCPOBOP) [79]. In the marine environment, dietary toxins (e.g., cyclopentenone prostaglandins) have been shown to be both inducers and substrates of GST enzymes in three marine molluscs [80–82].

In summary, it is likely, that xenobiotic-mediated *control* of detoxification pathway gene expression may evolve adaptively in response to the differing chemicals to which different animals are exposed to, with diet probably being the main route of exposure [66,67,77,83]. The next section will consider the role of a specific group of ligand-activated transcription factors, the XANRs, in xenobiotic-mediated control of gene expression and speculate how natural selection may influence the adaptive evolution of such XANR genes.

4. Xenobiotic Receptors: Functions, Structures, and Taxonomic Distribution

4.1. Vertebrate Pregnane X Receptor

Functional characterization of XANRs is most advanced in a few selected vertebrate taxa [6,84–86]. Mammalian genomes encode two XANRs: constitutive androstane receptor (CAR; NR notation: NR1I3) and pregnane X receptor (PXR; NR notation: NR1I2). Although both CAR and PXR regulate the transcription of genes involved in detoxification of endogenous and exogenous (*i.e.*, xenobiotic) chemicals [85–87], PXR is the better understood with respect to how its LBD structure relates to ligand-binding and subsequent activation [85]. Therefore, the following sections of this review will focus on PXR and its orthologues.

Mammalian PXR was originally identified in genomic sequence data and designated as an orphan NR as its ligand(s) were then unknown. In 1998, three groups independently reported mammalian PXR activation by both steroids and a range of xenobiotics resulting in three alternative receptor names—with pregnane X receptor (PXR) now the most widely used [88–90]. PXR appears to function much like a standard ligand-activated NR. After ligand-binding within the PXR LBD, the activated PXR protein forms a complex with retinoid X receptor (RXR) before translocating from the cell cytoplasm into the nucleus. The PXR/RXR heterodimer binds to appropriate DNA response elements, thereby influencing transcription of adjacent genes [91–94]. Many of the PXR regulated genes are involved in detoxification. Thus, PXR activation, following xenobiotic-binding to its LBD, provides a mechanistic link between the presence of xenobiotics in a cell and appropriate detoxification gene expression [95].

Vertebrate PXR ligands include a structurally diverse range of endogenously produced molecules (e.g., bile acids, steroid hormones, and vitamins) along with exogenously acquired chemicals (e.g., both synthetic and herbal drugs) [96–101]. Determination of the three dimensional structure of the human PXR protein has helped explain its striking permissiveness with respect to the differing structures of activating ligands. The classic model of NR function proposes that NR ligand specificity arises from the interaction between ligands and a ligand-binding domain (LBD) of the NR protein (Figure 2). In the majority of NRs the LBD cavities have well-defined shapes with restricted mobility, thereby ensuring specificity of ligand—NR LBD interactions [102,103]. In contrast, the human PXR LBD is both larger than is typical of NRs and also displays significant flexibility during ligand binding allowing it to accommodate a wider range of ligand sizes and structures [104,105].

Figure 2. Schematic summary of the generic nuclear receptor (NR) structure. The overall structure of ligand-activated NRs is conserved through evolution with five key structural domains: *N*-terminal transcription activation domain (activation function-1, AF-1), DNA-binding domain (DBD), flexible hinge region (Hinge), and ligand-binding domain (LBD), which includes a *C*-terminal activation domain (activation function-2, AF-2). The DBD interacts with DNA while the structurally separate LBD interacts with ligands [106].

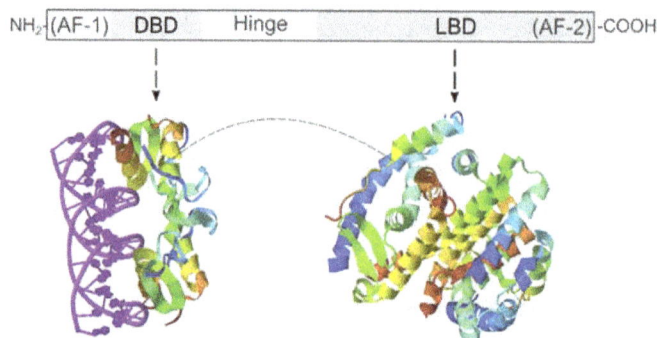

Vertebrate PXRs display greater inter-taxa variation in LBD sequences than is typical of NRs, along with some evidence of positive selection within the LBD [4,107–109]. From this observation it has been speculated that such inter-taxa PXR LBD sequence differences may reflect adaptive evolutionary changes which enhance PXR binding of those exogenous dietary bioactives/toxins typically encountered by an organism within its particular ecological niche [4,5,107,108,110,111]. Notwithstanding such speculation, it has been shown experimentally that the differential activation of human and rat PXRs by some ligands can be attributed, in part, to specific residues within the PXR LBD [88,89,112–116]. For example, rifampicin is an effective activator of human PXR but has little activity on rat PXR [112] and this difference can be attributed to differences between the two PXR LBDs at a single position: human PXR Leu$_{308}$/rat PXR Phe$_{305}$ [117].

4.2. Non-Marine Invertebrate XANRs

Although most advances in our understanding of XANR function and evolution have been within the *Vertebrata*, there has been some progress in characterizing invertebrate XANRs, building on the conceptual foundations provided by the vertebrate PXR studies. It should be noted that phylogenetic approaches to identifying XANR genes within the genomes of non-chordate invertebrate taxa face significant challenges, the biggest being the large evolutionary distances separating the functionally characterized vertebrate query sequences typically used to search invertebrate genomes. The scale of this challenge is reinforced by considering that the split between the deuterostome (includes vertebrates) and protostome (includes most invertebrates) lineages occurred approximately one billion years ago [118–120]. Fortunately, model invertebrate organisms within two protostome phyla, *Arthropoda* (*D. melanogaster*) and *Nematoda* (*C. elegans*), provide the experimental route of using mutant animal phenotypes to assess the functions of putative XANRs initially identified on the basis of sequence homologies [121,122]. A PXR/NR1I-like homologue, termed hormone receptor-like 96 (*HR96*), identified in the *D. melanogaster* genome was found to be selectively expressed in tissues involved in the metabolism of xenobiotics [121,123]. *D. melanogaster* flies homozygous for Dhr96 null alleles displayed increased sensitivity to both phenobarbital and the pesticide 1,1,1-trichloro-2,2-di(4-chlorophenyl) ethane (DDT) along with defects in phenobarbital induction of gene expression [121]. These findings are consistent with experiments, using a combination of RNA interference treatments and *Cyp6d1* promoter reporter assays, that indicate a role for *DHR96* in mediating phenobarbital associated gene induction in *Drosophila* Schneider (S2) cells [123]. In summary, Dhr96 represents a strong candidate as a *bona fide* arthropod XANR and it may also have roles in cholesterol homeostasis and lipid metabolism [124,125]. A DHR96 orthologue from a freshwater aquatic arthropod (*Daphnia pulex*), DappuHR96, has been shown to be activated by a structurally diverse range of both endogenously produced compounds and xenobiotics, consistent with a role as a lipid and/or xenobiotic sensor [126]. Probable *DHR96* orthologues have also been identified in a range of other arthropod genomes *albeit* with no reports of their functional characterization: beetles [*Tribolium castaneum*] [127], ants [*Camponotus floridabus*] [128], honey bee [*Apis mellifera*] [129], and fall armyworm [*Spodoptera frugiperda*] [130].

The genome of the model nematode *C. elegans* (phylum *Nematoda*) encodes an exceptionally large number of NRs (~284) [131]. Responses of mutant *C. elegans* strains to toxin exposure indicates that one of these NRs, denoted NHR-8, a homologue of *D. melanogaster* HR96, is involved in the regulation of detoxification enzyme induction which is consistent with NHR-8 functioning as a XANR [122].

From a phylogenetic perspective one useful generalization emerges. To date, all arthropod and nematode putative, and partially functionally verified, XANRs are placed in the NR1J group of the NR super-family [126]. NR1J forms a sister group of the NR1I sub-family, which contains all the chordate XANRs. This grouping is suggestive of a shared ancestral XANR gene being present in the genome of a common ancestor preceding the divergence of the protostome-deuterostome lineage [132].

4.3. Marine Invertebrate XANRs

Over a decade ago, phylogenetic analyses identified two orthologues of vertebrate PXR encoded in the genome of a marine invertebrate chordate; the solitary tunicate *Ciona intestinalis* [133–137]. As the functional characterization of one of these *C. intestinalis* putative XANRs is central to this review, the associated experiments will be described in more detail in Section 5. More recently, two putative PXR/NR1I orthologues have been detected in the genomic sequence of the colonial tunicate *Botryllus schlosseri* [138], while the genome of the pelagic tunicate *Oikopleura dioica* encodes as many as six NR1I clade genes [139].

Although NR encoding genes can be confidently identified in a growing number of publicly available non-chordate marine invertebrate genomic sequences [140], recognizing *bona fide* XANRs within the NR repertoire remains problematic. To date, all functionally characterised XANRs, both deuterostome and protostome, have been placed in the NR1 group of the NR super-family (deuterostome: NR1I; protostome: NR1J) [89,114,121,123,126,141–143]. Therefore, at present, any predicted marine invertebrate NRs placed in the NR1 group may be regarded as potential XANRs. However, such phylogeny-based designations are always highly tentative and only functional data can lead to the confident assignment of a NR1 protein as a functional XANR (Section 5).

Despite the evident phylogenetic/bioinformatic challenges, candidate XANR genes have been identified in sequence data derived from a number of non-chordate marine invertebrates. For example, at least one of the two NRs identified in a demosponge (*Amphimedon queenslandica*, Phylum *Porifera*) genome displays functional characteristics consistent with a role in detecting xenobiotics [144,145]. The genome of the starlet sea anemone (*Nematostella vectensis*, Phylum *Cnidaria*) encodes three NR genes equally related to sub-families NR1 and NR4, while the genome of the Pacific oyster (*Crassostrea gigas*, Phylum *Bivalvia*) encodes as many as 23 NR genes placed in the NR1 sub-family [146,147]. In contrast, the genome of a marine deuterostome, the sea urchin (*Strongylocentrotus purpuratus*, Phylum *Echinodermata*), appears to lack NR1I sub-family genes although three NR1H genes were identified [148]. The apparent absence of NR1I subfamily genes from the *S. purpuratus* genome should be treated with caution as it may simply reflect an incomplete data-set [148].

We suggest that functional characterization of at least one molluscan XANR would be a very useful advance, as both filter-feeding (*Bivalvia*) and grazing (*Gastropoda*) marine molluscs are expected to be exposed to a wide range of bioactive chemicals through their diet (Figure 1) [7,149]. It is also worth noting that molluscs provide a cautionary tale warning against assumptions that sequence homology/orthology necessarily predict shared function. For example, the molluscan homologue of the vertebrate steroid receptors is a constitutive, rather than a ligand-activated, transcription activation factor [150].

In summary, although significant bioinformatic challenges remain, it is clear that the ever increasing nucleotide sequence databases provide an informational resource in which putative marine invertebrate XANRs can be identified with varying degrees of confidence. However, functional characterization of such putative XANRs is always needed, as will be described in the next section.

5. Marine Invertebrate Putative XANR LBDs are Activated by Known Marine Bioactive Compounds

In 2002, the solitary tunicate *Ciona intestinalis* was the first marine invertebrate to have an assembled and annotated genome sequence published [137,151]. Analysis of the *C. intestinalis* genomic sequence, in combination with *C. intestinalis* expressed sequence tag (EST) databases, revealed two genes that phylogenetic analyses placed as orthologous to vertebrate NR1I genes. These two *C. intestinalis* NR1I-like genes were denoted "VDR/PXR" reflecting their putative orthology with both the vertebrate PXR and vitamin D receptor (VDR) genes [137,152]. Hereafter, these two *C. intestinalis* genes will be abbreviated as *Ci*VDR/PXRα (GenBank accession number: **NM 001078379**) and *Ci*VDR/PXRβ (**NM 001044366**) [153,154]. At the time of writing there is no functional data published regarding *Ci*VDR/PXRβ so it will not be discussed further here.

Functional characterization of *Ci*VDR/PXRα began with its LBD being expressed, as part of a chimeric protein, in a mammalian cell line. Briefly, the *Ci*VDR/PXRα LBD was joined to the generic GAL4 DNA-binding domain (GAL4-DBD) and the resulting chimeric protein was shown to mediate ligand-dependent expression of a luciferase reporter gene in mammalian cells [153,155]. Using this mammalian cell line bioassay, an extensive range of both natural and synthetic chemicals ($n = 166$) were screened for their activity [153,155] and three putative *Ci*VDR/PXRα LBD agonists were identified (6-formylindolo-[3,2-b]carbazole: $EC_{50} = 0.86$ μM; *n*-butyl-*p*-aminobenzoate: $EC_{50} = 16.5$ μM; carbamazepine: $EC_{50} > 10.0$ μM). Based on these results a pharmacophore model was tentatively defined which consisted of a planar structure with at least one off-center hydrogen bond acceptor flanked by two hydrophobic regions [153,156]. Note that none of the three CiVDR/PXRα LBD agonists identified were strikingly potent, having EC$_{50}$ values in the μM range, nor did it seem plausible that these chemicals would have been encountered by *C. intestinalis* over evolutionary time.

Pursuing the hypothesis that the natural ligands of CiVDR/PXRα include marine bioactive compounds frequently present in a marine filter-feeder's diet, four microalgal biotoxins (okadaic acid, pectenotoxin-2 (PTX-2), gymnodimine, and yessotoxin) were tested for activation of the CiVDR/PXRα LBD [156] (Figure 3). The four microalgal biotoxins investigated have diverse structures and all came with the caveat that their toxicity towards intact tunicate animals was, and still is, unknown. Of the four biotoxins tested, okadaic acid (EC$_{50}$ = 18.2 nM) and PTX-2 (EC$_{50}$ = 37.0 nM) activated the bioassay, while gymnodimine and yessotoxin were inactive [156] (Figure 4). Interestingly, the EC$_{50}$ values for okadaic acid and PTX-2 are in the low-to-mid nM range making these ligands two to three orders of magnitude more potent than the three synthetic compounds previously found to be active in the CiVDR/PXRα LBD-based bioassay [153].

Figure 3. Structures of four microalgal biotoxins tested for activation of *C. intestinalis* VDR/PXRα LBD-based bioassays. Pectenotoxin-2 and okadaic acid activated bioassays that used the *C. intestinalis* VDR/PXRα LBD as the sensor element while yessotoxin and gymnodimine did not. (**A**) pectenotoxin-2 (CID: 6437385); (**B**) okadaic acid (CID: 446512); (**C**) yessotoxin (CID: 6440821); and (**D**) gymnodimine (CID: 11649137). *Abbreviations*: CID, PubChem compound accession identifier [157].

Active in *Ciona intestinalis* VDR/PXRα LBD-based bioassays

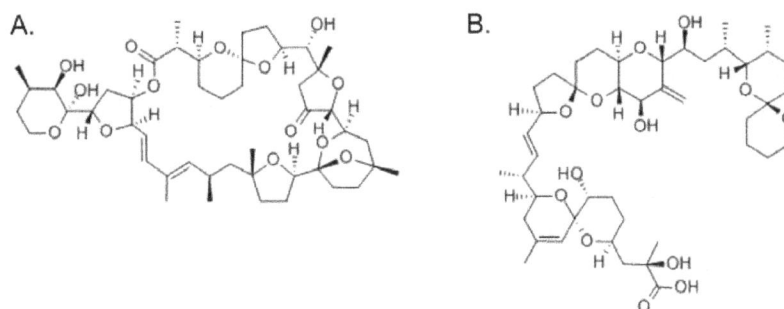

Inactive in *Ciona intestinalis* VDR/PXRα LBD-based bioassays

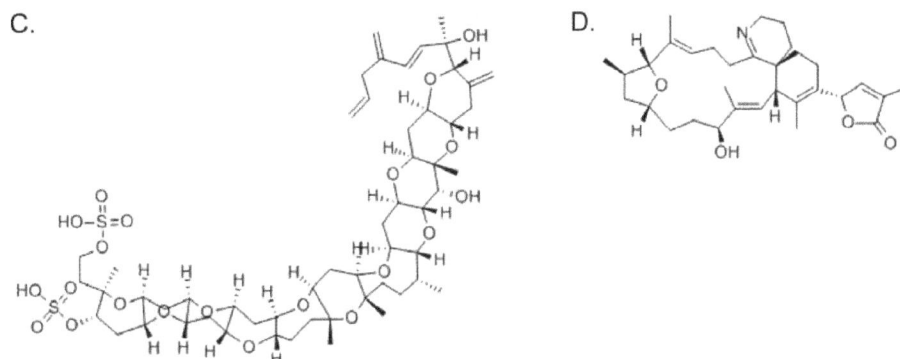

Figure 4. Microalgal biotoxin concentration-dependent response curves of luciferase expression induction by cell lines transfected with GAL4-DBD-CiVDR/PXRαLBD fusion genes. The doubly-labelled ordinate axes indicates fold induction compared to vehicle control (left axis) and efficacy relative to 5.0 mM 6-formylindolo-[3,2-b] carbazole (adapted from Fidler *et al.* [156], with permission from © 2014 Elsevier).

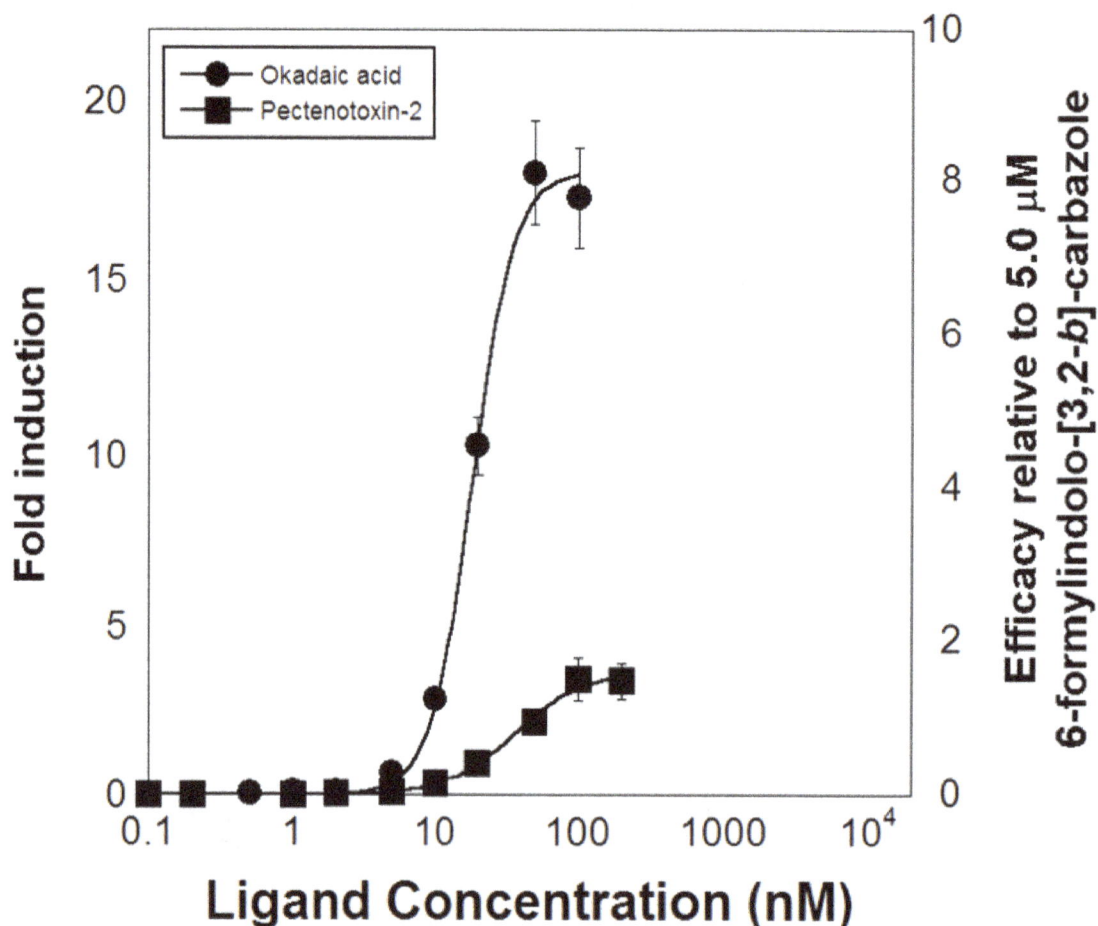

In summary, the combined studies of Ekins *et al.* (2008) and Fidler *et al.* (2012) established that the CiVDR/PXRα LBD displayed xenobiotic/ligand-binding characteristics consistent with CiVDR/PXRα having a role in detecting bioactive marine chemicals naturally encountered by filter-feeding marine invertebrates through their diet [153,156]. Interestingly, the CiVDR/PXRα LBD appears to have rather narrow ligand selectivity when compared to vertebrate, particularly human, PXRs [153,156]. It is possible that such tunicate VDR/PXR LBD ligand selectivity may reflect tunicate genomes encoding multiple VDR/PXR paralogues, with each paralogue subtype perhaps binding a differing range of ligand structures [138,139]. It should also be remembered that, despite the insights obtained from CiVDR/PXRα LBD functioning in mammalian cell lines, the actual role of VDR/PXR orthologues in intact, living tunicates has not been determined and this represents an important area of future research. Nonetheless, the critical point is that the work of Ekins *et al.* (2008) and Fidler *et al.* (2012) firmly established the feasibility of using marine filter-feeder XANR LBDs as "sensors" in bioassays for marine bioactives. How this basic concept might be developed into cost-effective, high throughput bioassays will be considered next.

6. Development of High-Throughput Bioassays Based on Marine Invertebrate XANR LBDs

6.1. XANR LBD-Based Bioassays: Technical Considerations and Challenges

As outlined in Section 5, it is established that marine invertebrate XANR LBDs can be utilized as "sensors" in bioassays for marine bioactive chemicals. However, significant technical challenges exist for expanding this simple insight into economically and technically viable bioassays for high-throughput screening of marine compounds. Fortunately, existing NR LBD-based assays provide a strong foundation to build on. This section outlines how well-established NR LBD-based approaches could be applied to marine invertebrate XANR LBDs [158,159].

6.1.1. NR LBD Bioassays Using Recombinant Yeast

Mammalian cell line-based bioassays, as described in Section 5, have the advantage that they provide a cellular/biochemical milieu shared by metazoan cells, which is expected to assist correct folding/functioning of proteins expressed from heterologous genes. However, mammalian cell lines do have significant limitations as they require costly, highly specialized culturing facilities and associated laboratory skills. In contrast, baker's yeast (*Saccharomyces cerevisiae*) provides a well-established eukaryotic expression system that is inexpensive and suitable for standard microbiology laboratories. Furthermore, the yeast's nutrient requirements are easily met in a 96-well plate format making *S. cerevisiae*-based bioassays well-suited to high-throughput screening formats [160–162]. In addition, *S. cerevisiae* cells do not contain endogenous NRs to potentially interfere with bioassays based on introduced NRs [163,164]. Finally, *S. cerevisiae* offers the possibility of directed evolution of NRs whereby NR LBD variant sequences, generated by *in vitro* mutagenesis, can be selected for enhancement of growth rates in the presence of a cognate ligand [165]. For example, Chen and Zhao (2003) combined random *in vitro* mutagenesis together with directed evolution to generate novel variants of the human estrogen receptor alpha (ERα) LBD that had significantly modified ligand-binding properties [165].

Despite the clear attractiveness of *S. cerevisiae* cells for NR LBD-based bioassays, there are aspects of NR functioning that require consideration when designing any associated high-throughput bioassays. Bioassay design needs to address how ligand-induced conformational changes in an introduced metazoan NR LBD will be transduced into a quantifiable output signal. Particular consideration needs to be given to the step in the NR signal transduction pathway selected for activation of the "output signal". Two ligand-dependent steps in a NR signal transduction pathway have been utilized: (i) ligand-dependent binding of the NR to co-activator protein(s) [166] or (ii) ligand-dependent binding of the NR/co-activator complex to a specific promoter control region [167]. Indeed approach (i) has been successfully used for detecting ligand-dependent interactions between the human PXR and its co-activator protein, human steroid co-activator-1 (hSRC-1) [159,168]. Despite this success with human PXR, it is clear that this approach requires extensive knowledge of a given NR's co-activator protein, knowledge, which generally will not be available for most marine invertebrate XANRs. Approach (ii) utilizes binding of ligand-activated NRs to control sequences in reporter gene promoters as the mechanism to generate an "output signal" [169,170]. For native, full-length marine invertebrate XANRs to be used in this approach, both the XANR's co-activators (if any) and the sequences of the cognate DNA elements to which the XANR DBD binds need to be known. Again, for marine invertebrate XANRs such specialized knowledge is unlikely to be available. Even when putative response elements for metazoan XANRs have been identified in marine invertebrate genomes [171], these may not function as required in *S. cerevisiae* cells. Fortunately, such knowledge gaps can be bypassed by exploiting the highly modular structure of NRs (Figure 2). Basically a chimeric protein can be generated in which the XANR LBD is fused to the generic GAL4-DBD, which is native to yeast cells, removing any need for knowledge of the natural heterodimer partners of the XANR or the DNA sequence elements to which the XANR binds through its native DBD.

Following binding to well-characterized DNA control elements, via the GAL4-DBD, the XANR's ligand-dependent activation domain (AF-2) (Figure 2) needs to function within the nuclear milieu of yeast cells. As previous studies have shown that the AF-2 domains of some vertebrate NRs do not function in yeast cells [172,173], a generic transcription activation domain (AD) from the *Herpes simplex* virion protein 16 (VP16) can be added to the C-terminus of the chimeric proteins [173]. In summary, a fusion gene can be generated encoding a chimeric protein that contains the GAL4-DBD, the XANR LBD, and the VP16-AD, with the ligand-binding characteristics of the chimeric protein determined by the XANR LBD.

The reporter gene selected to generate the "output" signal from yeast-based bioassays must combine low background with a clear response signal following activation of the NR LBD. In addition, it is highly desirable that the assay used to quantify reporter gene expression is non-lethal, allowing repeated measurements to be taken over time. Three types of NR-dependent reporter gene assays have been used in recombinant yeast: the *Escherichia coli lacZ* gene, encoding the enzyme β-galactosidase [174], yeast-enhanced green fluorescence protein (yEGFP) [163,175,176], and the luciferase gene [177]. Although the luciferase and yEGFP reporter assays have been shown to be somewhat more sensitive than *lacZ* [178] both have associated complications. For example, luciferase assays require the use of expensive substrates and involve cell lysis [174,179] which can be problematic either due to released cellular proteases [174] or incomplete cell lysis [179]. Although yEGFP assays do not require the addition of substrates, the assays are characterized by a high natural background of green fluorescence [163]. In contrast, *lacZ* assays are inexpensive and, when based on the chromogenic substrate (chlorophenol red-β-D-galactopyranoside, CPRG), do not require cell lysis [180]. Such non-lethal measurement of β-galactosidase activity is useful as it means that repeated measurements can be taken over time. This is a significant advantage because the time course of *lacZ* gene transcription induction will vary between ligands due to differences in parameters such as membrane permeability and solubility in the cytoplasm [160].

Notwithstanding the technical challenges many metazoan NRs have been successfully expressed in *S. cerevisiae* in combination with reporter genes [162,166,168,170,181–183]. Furthermore, some of the resulting yeast strains have found application in screening environmental samples for bioactivities—particularly for estrogenic activity [176,180,183–187]. Thus, it is well-established that *S. cerevisiae* provides a suitable expression system for bioassays in which a NR LBD acts as the sensor element that interacts with bioactive chemicals to be detected [183,185,187].

Despite the clear merits of recombinant yeast-based NR bioassays it remains true that all cell-based bioassays face limitations intrinsic to living cells. Such limitations include test compounds being unable to cross cell membranes or being directly toxic to the cells themselves. To address such limitations biosensor techniques have been developed to directly measure physical interactions between macromolecules and their potential ligands.

6.1.2. Biosensors for High-Throughput Screening for NR LBD Ligands

During the past two decades biosensors, which measure a range of physiochemical changes associated with interactions between molecules, have been developed within both industry and academia [188,189]. The main advantage of such biosensors is their cell-free nature thereby removing some of the limitations associated with cell-based bioassays, such as the need for test compounds to cross cell membranes [160]. Biosensors typically consist of a macromolecule immobilized on a surface via either covalent or strong non-covalent bonds [190]. An important consideration is that such attachments should not significantly influence the natural structure of the macromolecule or change its functioning in unpredictable ways [190]. Following immobilization, a wide range of techniques exist to detect and quantify interactions between the immobilized macromolecules and potential ligands—including calorimetric, acoustic, electrical, magnetic, and optical sensing techniques [189].

Numerous NR LBD-based biosensors have been developed, principally in the context of drug development [166,190]. Among the established xenobiotic receptors, the human PXR LBD has been successfully used as the sensor element in a number of differing biosensor formats [111,191,192]. Such biosensors have confirmed previously known human PXR ligands such as hyperforin, clotrimazole, ginkgolide A, SR12813, and 5b-pregnane-3,20-dione [111,112,191]. The successful development of human PXR LBD-based biosensors supports the theoretical feasibility of using marine invertebrate XANR LBDs in biosensor formats to screen for marine bioactive compounds. Furthermore, if routine production of correctly folded and soluble marine invertebrate XANR LBDs can be mastered, then they could be used in affinity chromatography to identify and isolate novel XANR ligands, as has recently been reported for the human PXR LBD [193].

6.2. XANR LBD-Based Assays: Biological, Ecological, and Evolutionary Considerations

The application of marine invertebrate XANR LBDs in high-throughput bioassays/biosensors entails some biological, ecological, and evolutionary considerations. The first point to emphasize is the vast taxonomic diversity of marine invertebrates, along with the myriad of ecological niches they occupy. It is expected that this diversity and complexity will be paralleled by the diversity and complexity of the bioactive xenobiotics to which these animals are exposed to during their life-cycles. If dietary bioactive xenobiotics do act as selective agents in shaping the structure of marine invertebrate XANR LBDs, then there exists an effectively unlimited supply of "sensors" in the sea that have been pre-molded by natural selection to facilitate detection of marine bioactive compounds.

If this perspective is correct, then a major decision confronting developers of marine invertebrate XANRs-based bioassays is how to select, from the virtually unlimited options, the marine invertebrate XANRs to use. We suggest that three considerations should both guide and restrict this decision. The first is the implicit assumption that a bioactive chemical that has acted as a significant selective pressure driving molecular evolution of a given marine invertebrate's XANR LBD may also be active on human cells/tissues. In this context it could be argued that the more closely related an organism is to humans, the more similar would be their susceptibility to marine bioactive chemicals. Whilst doubtless a simplification, this idea does suggest that selecting XANR LBDs from marine invertebrate taxa within the phylum *Chordata* (tunicates, *Urochordata*; lancets, *Cephalochordata*; acorn worms, *Hemichordata*) would be a useful strategy. However, it should be emphasized that there is no *a priori* reason why taxon selection should be restricted to the *Chordata*. For example, at least one marine microalgal biotoxin has acted as a selective pressure in the evolution of bivalves (phylum *Mollusca*) while also being toxic to vertebrates [194,195]. In addition, selections may be restricted simply because promising marine bioactives may exist in contexts/ecological niches to which no marine invertebrate chordate is adapted. This highlights a second consideration in selecting marine invertebrate XANRs for bioassays. The ecological niche occupied by the taxa from which XANR LBDs could be isolated requires careful consideration as such niches restrict the xenobiotics influencing XANR LBD function and evolution. For example, filter-feeding bivalves use a somewhat different mechanism for filtering seawater than do filter-feeding tunicates—and therefore these two groups of filter-feeders ingest somewhat differing profiles/size-ranges of marine microorganisms [196]. It is also important to consider that some bioactive chemicals may be produced by marine organisms that adhere to hard surfaces. For example, benthic microalgae, such as the dinoflagellate *Gambierdiscus toxicus* can produce highly toxic compounds (e.g., ciguatera-associated toxins) [197]. To detect such toxins, XANR LBDs from surface-grazing animals would probably be more suitable for the bioassay than XANR LBDs from filter-feeding animals. A third consideration guiding XANR LBD selection is simply the genomic resources available. Clearly, when a specific marine invertebrate taxon has been selected on evolutionary and ecological grounds, then the required genomic datasets can be generated for increasingly realistic costs. Nonetheless, as discussed earlier in this review, bioinformatic challenges exist when designating a NR as a putative XANR based solely on homology/phylogeny. As a generalization, more reliable predictions of *bona fide* XANRs are likely within those phyla, particularly *Chordata* and *Arthropoda*, for which a number of functionally characterized XANRs exist.

7. Future Prospects for Marine Invertebrate XANR LBD-Based Bioassays

The discovery of useful natural marine bioactive compounds, along with the subsequent development of derived pharmaceuticals, faces immense technical challenges [198]. Consequently, despite the enormous number of structurally unique bioactive marine natural products that are now known, to date the associated pharmaceutical pipeline comprises only eight approved drugs, along with twelve natural marine products (or derivatives thereof) in different phases of clinical testing [198,199]. Obviously the natural biological activities of potential drug lead compounds influence their potential medical applications, so bioassay design is a major limiting factor in the detection of useful bioactives [198,200]. In this context we see potential for XANR LBD-based bioassays as their specificities rest on natural evolutionary processes that may have molded the XANR LBD's structure and its associated ligand-binding properties. Thus, in a sense, these evolutionary processes can provide "creative input" into the bioassay design. Due to the highly modular structure of NRs the XANR LBDs can be combined with generic DNA-binding and transcription-activating domains in various cell-based bioassays or can be used as purified proteins in biosensors. By selecting, on the basis of taxonomy and ecology, the organism to source the XANR LBD from it may be possible to tailor bioassays to search for bioactive compounds from differing sources. As an example, LBDs of potential XANRs identified in the genomes of marine filter-feeding organisms like tunicates and bivalves could be used to test for bioactive compounds associated with the myriad of microorganisms that make up the diet of marine filter-feeders (Figure 1) [7,149].

8. Conclusions

This review began with the assertion that, from both an ecological and evolutionary perspective, many bioassays currently used to screen for marine bioactive chemicals are somewhat "arbitrary", in the sense that they bring together chemicals and biological detection systems that would rarely, if ever, be combined in nature. To address this deficiency, we have proposed that members of a specific group of ligand-activated transcription factors—marine invertebrate xenobiotic-activated nuclear receptors (XANRs)—provide a source of bioassay sensor elements that have been pre-molded by natural selection for detecting bioactive chemicals present in marine invertebrate diets. As a proof-of-concept we outlined recent work showing that mammalian cell lines expressing tunicate XANR LBDs, coupled with an appropriate reporter gene, can detect established microalgal biotoxins. Based on such success with mammalian cell lines we suggest that recombinant yeast strains expressing XANR LBDs may provide low-cost, high-throughput bioassays. Alternatively, it may be possible to entirely remove the need for live cells and adapt biosensor technologies to look for marine chemicals that directly bind to XANR LBDs. Whatever the assay format and technology used, marine invertebrate genomes, each with its own ecological niche and evolutionary history, represent an increasingly accessible informational resource of XANRs that can be harnessed to identify the chemical treasures that are undoubtedly hidden in the sea [198,201].

Acknowledgments

This work was funded by the New Zealand Ministry of Business, Innovation and Employment (Contract No. CAWX1001). We are grateful to Chris Woods (National Institute of Water and Atmospheric Research, New Zealand) and Rod Asher (Cawthron Institute, New Zealand) for providing photographs used in Figure 1.

References

1. Denison, M.S.; Nagy, S.R. Activation of the aryl hydrocarbon receptor by structurally diverse exogenous and endogenous chemicals. *Annu. Rev. Pharmacol. Toxicol.* **2003**, *43*, 309–334.

2. Kewley, R.J.; Whitelaw, M.L.; Chapman-Smith, A. The mammalian basic helix-loop-helix/PAS family of transcriptional regulators. *Int. J. Biochem. Cell Biol.* **2004**, *36*, 189–204.

3. Fujii-Kuriyama, Y.; Kawajiri, K. Molecular mechanisms of the physiological functions of the aryl hydrocarbon (dioxin) receptor, a multifunctional regulator that senses and responds to environmental stimuli. *Proc. Jpn. Acad. Ser. B-Phys. Biol. Sci.* **2010**, *86*, 40–53.

4. Moore, L.B.; Maglich, J.M.; McKee, D.D.; Wisely, B.; Willson, T.M.; Kliewer, S.A.; Lambert, M.H.; Moore, J.T. Pregnane X receptor (PXR), constitutive androstane receptor (CAR), and benzoate X receptor (BXR) define three pharmacologically distinct classes of nuclear receptors. *Mol. Endocrinol.* **2002**, *16*, 977–986.

5.   Krasowski, M.D.; Ni, A.; Hagey, L.R.; Ekins, S. Evolution of promiscuous nuclear hormone receptors: LXR, FXR, VDR, PXR, and CAR. *Mol. Cell. Endocrinol.* **2011**, *334*, 39–48.

6.   Reschly, E.J.; Krasowski, M.D. Evolution and function of the NR1I nuclear hormone receptor subfamily (VDR, PXR, and CAR) with respect to metabolism of xenobiotics and endogenous compounds. *Curr. Drug Metab.* **2006**, *7*, 349–365.

7.   Echevarria, M.; Naar, J.P.; Tomas, C.; Pawlik, J.R. Effects of *Karenia brevis* on clearance rates and bioaccumulation of brevetoxins in benthic suspension feeding invertebrates. *Aquat. Toxicol.* **2012**, *106*, 85–94.

8.   Haberkorn, H.; Tran, D.; Massabuau, J.C.; Ciret, P.; Savar, V.; Soudant, P. Relationship between valve activity, microalgae concentration in the water and toxin accumulation in the digestive gland of the Pacific oyster *Crassostrea gigas* exposed to *Alexandrium minutum*. *Mar. Pollut. Bull.* **2011**, *62*, 1191–1197.

9.   Forbey, J.S.; Dearing, M.D.; Gross, E.M.; Orians, C.M.; Sotka, E.E.; Foley, W.J. A Pharm-Ecological perspective of terrestrial and aquatic plant-herbivore interactions. *J. Chem. Ecol.* **2013**, *39*, 465–480.

10. Raubenheimer, D.; Simpson, S.J. Nutritional PharmEcology: Doses, nutrients, toxins, and medicines. *Integr. Comp. Biol.* **2009**, *49*, 329–337.

11. Deng, H.; Kerppola, T.K. Regulation of Drosophila metamorphosis by xenobiotic response regulators. *PLoS Genet.* **2013**, *9*, e1003263.

12. Ortiz-Ramirez, F.A.; Vallim, M.A.; Cavalcanti, D.N.; Teixeira, V.L. Effects of the secondary metabolites from *Canistrocarpus cervicornis* (Dictyotales, Phaeophyceae) on fertilization and early development of the sea urchin *Lytechinus variegatus*. *Lat. Am. J. Aquat. Res.* **2013**, *41*, 296–304.

13. Targett, N.M.; Arnold, T.M. Effects of secondary metabolites on digestion in marine herbivores. In *Marine Chemical Ecology*; CRC Press: Boca Raton, FL, USA, 2001; pp. 391–411.

14. Hay, M.E. Marine chemical ecology: Chemical signals and cues structure marine populations, communities, and ecosystems. *Annu. Rev. Mar. Sci.* **2009**, *1*, 193–212.

15. Dearing, M.D.; Foley, W.J.; McLean, S. The influence of plant secondary metabolites on the nutritional ecology of herbivorous terrestrial vertebrates. *Annu. Rev. Ecol. Evol. Syst.* **2005**, *36*, 169–189.

16. Marsh, K.J.; Wallis, I.R.; Andrew, R.L.; Foley, W.J. The detoxification limitation hypothesis: Where did it come from and where is it going? *J. Chem. Ecol.* **2006**, *32*, 1247–1266.

17. Slattery, M.; Avila, C.; Starmer, J.; Paul, V.J. A sequestered soft coral diterpene in the aeolid nudibranch *Phyllodesmium guamensis*. *J. Exp. Mar. Biol. Ecol.* **1998**, *226*, 33–49.

18. Glendinning, J.I. Is the bitter rejection response always adaptive? *Physiol. Behav.* **1994**, *56*, 1217–1227.

19. Chandrashekar, J.; Mueller, K.L.; Hoon, M.A.; Adler, E.; Feng, L.X.; Guo, W.; Zuker, C.S.; Ryba, N.J.P. T2Rs function as bitter taste receptors. *Cell* **2000**, *100*, 703–711.

20. Dong, D.; Jones, G.; Zhang, S. Dynamic evolution of bitter taste receptor genes in vertebrates. *BMC Evol. Biol.* **2009**, *9*, doi:10.1186/1471-2148-9-12.

21. Li, D.; Zhang, J. Diet shapes the evolution of the vertebrate bitter taste receptor gene repertoire. *Mol. Biol. Evol.* **2013**, *7*, 7, doi:10.1093/molbev/mst219.

22. Sugawara, T.; Go, Y.; Udono, T.; Morimura, N.; Tomonaga, M.; Hirai, H.; Imai, H. Diversification of bitter taste receptor gene family in western chimpanzees. *Mol. Biol. Evol.* **2011**, *28*, 921–931.

23. Miller, A.M.; Pawlik, J.R. Do coral reef fish learn to avoid unpalatable prey using visual cues? *Anim. Behav.* **2013**, *85*, 339–347.

24. Long, J.D.; Hay, M.E. Fishes learn aversions to a nudibranch's chemical defense. *Mar. Ecol. Prog. Ser.* **2006**, *307*, 199–208.

25. Ritson-Williams, R.; Paul, V.J. Marine benthic invertebrates use multimodal cues for defense against reef fish. *Mar. Ecol. Prog. Ser.* **2007**, *340*, 29–39.

26. Hegaret, H.; Wikfors, G.H.; Shumway, S.E. Diverse feeding responses of five species of bivalve mollusc when exposed to three species of harmful algae. *J. Shellfish Res.* **2007**, *26*, 549–559.

27. Fdil, M.A.; Mouabad, A.; Outzourhit, A.; Benhra, A.; Maarouf, A.; Pihan, J.C. Valve movement response of the mussel *Mytilus galloprovincialis* to metals (Cu, Hg, Cd and Zn) and phosphate industry effluents from Moroccan Atlantic coast. *Ecotoxicology* **2006**, *15*, 477–486.

28. Wildish, D.; Lassus, P.; Martin, J.; Saulnier, A.; Bardouil, M. Effect of the PSP-causing dinoflagellate, *Alexandrium sp.* on the initial feeding response of *Crassostrea gigas*. *Aquat. Living Resour.* **1998**, *11*, 35–43.

29. Glendinning, J.I. How do predators cope with chemically defended foods? *Biol. Bull.* **2007**, *213*, 252–266.

30. Manfrin, C.; de Moro, G.; Torboli, V.; Venier, P.; Pallavicini, A.; Gerdol, M. Physiological and molecular responses of bivalves to toxic dinoflagellates. *Invertebr. Surviv. J.* **2012**, *9*, 184–199.

31. Fernandez-Reiriz, M.J.; Navarro, J.M.; Contreras, A.M.; Labarta, U. Trophic interactions between the toxic dinoflagellate *Alexandrium catenella* and *Mytilus chilensis*: Feeding and digestive behaviour to long-term exposure. *Aquat. Toxicol.* **2008**, *87*, 245–251.

32. Sotka, E.E.; Gantz, J. Preliminary evidence that the feeding rates of generalist marine herbivores are limited by detoxification rates. *Chemoecology* **2013**, *23*, 233–240.

33. Sotka, E.E.; Forbey, J.; Horn, M. The emerging role of pharmacology in understanding consumer-prey interactions in marine and freshwater systems. *Integr. Comp. Biol.* **2009**, *49*, 291–313.

34. Zanger, U.M.; Schwab, M. Cytochrome P450 enzymes in drug metabolism: Regulation of gene expression, enzyme activities, and impact of genetic variation. *Pharmacol. Ther.* **2013**, *138*, 103–141.

35. Oakley, A. Glutathione transferases: A structural perspective. *Drug Metab. Rev.* **2011**, *43*, 138–151.

36. Li, X.C.; Schuler, M.A.; Berenbaum, M.R. Molecular mechanisms of metabolic resistance to synthetic and natural xenobiotics. *Annu. Rev. Entomol.* **2007**, *52*, 231–253.

37. Nelson, D.R.; Goldstone, J.V.; Stegeman, J.J. The cytochrome P450 genesis locus: The origin and evolution of animal cytochrome P450s. *Philos. Trans. R. Soc. B-Biol. Sci.* **2013**, *368*, doi: 10.1098/rstb.2012.0474.

38. Testa, B.; Pedretti, A.; Vistoli, G. Foundation review: Reactions and enzymes in the metabolism of drugs and other xenobiotics. *Drug Discov. Today* **2012**, *17*, 549–560.

39. Sladek, F.M. What are nuclear receptor ligands? *Mol. Cell. Endocrinol.* **2011**, *334*, 3–13.

40. Baker, M.E. Xenobiotics and the evolution of multicellular animals: Emergence and diversification of ligand-activated transcription factors. *Integr. Comp. Biol.* **2005**, *45*, 172–178.

41. Iyanagi, T. Molecular mechanism of phase I and phase II drug-metabolizing enzymes: Implications for detoxification, In *International Review of Cytology—A Survey of Cell Biology*, 1st ed.; Jeon, K.W., Ed.; Elsevier Academic Press Inc: San Diego, CA, USA, 2007; Volume 260, pp. 35–112.

42. Yang, Y.M.; Noh, K.; Han, C.Y.; Kim, S.G. Transactivation of genes encoding for phase II enzymes and phase III transporters by phytochemical antioxidants. *Molecules* **2010**, *15*, 6332–6348.

43. Sotka, E.E.; Whalen, K.E. Herbivore offense in the sea: The detoxification and transport of secondary metabolites. In *Algal Chemical Ecology*, XVIII ed.; Amsler, C.D., Ed.; Springer: Berlin/Heidelberg, Germany, 2008; pp. 203–228.

44. Court, M.H. Feline drug metabolism and disposition pharmacokinetic evidence for species differences and molecular mechanisms. *Vet. Clin. N. Am. Small Anim. Pract.* **2013**, *43*, 1039–1054.

45. Shrestha, B.; Reed, J.M.; Starks, P.T.; Kaufman, G.E.; Goldstone, J.V.; Roelke, M.E.; O'Brien, S.J.; Koepfli, K.-P.; Frank, L.G.; Court, M.H.; *et al.* Evolution of a major drug metabolizing enzyme defect in the domestic cat and other Felidae: Phylogenetic timing and the role of hypercarnivory. *PLoS One* **2011**, *6*, e18046.

46. Morris, J.G. Idiosyncratic nutrient requirements of cats appear to be diet-induced evolutionary adaptations. *Nutr. Res. Rev.* **2002**, *15*, 153–168.

47. James, M.O.; Ambadapadi, S. Interactions of cytosolic sulfotransferases with xenobiotics. *Drug Metab. Rev.* **2013**, *45*, 401–414.

48. Pinto, N.; Dolan, M.E. Clinically relevant genetic variations in drug metabolizing enzymes. *Curr. Drug Metab.* **2011**, *12*, 487–497.

49. Zanger, U.M.; Turpeinen, M.; Klein, K.; Schwab, M. Functional pharmacogenetics/genomics of human cytochromes P450 involved in drug biotransformation. *Anal. Bioanal. Chem.* **2008**, *392*, 1093–1108.

50. Yamazaki, H. Roles of human cytochrome P450 enzymes involved in drug metabolism and toxicological studies. *J. Pharm. Soc. Jpn.* **2000**, *120*, 1347–1357.

51. Pavek, P.; Dvorak, Z. Xenobiotic-induced transcriptional regulation of xenobiotic metabolizing enzymes of the cytochrome P450 superfamily in human extrahepatic tissues. *Curr. Drug Metab.* **2008**, *9*, 129–143.

52. Thummel, K.E.; Wilkinson, G.R. In vitro and *in vivo* drug interactions involving human CYP3A. *Annu. Rev. Pharm. Toxicol.* **1998**, *38*, 389–430.

53. Dobrinas, M.; Cornuz, J.; Pedrido, L.; Eap, C.B. Influence of cytochrome P450 oxidoreductase genetic polymorphisms on CYP1A2 activity and inducibility by smoking. *Pharm. Genomics* **2012**, *22*, 143–151.

54. Martignoni, M.; Groothuis, G.M.M.; de Kanter, R. Species differences between mouse, rat, dog, monkey and human CYP-mediated drug metabolism, inhibition and induction. *Expert Opin. Drug Metab. Toxicol.* **2006**, *2*, 875–894.

55. Vignati, L.A.; Bogni, A.; Grossi, P.; Monshouwer, M. A human and mouse pregnane X receptor reporter gene assay in combination with cytotoxicity measurements as a tool to evaluate species-specific CYP3A induction. *Toxicology* **2004**, *199*, 23–33.

56. Martignoni, M.; de Kanter, R.; Grossi, P.; Mahnke, A.; Saturno, G.; Monshouwer, M. An *in vivo* and *in vitro* comparison of CYP induction in rat liver and intestine using slices and quantitative RT-PCR. *Chem.-Biol. Interact.* **2004**, *151*, 1–11.

57. Lu, C.; Li, A.P. Species comparison in P450 induction: Effects of dexamethasone, omeprazole, and rifampin on P450 isoforms 1A and 3A in primary cultured hepatocytes from man, Sprague-Dawley rat, minipig, and beagle dog. *Chem. Biol. Interact.* **2001**, *134*, 271–281.

58. Kocarek, T.A.; Schuetz, E.G.; Strom, S.C.; Fisher, R.A.; Guzelian, P.S. Comparative analysis of cytochrome P4503A induction in primary cultures of rat, rabbit, and human hepatocytes. *Drug Metab. Dispos.* **1995**, *23*, 415–421.

59. Rewitz, K.F.; Styrishave, B.; Lobner-Olesen, A.; Andersen, O. Marine invertebrate cytochrome P450: Emerging insights from vertebrate and insect analogies. *Comp. Biochem. Physiol. C Toxicol. Pharmacol.* **2006**, *143*, 363–381.

60. Schuler, M.A.; Berenbaum, M.R. Structure and function of cytochrome P450S in insect adaptation to natural and synthetic toxins: Insights gained from molecular modeling. *J. Chem. Ecol.* **2013**, *39*, 1232–1245.

61. Feyereisen, R. Arthropod CYPomes illustrate the tempo and mode in P450 evolution. *Biochim. Biophys. Acta (BBA) Proteins Proteomics* **2011**, *1814*, 19–28.

62. Fisher, T.; Crane, M.; Callaghan, A. Induction of cytochrome P-450 activity in individual *Chironomus riparius* (Meigen) larvae exposed to xenobiotics. *Ecotoxicol. Environ. Saf.* **2003**, *54*, 1–6.

63. Kasai, S.; Scott, J.G. Expression and regulation of CYP6D3 in the house fly, *Musca domestica* (L.). *Insect Biochem. Mol. Biol.* **2001**, *32*, 1–8.

64. Tomita, T.; Itokawa, K.; Komagata, O.; Kasai, S. Overexpression of cytochrome P450 genes in insecticide-resistant mosquitoes. *J. Pestic. Sci.* **2010**, *35*, 562–568.

65. Natsuhara, K.; Shimada, K.; Tanaka, T.; Miyata, T. Phenobarbital induction of permethrin detoxification and phenobarbital metabolism in susceptible and resistant strains of the beet armyworm *Spodoptera exigua* (Hubner). *Pestic. Biochem. Physiol.* **2004**, *79*, 33–41.

66. Johnson, R.M.; Mao, W.; Pollock, H.S.; Niu, G.; Schuler, M.A.; Berenbaum, M.R. Ecologically appropriate xenobiotics induce cytochrome P450s in *Apis mellifera. PLoS One* **2012**, *7*, e31051, doi:10.1371/journal.pone.0031051.

67. Niu, G.; Johnson, R.M.; Berenbaum, M.R. Toxicity of mycotoxins to honeybees and its amelioration by propolis. *Apidologie* **2011**, *42*, 79–87.

68. Morra, R.; Kuruganti, S.; Lam, V.; Lucchesi, J.C.; Ganguly, R. Functional analysis of the *cis-acting* elements responsible for the induction of the Cyp6a8 and Cyp6g1 genes of *Drosophila melanogaster* by DDT, phenobarbital and caffeine. *Insect Mol. Biol.* **2010**, *19*, 121–130.

69. Maitra, S.; Dombrowski, S.M.; Waters, L.C.; Ganguly, R. Three second chromosome-linked clustered Cyp6 genes show differential constitutive and barbital-induced expression in DDT-resistant and susceptible strains of *Drosophila melanogaster. Gene* **1996**, *180*, 165–171.

70. Misra, J.R.; Horner, M.A.; Lam, G.; Thummel, C.S. Transcriptional regulation of xenobiotic detoxification in Drosophila. *Genes Dev.* **2011**, *25*, 1796–1806.

71. Danielson, P.B.; MacIntyre, R.J.; Fogleman, J.C. Molecular cloning of a family of xenobiotic-inducible drosophilid cytochrome P450s: Evidence for involvement in host-plant allelochemical resistance. *PNAS* **1997**, *94*, 10797–10802.

72. Menzel, R.; Bogaert, T.; Achazi, R. A systematic gene expression screen of *Caenorhabditis elegans* cytochrome P450 genes reveals CYP35 as strongly xenobiotic inducible. *Arch. Biochem. Biophys.* **2001**, *395*, 158–168.

73. Menzel, R.; Rodel, M.; Kulas, J.; Steinberg, C.E.W. CYP35: Xenobiotically induced gene expression in the nematode *Caenorhabditis elegans*. *Arch. Biochem. Biophys.* **2005**, *438*, 93–102.

74. Schaefer, P.; Mueller, M.; Krueger, A.; Steinberg, C.E.W.; Menzel, R. Cytochrome P450-dependent metabolism of PCB52 in the nematode *Caenorhabditis elegans*. *Arch. Biochem. Biophys.* **2009**, *488*, 60–68.

75. Chakrapani, B.P.; Kumar, S.; Subramaniam, J.R. Development and evaluation of an *in vivo* assay in *Caenorhabditis elegans* for screening of compounds for their effect on cytochrome P450 expression. *J. Biosci.* **2008**, *33*, 269–277.

76. Zheng, S.L.; Chen, B.; Qiu, X.Y.; Lin, K.L.; Yu, X.G. Three novel cytochrome P450 genes identified in the marine polychaete *Perinereis nuntia* and their transcriptional response to xenobiotics. *Aquat. Toxicol.* **2013**, *134*, 11–22.

77. Whalen, K.E.; Starczak, V.R.; Nelson, D.R.; Goldstone, J.V.; Hahn, M.E. Cytochrome P450 diversity and induction by gorgonian allelochemicals in the marine gastropod *Cyphoma gibbosum*. *BMC Ecol.* **2010**, *10*, 24, doi:10.1186/1472-6785-10-24.

78. Bousova, I.; Skalova, L. Inhibition and induction of glutathione S-transferases by flavonoids: Possible pharmacological and toxicological consequences. *Drug Metab. Rev.* **2012**, *44*, 267–286.

79. Maglich, J.M.; Stoltz, C.M.; Goodwin, B.; Hawkins-Brown, D.; Moore, J.T.; Kliewer, S.A. Nuclear pregnane X receptor and constitutive androstane receptor regulate overlapping but distinct sets of genes involved in xenobiotic detoxification. *Mol. Pharmacol.* **2002**, *62*, 638–646.

80. Whalen, K.E.; Morin, D.; Lin, C.Y.; Tjeerdema, R.S.; Goldstone, J.V.; Hahn, M.E. Proteomic identification, cDNA cloning and enzymatic activity of glutathione S-transferases from the generalist marine gastropod, *Cyphoma gibbosum*. *Arch. Biochem. Biophys.* **2008**, *478*, 7–17.

81. Kuhajek, J.M.; Schlenk, D. Effects of the brominated phenol, lanosol, on cytochrome P-450 and glutathione transferase activities in *Haliotis rufescens* and *Katharina tunicata*. *Comp. Biochem. Physiol. C Toxicol. Pharmacol.* **2003**, *134*, 473–479.

82. Whalen, K.E.; Lane, A.L.; Kubanek, J.; Hahn, M.E. Biochemical warfare on the reef: The role of glutathione transferases in consumer tolerance of dietary prostaglandins. *PLoS One* **2010**, *5*, 0008537.

83. Mao, W.; Rupasinghe, S.G.; Johnson, R.M.; Zangerl, A.R.; Schuler, M.A.; Berenbaum, M.R. Quercetin-metabolizing CYP6AS enzymes of the pollinator *Apis mellifera* (Hymenoptera: Apidae). *Comp. Biochem. Physiol. B Biochem. Mol. Biol.* **2009**, *154*, 427–434.

84. Bainy, A.C.D.; Kubota, A.; Goldstone, J.V.; Lille-Langoy, R.; Karchner, S.I.; Celander, M.C.; Hahn, M.E.; Goksoyr, A.; Stegeman, J.J. Functional characterization of a full length pregnane X receptor, expression *in vivo*, and identification of PXR alleles, in Zebrafish (*Danio rerio*). *Aquat. Toxicol.* **2013**, *142–143*, 447–457.

85. Chai, X.; Zeng, S.; Xie, W. Nuclear receptors PXR and CAR: Implications for drug metabolism regulation, pharmacogenomics and beyond. *Expert Opin. Drug Metab. Toxicol.* **2013**, *9*, 253–266.

86. Xie, W.; Chiang, J.Y.L. Nuclear receptors in drug metabolism and beyond. *Drug Metab. Rev.* **2013**, *45*, 1–2.

87. NR1I2 nuclear receptor subfamily 1, group I, member 2. Available online: http://www.ncbi.nlm.nih.gov/gene/8856 (accessed on 26 February 2014).

88. Kliewer, S.A.; Moore, J.T.; Wade, L.; Staudinger, J.L.; Watson, M.A.; Jones, S.A.; McKee, D.D.; Oliver, B.B.; Willson, T.M.; Zetterstrom, R.H.; *et al.* An orphan nuclear receptor activated by pregnanes defines a novel steroid signaling pathway. *Cell* **1998**, *92*, 73–82.

89. Blumberg, B.; Sabbagh, W.; Juguilon, H.; Bolado, J.; van Meter, C.M.; Ono, E.S.; Evans, R.M. SXR, a novel steroid and xenobiotic-sensing nuclear receptor. *Genes Dev.* **1998**, *12*, 3195–3205.

90. Bertilsson, G.; Heidrich, J.; Svensson, K.; Asman, M.; Jendeberg, L.; Sydow-Backman, M.; Ohlsson, R.; Postlind, H.; Blomquist, P.; Berkenstam, A.; *et al.* Identification of a human nuclear receptor defines a new signaling pathway for CYP3A induction. *PNAS* **1998**, *95*, 12208–12213.

91. Orans, J.; Teotico, D.G.; Redinbo, M.R. The nuclear xenobiotic receptor pregnane X receptor: Recent insights and new challenges. *Mol. Endocrinol.* **2005**, *19*, 2891–2900.

92. Li, T.G.; Chiang, J.Y.L. Mechanism of rifampicin and pregnane X receptor inhibition of human cholesterol 7 alpha-hydroxylase gene transcription. *Am. J. Physiol. Gastrointest. Liver Physiol.* **2005**, *288*, G74–G84.

93. McKenna, N.J.; Lanz, R.B.; O'Malley, B.W. Nuclear receptor coregulators: Cellular and molecular biology. *Endocr. Rev.* **1999**, *20*, 321–344.

94. Teotico, D.G.; Frazier, M.L.; Ding, F.; Dokholyan, N.V.; Temple, B.R.S.; Redinbo, M.R. Active Nuclear Receptors Exhibit Highly Correlated AF-2 Domain Motions. *PLoS Comput. Biol.* **2008**, *4*, e1000111, doi:10.1371/journal.pcbi.1000111.

95. Tolson, A.H.; Wang, H.B. Regulation of drug-metabolizing enzymes by xenobiotic receptors: PXR and CAR. *Adv. Drug Deliv. Rev.* **2010**, *62*, 1238–1249.

96. Biswas, A.; Mani, S.; Redinbo, M.R.; Krasowski, M.D.; Li, H.; Ekins, S. Elucidating the "Jekyll and Hyde" nature of PXR: The case for discovering antagonists or allosteric antagonists. *Pharm. Res.* **2009**, *26*, 1807–1815.

97. Chang, T.K.; Waxman, D.J. Synthetic drugs and natural products as modulators of constitutive androstane receptor (CAR) and pregnane X receptor (PXR). *Drug Metab. Rev.* **2006**, *38*, 51–73.

98. Hernandez, J.P.; Mota, L.C.; Baldwin, W.S. Activation of CAR and PXR by Dietary, Environmental and Occupational Chemicals Alters Drug Metabolism, Intermediary Metabolism, and Cell Proliferation. *Curr. Pharm. Pers. Med.* **2009**, *7*, 81–105.

99. Manez, S. A fresh insight into the interaction of natural products with pregnane X receptor. *Nat. Prod. Commun.* **2008**, *3*, 2123–2128.

100. Staudinger, J.L.; Ding, X.; Lichti, K. Pregnane X receptor and natural products: Beyond drug-drug interactions. *Expert Opin. Drug Metab. Toxicol.* **2006**, *2*, 847–857.

101. Zhou, C.; Verma, S.; Blumberg, B. The steroid and xenobiotic receptor (SXR), beyond xenobiotic metabolism. *Nucl. Recept. Signal.* **2009**, *7*, e001.

102. Harms, M.J.; Eick, G.N.; Goswami, D.; Colucci, J.K.; Griffin, P.R.; Ortlund, E.A.; Thornton, J.W. Biophysical mechanisms for large-effect mutations in the evolution of steroid hormone receptors. *PNAS* **2013**, *110*, 11475–11480.

103. Eick, G.N.; Colucci, J.K.; Harms, M.J.; Ortlund, E.A.; Thornton, J.W. Evolution of minimal specificity and promiscuity in steroid hormone receptors. *PLoS Genet.* **2012**, *8*, e1003072, doi:10.1371/journal.pgen.1003072.

104. Wu, B.; Li, S.; Dong, D. 3D structures and ligand specificities of nuclear xenobiotic receptors CAR, PXR and VDR. *Drug Discov. Today* **2013**, *18*, 574–581.

105. Wallace, B.D.; Redinbo, M.R. Xenobiotic-sensing nuclear receptors involved in drug metabolism: A structural perspective. *Drug Metab. Rev.* **2013**, *45*, 79–100.

106. Mangelsdorf, D.J.; Thummel, C.; Beato, M.; Herrlich, P.; Schutz, G.; Umesono, K.; Blumberg, B.; Kastner, P.; Mark, M.; Chambon, P.; *et al.* The nuclear receptor superfamily: The second decade. *Cell* **1995**, *83*, 835–839.

107. Krasowski, M.D.; Yasuda, K.; Hagey, L.R.; Schuetz, E.G. Evolution of the pregnane X receptor: Adaptation to cross-species differences in biliary bile salts. *Mol. Endocrinol.* **2005**, *19*, 1720–1739.

108. Krasowski, M.D.; Yasuda, K.; Hagey, L.R.; Schuetz, E.G. Evolutionary selection across the nuclear hormone receptor superfamily with a focus on the NR1I subfamily (vitamin D, pregnane X, and constitutive androstane receptors). *Nucl. Recept.* **2005**, *3*, 2, doi:10.1186/1478-1336-3-2.

109. Zhang, Z.D.; Burch, P.E.; Cooney, A.J.; Lanz, R.B.; Pereira, F.A.; Wu, J.Q.; Gibbs, R.A.; Weinstock, G.; Wheeler, D.A. Genomic analysis of the nuclear receptor family: New insights into structure, regulation, and evolution from the rat genome. *Genome Res.* **2004**, *14*, 580–590.

110. Krasowski, M.D.; Ai, N.; Hagey, L.R.; Kollitz, E.M.; Kullman, S.W.; Reschly, E.J.; Ekins, S. The evolution of farnesoid X, vitamin D, and pregnane X receptors: Insights from the green-spotted pufferfish (*Tetraodon nigriviridis*) and other non-mammalian species. *BMC Biochem.* **2011**, *12*, doi:10.1186/1471-2091-12-5.

111. Moore, L.B.; Parks, D.J.; Jones, S.A.; Bledsoe, R.K.; Consler, T.G.; Stimmel, J.B.; Goodwin, B.; Liddle, C.; Blanchard, S.G.; Willson, T.M.; *et al.* Orphan nuclear receptors constitutive androstane receptor and pregnane X receptor share xenobiotic and steroid ligands. *J. Biol. Chem.* **2000**, *275*, 15122–15127.

112. Jones, S.A.; Moore, L.B.; Shenk, J.L.; Wisely, G.B.; Hamilton, G.A.; McKee, D.D.; Tomkinson, N.C.O.; LeCluyse, E.L.; Lambert, M.H.; Willson, T.M.; *et al.* The pregnane X receptor: A promiscuous xenobiotic receptor that has diverged during evolution. *Mol. Endocrinol.* **2000**, *14*, 27–39.

113. LeCluyse, E.L. Pregnane X receptor: Molecular basis for species differences in CYP3A induction by xenobiotics. *Chem. Biol. Interact.* **2001**, *134*, 283–289.

114. Lehmann, J.M.; McKee, D.D.; Watson, M.A.; Willson, T.M.; Moore, J.T.; Kliewer, S.A. The human orphan nuclear receptor PXR is activated by compounds that regulate CYP3A4 gene expression and cause drug interactions. *J. Clin. Investig.* **1998**, *102*, 1016–1023.

115. Xie, W.; Evans, R.M. Orphan nuclear receptors: The exotics of xenobiotics. *J. Biol. Chem.* **2001**, *276*, 37739–37742.

116. Xie, W.; Barwick, J.L.; Downes, M.; Blumberg, B.; Simon, C.M.; Nelson, M.C.; Neuschwander-Tetri, B.A.; Bruntk, E.M.; Guzelian, P.S.; Evans, R.M.; *et al.* Humanized xenobiotic response in mice expressing nuclear receptor SXR. *Nature* **2000**, *406*, 435–439.

117. Tirona, R.G.; Leake, B.F.; Podust, L.M.; Kim, R.B. Identification of amino acids in rat pregnane X receptor that determine species-specific activation. *Mol. Pharmacol.* **2004**, *65*, 36–44.

118. Wang, D.Y.C.; Kumar, S.; Hedges, S.B. Divergence time estimates for the early history of animal phyla and the origin of plants, animals and fungi. *PNAS* **1999**, *266*, 163–171.

119. Erwin, D.H.; Davidson, E.H. The last common bilaterian ancestor. *Development* **2002**, *129*, 3021–3032.

120. Hedges, S.B.; Dudley, J.; Kumar, S. TimeTree: A public knowledge-base of divergence times among organisms. *Bioinformatics* **2006**, *22*, 2971–2972.

121. King-Jones, K.; Horner, M.A.; Lam, G.; Thummel, C.S. The DHR96 nuclear receptor regulates xenobiotic responses in Drosophila. *Cell Metab.* **2006**, *4*, 37–48.

122. Lindblom, T.H.; Pierce, G.J.; Sluder, A.E. A *C. elegans* orphan nuclear receptor contributes to xenobiotic resistance. *Curr. Biol.* **2001**, *11*, 864–868.

123. Lin, G.G.H.; Kozaki, T.; Scott, J.G. Hormone receptor-like in 96 and Broad-Complex modulate phenobarbital induced transcription of cytochrome P450 CYP6D1 in Drosophila S2 cells. *Insect Mol. Biol.* **2011**, *20*, 87–95.

124. Palanker, L.; Necakov, A.S.; Sampson, H.M.; Ni, R.; Hu, C.H.; Thummel, C.S.; Krause, H.M. Dynamic regulation of Drosophila nuclear receptor activity *in vivo*. *Development* **2006**, *133*, 3549–3562.

125. Sieber, M.H.; Thummel, C.S. Coordination of triacylglycerol and cholesterol homeostasis by DHR96 and the Drosophila LipA homolog magro. *Cell Metab.* **2012**, *15*, 122–127.

126. Karimullina, E.; Li, Y.; Ginjupalli, G.K.; Baldwin, W.S. Daphnia HR96 is a promiscuous xenobiotic and endobiotic nuclear receptor. *Aquat. Toxicol.* **2012**, *116*, 69–78.

127. Xu, J.; Tan, A.; Palli, S.R. The function of nuclear receptors in regulation of female reproduction and embryogenesis in the red flour beetle, *Tribolium castaneum*. *J. Insect Physiol.* **2010**, *56*, 1471–1480.

128. Bonasio, R.; Zhang, G.; Ye, C.; Mutti, N.S.; Fang, X.; Qin, N.; Donahue, G.; Yang, P.; Li, Q.; Li, C.; *et al.* Genomic comparison of the ants *Camponotus floridanus* and *Harpegnathos saltator*. *Science* **2010**, *329*, 1068–1071.

129. Velarde, R.A.; Robinson, G.E.; Fahrbach, S.E. Nuclear receptors of the honey bee: Annotation and expression in the adult brain. *Insect Mol. Biol.* **2006**, *15*, 583–595.

130. Giraudo, M.; Audant, P.; Feyereisen, R.; le Goff, G. Nuclear receptors HR96 and ultraspiracle from the fall armyworm (*Spodoptera frugiperda*), developmental expression and induction by xenobiotics. *J. Insect Physiol.* **2013**, *59*, 560–568.

131. Sluder, A.E.; Mathews, S.W.; Hough, D.; Yin, V.P.; Maina, C.V. The nuclear receptor superfamily has undergone extensive proliferation and diversification in nematodes. *Genome Res.* **1999**, *9*, 103–120.

132. Bertrand, W.; Brunet, F.G.; Escriva, H.; Parmentier, G.; Laudet, V.; Robinson-Rechavi, M. Evolutionary genomics of nuclear receptors: From twenty-five ancestral genes to derived endocrine systems. *Mol. Biol. Evol.* **2004**, *21*, 1923–1937.

133. Delsuc, F.; Brinkmann, H.; Chourrout, D.; Philippe, H. Tunicates and not cephalochordates are the closest living relatives of vertebrates. *Nature* **2006**, *439*, 965–968.

134. Delsuc, F.; Tsagkogeorga, G.; Lartillot, N.; Philippe, H. Additional molecular support for the new chordate phylogeny. *Genesis* **2008**, *46*, 592–604.

135. Singh, T.R.; Tsagkogeorga, G.; Delsuc, F.; Blanquart, S.; Shenkar, N.; Loya, Y.; Douzery, E.J.P.; Huchon, D. Tunicate mitogenomics and phylogenetics: Peculiarities of the *Herdmania momus* mitochondrial genome and support for the new chordate phylogeny. *BMC Genomics* **2009**, *10*, doi:10.1186/1471-2164-10-534.

136. Rowe, T. Chordate phylogeny and development, In *Assembling The Tree of Life*, 1st ed.; Cracraft, J., Donoghue, M.J., Eds.; Oxforf University Press: Oxford, UK, 2004; pp. 384–409.

137. Dehal, P.; Satou, Y.; Campbell, R.K.; Chapman, J.; Degnan, B.; de Tomaso, A.; Davidson, B.; di Gregorio, A.; Gelpke, M.; Goodstein, D.M.; *et al.* The draft genome of *Ciona intestinalis*: Insights into chordate and vertebrate origins. *Science* **2002**, *298*, 2157–2167.

138. Voskoboynik, A.; Neff, N.F.; Sahoo, D.; Newman, A.M.; Pushkarev, D.; Koh, W.; Passarelli, B.; Fan, H.C.; Mantalas, G.L.; Palmeri, K.J.; *et al.* The genome sequence of the colonial chordate, *Botryllus schlosseri*. *ELife* **2013**, *2*, doi:10.7554/eLife.00569.

139. Denoeud, F.; Henriet, S.; Mungpakdee, S.; Aury, J.-M.; da Silva, C.; Brinkmann, H.; Mikhaleva, J.; Olsen, L.C.; Jubin, C.; Canestro, C.; *et al.* Plasticity of animal genome architecture unmasked by rapid evolution of a pelagic tunicate. *Science* **2010**, *330*, 1381–1385.

140. Bracken-Grissom, H.; Collins, A.G.; Collins, T.; Crandall, K.; Distel, D.; Dunn, C.; Giribet, G.; Haddock, S.; Knowlton, N.; Martindale, M.; *et al.* The Global Invertebrate Genomics Alliance (GIGA): Developing community resources to study diverse invertebrate genomes. *J. Hered.* **2014**, *105*, 1–18.

141. Dussault, I.; Forman, B.M. The nuclear receptor PXR: A master regulator of "homeland" defense. *Crit. Rev. Eukaryot. Gene Expr.* **2002**, *12*, 53–64.

142. Kliewer, S.; Goodwin, B.; Willson, T. The nuclear pregnane X receptor: A key regulator of xenobiotic metabolism. *Endocr. Rev.* **2002**, *23*, 687–702.

143. Timsit, Y.E.; Negishi, M. CAR and PXR: The xenobiotic-sensing receptors. *Steroids* **2007**, *72*, 231–246.

144. Bridgham, J.T.; Eick, G.N.; Larroux, C.; Deshpande, K.; Harms, M.J.; Gauthier, M.E.A.; Ortlund, E.A.; Degnan, B.M.; Thornton, J.W. Protein evolution by molecular tinkering: Diversification of the nuclear receptor superfamily from a ligand-dependent ancestor. *PLoS Biol.* **2010**, *8*, e1000497, doi:10.1371/journal.pbio.1000497.

145. Srivastava, M.; Simakov, O.; Chapman, J.; Fahey, B.; Gauthier, M.E.A.; Mitros, T.; Richards, G.S.; Conaco, C.; Dacre, M.; Hellsten, U.; *et al.* The *Amphimedon queenslandica* genome and the evolution of animal complexity. *Nature* **2010**, *466*, 720–726.

146. Reitzel, A.M.; Tarrant, A.M. Nuclear receptor complement of the cnidarian *Nematostella vectensis*: Phylogenetic relationships and developmental expression patterns. *BMC Evol. Biol.* **2009**, *9*, 230, doi:10.1186/1471-2148-9-230.

147. Vogeler, S.; Galloway, T.; Lyons, B.; Bean, T. The nuclear receptor gene family in the Pacific oyster, *Crassostrea gigas*, contains a novel subfamily group. *BMC Genomics* **2014**, *15*, 369, doi:10.1186/1471-2164-15-369.

148. Sodergren, E.; Weinstock, G.M.; Davidson, E.H.; Cameron, R.A.; Gibbs, R.A.; Angerer, R.C.; Angerer, L.M.; Arnone, M.I.; Burgess, D.R.; Burke, R.D.; *et al*. The genome of the sea urchin *Strongylocentrotus purpuratus*. *Science* **2006**, *314*, 941–952.

149. Gazulha, V.; Mansur, M.C.D.; Cybis, L.F.; Azevedo, S.M.F.O. Feeding behavior of the invasive bivalve *Limnoperna fortunei* (Dunker, 1857) under exposure to toxic cyanobacteria *Microcystis aeruginosa*. *Braz. J. Biol.* **2012**, *72*, 41–49.

150. Bridgham, J.T.; Keay, J.; Ortlund, E.A.; Thornton, J.W. Vestigialization of an allosteric switch: Genetic and structural mechanisms for the evolution of constitutive activity in a steroid hormone receptor. *PLoS Genet.* **2014**, *10*, e1004058, doi:10.1371/journal.pgen.1004058.

151. Satou, Y.; Yamada, L.; Mochizuki, Y.; Takatori, N.; Kawashima, T.; Sasaki, A.; Hamaguchi, M.; Awazu, S.; Yagi, K.; Sasakura, Y.; *et al*. A cDNA resource from the basal chordate *Ciona intestinalis*. *Genesis* **2002**, *33*, 153–154.

152. Yagi, K.; Satou, Y.; Mazet, F.; Shimeld, S.M.; Degnan, B.; Rokhsar, D.; Levine, M.; Kohara, Y.; Satoh, N. A genomewide survey of developmentally relevant genes in *Ciona intestinalis*—III. Genes for Fox, ETS, nuclear receptors and NF kappa B. *Dev. Genes Evol.* **2003**, *213*, 235–244.

153. Ekins, S.; Reschly, E.J.; Hagey, L.R.; Krasowski, M.D. Evolution of pharmacologic specificity in the pregnane X receptor. *BMC Evol. Biol.* **2008**, *8*, doi:10.1186/1471-2148-8-103.

154. Satou, Y.; Kawashima, T.; Shoguchi, E.; Nakayama, A.; Satoh, N. An integrated database of the ascidian, *Ciona intestinalis*: Towards functional genomics. *Zool. Sci.* **2005**, *22*, 837–843.

155. Reschly, E.J.; Bainy, A.C.D.; Mattos, J.J.; Hagey, L.R.; Bahary, N.; Mada, S.R.; Ou, J.; Venkataramanan, R.; Krasowski, M.D. Functional evolution of the vitamin D and pregnane X receptors. *BMC Evol. Biol.* **2007**, *7*, 222, doi:10.1186/1471-2148-7-222.

156. Fidler, A.E.; Holland, P.T.; Reschly, E.J.; Ekins, S.; Krasowski, M.D. Activation of a tunicate (*Ciona intestinalis*) xenobiotic receptor orthologue by both natural toxins and synthetic toxicants. *Toxicon* **2012**, *59*, 365–372.

157. PubChem Compound Database. Available online : https://www.ncbi.nlm.nih.gov/pccompound?cmd=search (accessed on 7 September 2014).

158. Pinne, M.; Raucy, J.L. Advantages of cell-based high-volume screening assays to assess nuclear receptor activation during drug discovery. *Expert Opin. Drug Discov.* **2014**, *9*, 669–686.

159. Raucy, J.L.; Lasker, J.M. Cell-based systems to assess nuclear receptor activation and their use in drug development. *Drug Metab. Rev.* **2013**, *45*, 101–109.

160. Norcliffe, J.L.; Alvarez-Ruiz, E.; Martin-Plaza, J.J.; Steel, P.G.; Denny, P.W. The utility of yeast as a tool for cell-based, target-directed high-throughput screening. *Parasitology* **2013**, *141*, 8–16.

161. Hontzeas, N.; Hafer, K.; Schiestl, R.H. Development of a microtiter plate version of the yeast DEL assay amenable to high-throughput toxicity screening of chemical libraries. *Mutat. Res. Genet. Toxicol. Environ. Mutagen.* **2007**, *634*, 228–234.

162. Rajasarkka, J.; Virta, M. Miniaturization of a panel of high throughput yeast-cell-based nuclear receptor assays in 384-and 1536-well microplates. *Comb. Chem. High Throughput Screen.* **2011**, *14*, 47–54.

163. Bovee, T.F.H.; Helsdingen, R.J.R.; Hamers, A.R.M.; van Duursen, M.B.M.; Nielen, M.W.F.; Hoogenboom, R.L.A.P. A new highly specific and robust yeast androgen bioassay for the detection of agonists and antagonists. *Anal. Bioanal. Chem.* **2007**, *389*, 1549–1558.

164. An, W.F.; Tolliday, N. Cell-based assays for high-throughput screening. *Mol. Biotechnol.* **2010**, *45*, 180–186.

165. Chen, Z.L.; Zhao, H.M. A highly efficient and sensitive screening method for trans-activation activity of estrogen receptors. *Gene* **2003**, *306*, 127–134.

166. Raucy, J.L.; Lasker, J.M. Current *in vitro* high-throughput screening approaches to assess nuclear receptor activation. *Curr. Drug Metab.* **2010**, *11*, 806–814.

167. Chu, W.-L.; Shiizaki, K.; Kawanishi, M.; Kondo, M.; Yagi, T. Validation of a new yeast-based reporter assay consisting of human estrogen receptors α/β and coactivator SRC-1: Application for detection of estrogenic activity in environmental samples. *Environ. Toxicol.* **2009**, *24*, 513–521.

168. Li, H.; Redinbo, M.R.; Venkatesh, M.; Ekins, S.; Chaudhry, A.; Bloch, N.; Negassa, A.; Mukherjee, P.; Kalpana, G.; Mani, S.; *et al.* Novel yeast-based strategy unveils antagonist binding regions on the nuclear xenobiotic receptor PXR. *J. Biol. Chem.* **2013**, *288*, 13655–13668.

169. McEwan, I.J. Bakers yeast rises to the challenge: Reconstitution of mammalian steroid receptor signalling in *S. cerevisiae*. *Trends Genet.* **2001**, *17*, 239–243.

170. Fox, J.E.; Burow, M.E.; McLachlan, J.A.; Miller, C.A. Detecting ligands and dissecting nuclear receptor-signaling pathways using recombinant strains of the yeast *Saccharomyces cerevisiae*. *Nat. Protoc.* **2008**, *3*, 637–645.

171. Goldstone, J.V.; Goldstone, H.M.H.; Morrison, A.M.; Tarrant, A.; Kern, S.E.; Woodin, B.R.; Stegeman, J.J. Cytochrome P450 1 genes in early deuterostomes (tunicates and sea urchins) and vertebrates (chicken and frog): Origin and diversification of the CYP1 gene family. *Mol. Biol. Evol.* **2007**, *24*, 2619–2631.

172. Berry, M.; Metzger, D.; Chambon, P. Role of the 2 activating domains of the estrogen receptor in the cell-type and promoter-context dependent agonistic activity of the antiestrogen 4-hydroxytamoxifen. *EMBO J.* **1990**, *9*, 2811–2818.

173. Louvion, J.F.; Havauxcopf, B.; Picard, D. Fusion of GAL4-VP16 to a steroid-binding domain provides a tool for gratuitous induction of galactose-responsive genes in yeast. *Gene* **1993**, *131*, 129–134.

174. De Almeida, R.A.; Burgess, D.; Shema, R.; Motlekar, N.; Napper, A.D.; Diamond, S.L.; Pavitt, G.D. A *Saccharomyces cerevisiae* cell-based quantitative beta-galactosidase handling and assay compatible with robotic high-throughput screening. *Yeast* **2008**, *25*, 71–76.

175. Bovee, T.F.H.; Hendriksen, P.J.M.; Portier, L.; Wang, S.; Elliott, C.T.; van Egmond, H.P.; Nielen, M.W.F.; Peijnenburg, A.; Hoogenboom, L.A.P. Tailored microarray platform for the detection of marine toxins. *Environ. Sci. Technol.* **2011**, *45*, 8965–8973.

176. Chatterjee, S.; Kumar, V.; Majumder, C.B.; Roy, P. Screening of some anti-progestin endocrine disruptors using a recombinant yeast-based *in vitro* bioassay. *Toxicol. Vitr.* **2008**, *22*, 788–798.

177. Nordeen, S.K. Luciferase reporter gene vectors for analysis of promoters and enhancers. *Biotechniques* **1988**, *6*, 454–458.

178. Fan, F.; Wood, K.V. Bioluminescent assays for high-throughput screening. *Assay Drug Dev. Technol.* **2007**, *5*, 127–136.

179. Hancock, M.K.; Medina, M.N.; Smith, B.M.; Orth, A.P. Microplate orbital mixing improves high-throughput cell-based reporter assay readouts. *J. Biomol. Screen.* **2007**, *12*, 140–144.

180. Routledge, E.J.; Sumpter, J.P. Estrogenic activity of surfactants and some of their degradation products assessed using a recombinant yeast screen. *Environ. Toxicol. Chem.* **1996**, *15*, 241–248.

181. Miller, C.A.; Tan, X.B.; Wilson, M.; Bhattacharyya, S.; Ludwig, S. Single plasmids expressing human steroid hormone receptors and a reporter gene for use in yeast signaling assays. *Plasmid* **2010**, *63*, 73–78.

182. Chen, Q.; Chen, J.; Sun, T.; Shen, J.H.; Shen, X.; Jiang, H.L. A yeast two-hybrid technology-based system for the discovery of PPAR gamma agonist and antagonist. *Anal. Biochem.* **2004**, *335*, 253–259.

183. Balsiger, H.A.; de la Torre, R.; Lee, W.-Y.; Cox, M.B. A four-hour yeast bioassay for the direct measure of estrogenic activity in wastewater without sample extraction, concentration, or sterilization. *Sci. Total Environ.* **2010**, *408*, 1422–1429.

184. Collins, B.M.; McLachlan, J.A.; Arnold, S.F. The estrogenic and antiestrogenic activities of phytochemicals with human estrogen receptor expressed in yeast. *Steroids* **1997**, *62*, 365–372.

185. Gaido, K.W.; Leonard, L.S.; Lovell, S.; Gould, J.C.; Babai, D.; Portier, C.J.; McDonnell, D.P. Evaluation of chemicals with endocrine modulating activity in a yeast-based steroid hormone receptor gene transcription assay. *Toxicol. Appl. Pharmacol.* **1997**, *143*, 205–212.

186. Passos, A.L.; Pinto, P.I.; Power, D.M.; Canario, A.V. A yeast assay based on the gilthead sea bream (teleost fish) estrogen receptor beta for monitoring estrogen mimics. *Ecotoxicol. Environ. Saf.* **2009**, *72*, 1529–1537.

187. Chen, C.H.; Chou, P.H.; Kawanishi, M.; Yagi, T. Occurrence of xenobiotic ligands for retinoid X receptors and thyroid hormone receptors in the aquatic environment of Taiwan. *Mar. Pollut. Bull.* **2014**, *23*, doi:10.1016/j.marpolbul.2014.01.025.

188. Holdgate, G.A.; Anderson, M.; Edfeldt, F.; Geschwindner, S. Affinity-based, biophysical methods to detect and analyze ligand binding to recombinant proteins: Matching high information content with high throughput. *J. Struct. Biol.* **2010**, *172*, 142–157.

189. Senveli, S.U.; Tigli, O. Biosensors in the small scale: Methods and technology trends. *IET Nanobiotechnol.* **2013**, *7*, 7–21.

190. Fechner, P.; Gauglitz, G.; Gustafsson, J.-A. Nuclear receptors in analytics—A fruitful joint venture or a wasteful futility? *Trends Anal. Chem.* **2010**, *29*, 297–305.

191. Lin, W.; Liu, J.; Jeffries, C.; Yang, L.; Lu, Y.; Lee, R.E.; Chen, T. Development of BODIPY FL Vindoline as a novel and high-affinity pregnane X receptor fluorescent probe. *Bioconjug. Chem.* **2014**, *18*, 18.

192. Hill, K.L.; Dutta, P.; Long, Z.; Sepaniak, M.J. Microcantilever-based nanomechanical studies of the orphan nuclear receptor pregnane X receptor-ligand interactions. *J. Biomater. Nanobiotechnol.* **2011**, *3*, 133–142.

193. Dagnino, S.; Bellet, V.; Grimaldi, M.; Riu, A.; Ait-Aissa, S.; Cavailles, V.; Fenet, H.; Balaguer, P. Affinity purification using recombinant PXR as a tool to characterize environmental ligands. *Environ. Toxicol.* **2014**, *29*, 207–215.

194. Bricelj, V.M.; Connell, L.; Konoki, K.; MacQuarrie, S.P.; Scheuer, T.; Catterall, W.A.; Trainer, V.L. Sodium channel mutation leading to saxitoxin resistance in clams increases risk of PSP. *Nature* **2005**, *434*, 763–767.

195. Bricelj, V.M.; MacQuarrie, S.P.; Doane, J.A.E.; Connell, L.B. Evidence of selection for resistance to paralytic shellfish toxins during the early life history of soft-shell clam (*Mya arenaria*) populations. *Limnol. Oceanogr.* **2010**, *55*, 2463–2475.

196. Roje-Busatto, R.; Ujević, I. PSP toxins profile in ascidian *Microcosmus vulgaris* (Heller, 1877) after human poisoning in Croatia (Adriatic Sea). *Toxicon* **2014**, *79*, 28–36.

197. Parsons, M.L.; Settlemier, C.J.; Ballauer, J.M. An examination of the epiphytic nature of *Gambierdiscus toxicus*, a dinoflagellate involved in ciguatera fish poisoning. *Harmful Algae* **2011**, *10*, 598–605.

198. Martins, A.; Vieira, H.; Gaspar, H.; Santos, S. Marketed marine natural products in the pharmaceutical and cosmeceutical industries: Tips for success. *Mar. Drugs* **2014**, *12*, 1066–1101.

199. Mayer, A.M.S.; Glaser, K.B.; Cuevas, C.; Jacobs, R.S.; Kem, W.; Little, R.D.; McIntosh, J.M.; Newman, D.J.; Potts, B.C.; Shuster, D.E.; *et al.* The odyssey of marine pharmaceuticals: A current pipeline perspective. *Trends Pharmacol. Sci.* **2010**, *31*, 255–265.

200. Imhoff, J.F.; Labes, A.; Wiese, J. Bio-mining the microbial treasures of the ocean: New natural products. *Biotechnol. Adv.* **2011**, *29*, 468–482.

201. Blunt, J.W.; Copp, B.R.; Keyzers, R.A.; Munro, M.H.G.; Prinsep, M.R. Marine natural products. *Nat. Prod. Rep.* **2013**, *30*, 237–323.

10

Discovery of Novel Saponins from the Viscera of the Sea Cucumber *Holothuria lessoni*

Yadollah Bahrami [1,2,3,4], Wei Zhang [1,2,3] and Chris Franco [1,2,3,]*

[1] Department of Medical Biotechnology, School of Medicine, Flinders University, Adelaide 5001, SA 5042, Australia; E-Mails: yadollah.bahrami@flinders.edu.au (Y.B.); wei.zhang@flinders.edu.au (W.Z.)

[2] Centre for Marine Bioproducts Development, Flinders University, Adelaide 5001, SA 5042, Australia

[3] Australian Seafood Cooperative Research Centre, Mark Oliphant Building, Science Park, Adelaide 5001, SA 5042, Australia

[4] Medical Biology Research Center, Kermanshah University of Medical Sciences, Kermanshah 6714415185, Iran; E-Mail: ybahrami@mbrc.ac.ir

* Author to whom correspondence should be addressed; E-Mail: Chris.franco@flinders.edu.au

Abstract: Sea cucumbers, sometimes referred to as marine ginseng, produce numerous compounds with diverse functions and are potential sources of active ingredients for agricultural, nutraceutical, pharmaceutical and cosmeceutical products. We examined the viscera of an Australian sea cucumber *Holothuria lessoni* Massin *et al.* 2009, for novel bioactive compounds, with an emphasis on the triterpene glycosides, saponins. The viscera were extracted with 70% ethanol, and this extract was purified by a liquid-liquid partition process and column chromatography, followed by isobutanol extraction. The isobutanol saponin-enriched mixture was further purified by high performance centrifugal partition chromatography (HPCPC) with high purity and recovery. The resultant purified polar samples were analyzed using matrix-assisted laser desorption/ionization mass spectrometry (MALDI-MS)/MS and electrospray ionization mass spectrometry (ESI-MS)/MS to identify saponins and characterize their molecular structures. As a result, at least 39 new saponins were identified in the viscera of *H. lessoni* with a high structural diversity, and another 36 reported triterpene glycosides, containing different aglycones and sugar moieties. Viscera samples have provided a higher diversity and yield of compounds than observed from the body wall. The high structural diversity and novelty of saponins from *H. lessoni* with potential functional activities presents a great opportunity to exploit their applications for industrial, agricultural and pharmaceutical use.

Keywords: sea cucumber viscera; saponins; *Holothuria lessoni*; bioactive compounds; MALDI; mass spectrometry; ESI; HPCPC; triterpene glycosides; structure elucidation; marine invertebrate; *Echinodermata*; holothurian

---

1. Introduction

Holothurians are sedentary marine invertebrates, commonly known as sea cucumbers, trepang, bêche-de-mer, or gamat [1,2], belonging to the class Holothuroidea of the *Echinodermata* phylum. Sea cucumbers produce numerous compounds with diverse functions and are potential sources of agricultural or agrochemical, nutraceutical, pharmaceutical and cosmeceutical products [3–5]. It is for this reason they are called "marine ginseng" in Mandarin.

Even though sea cucumbers contain different types of natural compounds, saponins are their most important and abundant secondary metabolites [6–12]. Saponins are reported as the major bioactive compound in many effective traditional Chinese and Indian herbal medicines.

Sea cucumber saponins are known to have a wide range of medicinal properties due to their cardiovascular, immunomodulator, cytotoxic, anti-asthma, anti-eczema, anti-inflammatory, anti-arthritis, anti-oxidant, anti-diabetics, anti-bacterial, anti-viral, anti-cancer, anti-angiogenesis, anti-fungal, hemolytic, cytostatic, cholesterol-lowering, hypoglycemia and anti-dementia activities [4,7,13–24].

Saponins are amphipathic compounds that generally possess a triterpene or steroid backbone or aglycone. Triterpenoid saponins have aglycones that consist of 30 carbons, whereas steroidal saponins possess aglycones with 27 carbons, which are rare in nature [4].

Triterpene saponins belong to one of the most numerous and diverse groups of natural occurring products, which are produced in relatively high abundance. They are reported primarily as typical metabolites of terrestrial plants [25]. A few marine species belonging to the phylum *Echinodermata* [26] namely holothuroids (sea cucumbers) [7,10,13,27–33] and asteroids, and sponges from the phylum *Porifera* [13,34,35] produce saponins.

The majority of sea cucumber saponins, generally known as Holothurins, are usually triterpene glycosides, belonging to the holostane type group rather than nonholostane [36,37], which is comprised of a lanostane-3β-ol type aglycone containing a γ-18 (20)-lactone in the D-ring of tetracyclic triterpene (3β,20S-dihydroxy-5α-lanostano-18,20-lactone) [25] sometimes containing shortened side chains, and a carbohydrate moiety consisting of up to six monosaccharide units covalently connected to C-3 of the aglycone [7,8,13,37–42].

The sugar moiety of the sea cucumber saponins consists mainly of D-xylose, D-quinovose, 3-*O*-methyl-D-glucose, 3-*O*-methyl-D-xylose and D-glucose and sometimes 3-*O*-methyl-D-quinovose, 3-*O*-methyl-D-glucuronic acid and 6-*O*-acetyl-D-glucose [40,41,43–48]. In the oligosaccharide chain, the first monosaccharide unit is always a xylose, whereas either 3-*O*-methylglucose or 3-*O*-methylxylose is always the terminal sugar.

Although some identical saponins have been given different names by independent research groups [6] as they could be isomeric compounds, our comprehensive literature review showed that more than 250 triterpene glycosides have been reported from various species of sea cucumbers [7,13,18,25,29,41,44,49,50]. They are classified into four main structural categories based on their aglycone moieties; three holostane type glycoside group saponins containing a (1) 3β-hydroxyholost-9 (11)-ene aglycone skeleton; (2) saponins with a 3β-hydroxyholost-7-ene skeleton and (3) saponins with an aglycone moiety different to the other two holostane type aglycones (other holostane type aglycones); and (4) a nonholostane aglycone [25,38,46,51,52].

One of the most noteworthy characteristics of many of the saponins from marine organisms is the sulfation of aglycones or sugar moieties [4]. In sea cucumber saponins, sulfation of the oligosaccharide chain in the Xyl, Glc and MeGlc residues has been reported [38,40,46,53,54]. Most of them are mono-sulfated glycosides with few occurrences of di- and tri-sulfated glycosides. Saponin diversity can be further enhanced by the position of double bonds and lateral groups in the aglycone.

Triterpene glycosides have been considered a defense mechanism, as they are deleterious for most organisms [6–10,12,55–57]. In contrast, a recent study has shown that these repellent chemicals are also kairomones that attract the symbionts and are used as chemical "signals" [58]. However, in the sea cucumber, it has been suggested that saponins may also have two regulatory roles during reproduction: (1) to prevent oocyte maturation and (2) to act as a mediator of gametogenesis [18,59].

The wide range of biological properties and various physiological functions of sea cucumber extracts with high chemical structural diversity and the abundance of their metabolites have spurred researchers to study the ability of sea cucumbers to be used as an effective alternative source for potential future drugs. However, the large number of very similar saponin glycosides structures has led to difficulties in purification, and the complete structure elucidation of these molecules (especially isomers), has made it difficult to conduct tests to determine structure-activity relationships, which can lead to the development of new compounds with commercial applications [16]. Therefore, in order to overcome this problem, we employed High Performance Centrifugal Partition Chromatography (HPCPC) to successfully purify saponins in this study. HPCPC is more efficient in purifying large amounts of a given sample and also lower solvent consumption with high yields compared to other conventional chromatography methods.

This project aims to identify and characterize the novel bioactive compounds from the viscera (all internal organs other than the body wall) of an Australian sea cucumber *Holothuria lessoni* Massin *et al.* 2009 (golden sandfish) with an emphasis on saponins. *H. lessoni* was selected because it is a newly-identified Holothurian species, which is abundant in Australian waters. While only a few studies have compared the saponin contents of the body wall with that of the cuvierian tubules in other species [50,60–62], to our knowledge, no study has investigated the contribution of saponins of the body wall or the viscera of *Holothuria lessoni*. Sea cucumbers expel their internal organs as a defense mechanism called evisceration, a reaction that includes release of the respiratory tree, intestine, cuvierian tubules and gonads through the anal opening [50,58,61,63–68]. We hypothesize that the reason for this ingenious form of defense is because these organs contain high levels of compounds that repel predators [60,61,69,70]. Furthermore, the results of this project may identify the potential economic benefits of transforming viscera of the sea cucumber into high value co-products important to human health and industry.

Matrix-assisted laser desorption/ionization time-of-flight mass spectrometry (MALDI-ToF/MS) and electrospray ionization mass spectrometry (ESI-MS) techniques allow the "soft" ionization of large biomolecules, which has been a big challenge until recently [71]. Therefore, MALDI and ESI-MS, and MS/MS were performed to detect saponins and to elucidate their structures.

2. Results and Discussion

An effective method for the purification of saponins has been developed, and several saponins were isolated and purified from the viscera of *H. lessoni*. The enriched saponin mixtures of the viscera extract were successfully purified further by HPCPC, which is very efficient in purifying compounds with low polarity as well as in processing large amounts of sample. This method yielded saponins with higher than a 98% recovery of sample with high purities [72]. Purifying saponins from mixtures of saponins also helps to overcome the problem associated with identifying multiple saponins with liquid chromatography-tandem mass spectrometry (LC-MS) and ESI-MS. Mass spectrometry has been applied for the structure elucidation of saponins in both negative and positive ion modes [73–79]. In this study, identification of the saponin compounds was attempted by soft ionization MS techniques including MALDI and ESI in the positive mode. Previous studies have reported that the fragment ions of alkali metal adducts of saponins provide valuable structural information about the feature of the aglycone and the sequence and linkage site of the sugar residues [80]. Therefore, the MS analyses were conducted by introducing sodium ions to the samples. However, saponin spectra can also be detected without adding a sodium salt. Because of the high affinity of alkali cations for triterpene glycosides, all saponins detected in the positive ion mode spectra were predominantly singly charged sodium adducts of the molecules $[M + Na]^+$ [19,81]. The main fragmentation of saponins generated by cleavage of the glycosidic bond yielded oligosaccharide and monosaccharide fragments [19]. Other visible peaks and fragments were generated by the loss of other neutral moieties such as CO_2, H_2O or CO_2 coupled with H_2O.

The saponins obtained from the viscera of this tropical holothurian were profiled using MALDI-MS and ESI-MS. MALDI is referred to as a "soft" ionization technique, because the spectrum shows mostly intact, singly charged ions for the analyte molecules. However, in some cases, MALDI causes minimal fragmentation of analytes [71].

The chromatographic purification of isobutanol-soluble saponin-enriched fractions of *H. lessoni* viscera was monitored on pre-coated thin-layer chromatography (TLC) plates (Figure 1A) showing the presence of several bands. As a typical example, the TLC profile of HPCPC Fractions 52–61 of the isobutanol-saponin enriched fraction from the viscera of the *H. lessoni* sea cucumber is shown in Figure 1B. The centrifugal partition chromatography (CPC) technique not only allowed for the purification of saponins, but in some cases it could separate isomeric saponins e.g., separation of the isomers detected in the ion peak at *m/z* 1303.6, which will be discussed later.

Figure 1. The thin-layer chromatography (TLC) pattern of a saponin mixture (**A**) and the high performance centrifugal partition chromatography (HCPCP) fractions (**B**) from the purified extracts of the viscera of the *Holothuria lessoni* sea cucumber using the lower phase of CHCl$_3$-MeOH-H$_2$O (7:13:8) system. The numbers under each lane indicate the fraction number of fractions in the fraction collector. Here, only the fractions 52 to 61 of one analysis (of 110 fractions) are shown as a representative.

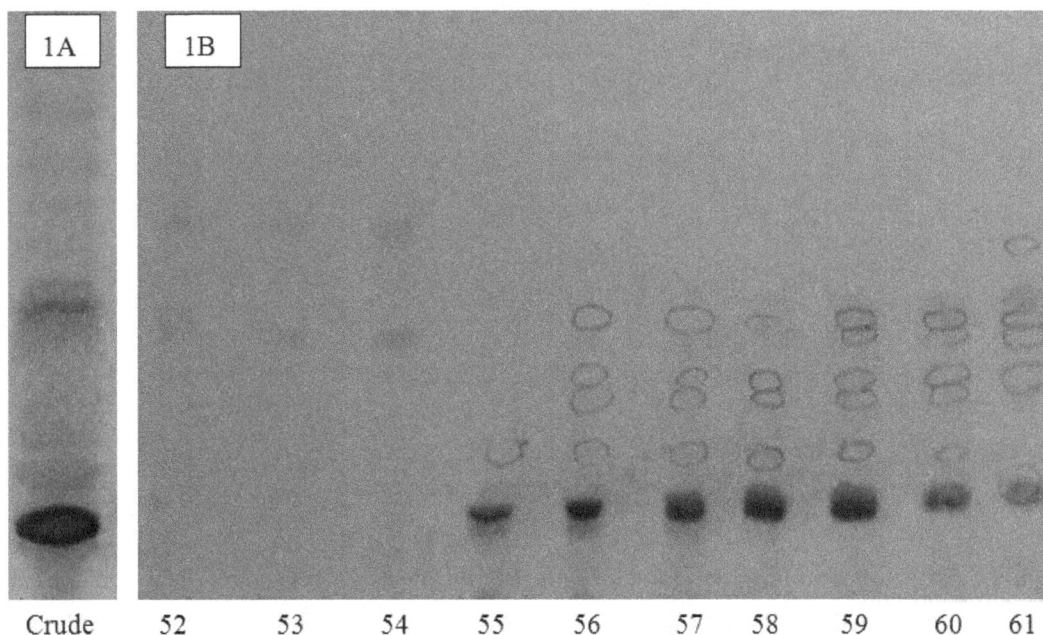

Mass spectrometry has been used extensively for the characterization of saponins and their structural confirmation. One of the powerful methods, which are widely used for the analysis of high molecular weight, non-volatile molecules is MALDI [82]. The appropriate HPCPC fractions were consequently pooled based on their TLC profiles and concentrated to dryness and analyzed by MALDI MS and MS/MS, and ESI MS/MS. In the positive ion mode, all detected ions were sodium-coordinated species such as [M + Na]$^+$ corresponding to sulfated and non-sulfated saponins [64]. The prominence of the parent ions [M + Na]$^+$ in MS spectra also enables the analysis of saponins in mixtures or fractions. The MALDI results indicate that the saponin fractions are quite pure, which is consistent with the TLC data. As a representative example, the full-scan MALDI mass spectrum of the saponin extract obtained from HPCPC Fraction 55 of the *H. lessoni* viscera is shown in Figure 2.

This spectrum displays the major intense peak detected at *m/z* 1243.4, which corresponds to Holothurin A, with an elemental composition of C$_{54}$H$_{85}$NaO$_{27}$S [M + Na]$^+$. Other visible peaks seem to correspond to the sugar moieties and aglycone ions generated by the losses of sugars and/or losses of water and/or carbon dioxide from cationized saponins upon MALDI ionization. These analyses show that this fraction contains one main saponin. Therefore, even though the HPCPC fractionation separated the saponin mixture, some saponin congeners, due to the similarity in their TLC migration, were detected in some of the pooled fractions. It was found that the total separation of the saponins was difficult within a single HPCPC run. However, this technique allowed the separation of a number of saponins, including some isomers (Figure 3).

Figure 2. The full-scan matrix-assisted laser desorption/ionization mass spectrometry (MALDI) mass spectrum of HPCPC Fraction 55 in the (+) ion mode.

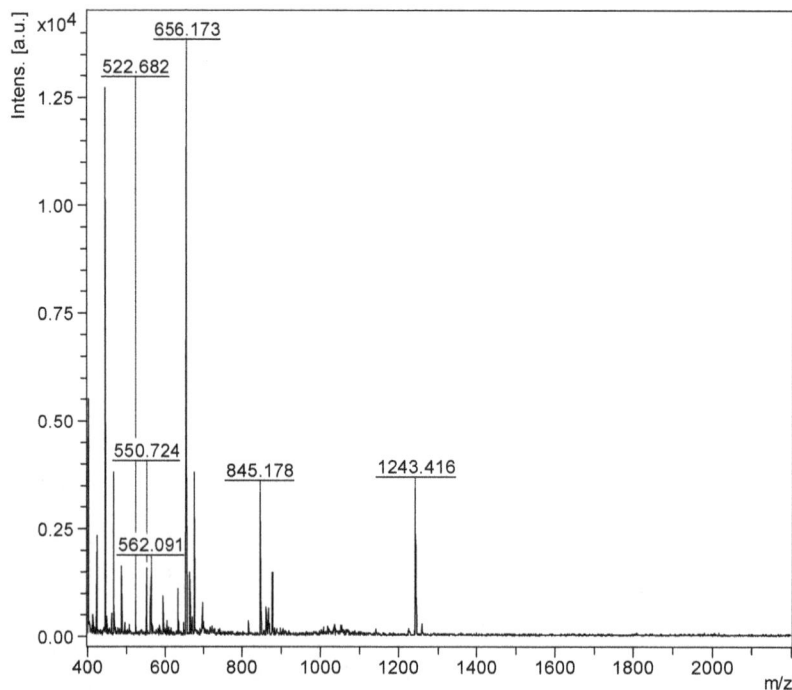

Figure 3. Schematic fragmentation patterns of the ion detected at m/z 1303.6; (**A**) Fraction 15; (**B**) Fraction 14 and (**C**) Fraction 12. Full and dotted arrows show the two main feasible fragmentation pathways. The predominant peak (**A** and **B**) at m/z 507 corresponds to the key sugar residue and aglycone moiety. The major abundant peak (**C**) at m/z 523 corresponds to both the key sugar residue and aglycone moiety. Abbreviations; G = Glc, MG = MeGlc, Q = Qui, X = Xyl.

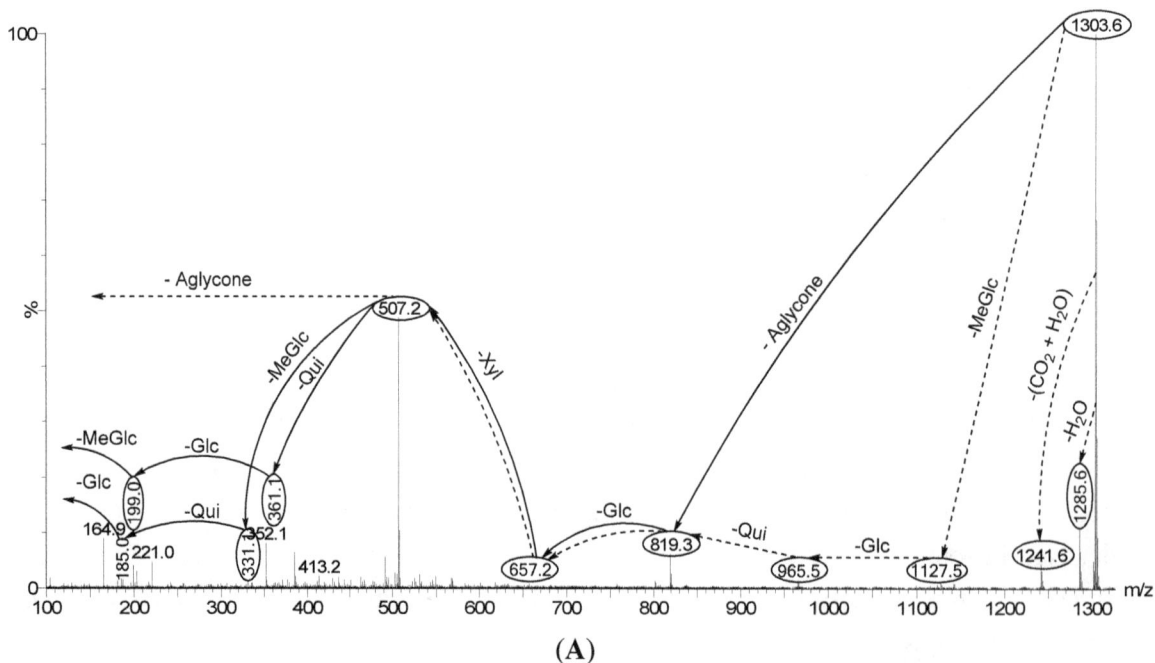

(**A**)

Figure 3. *Cont.*

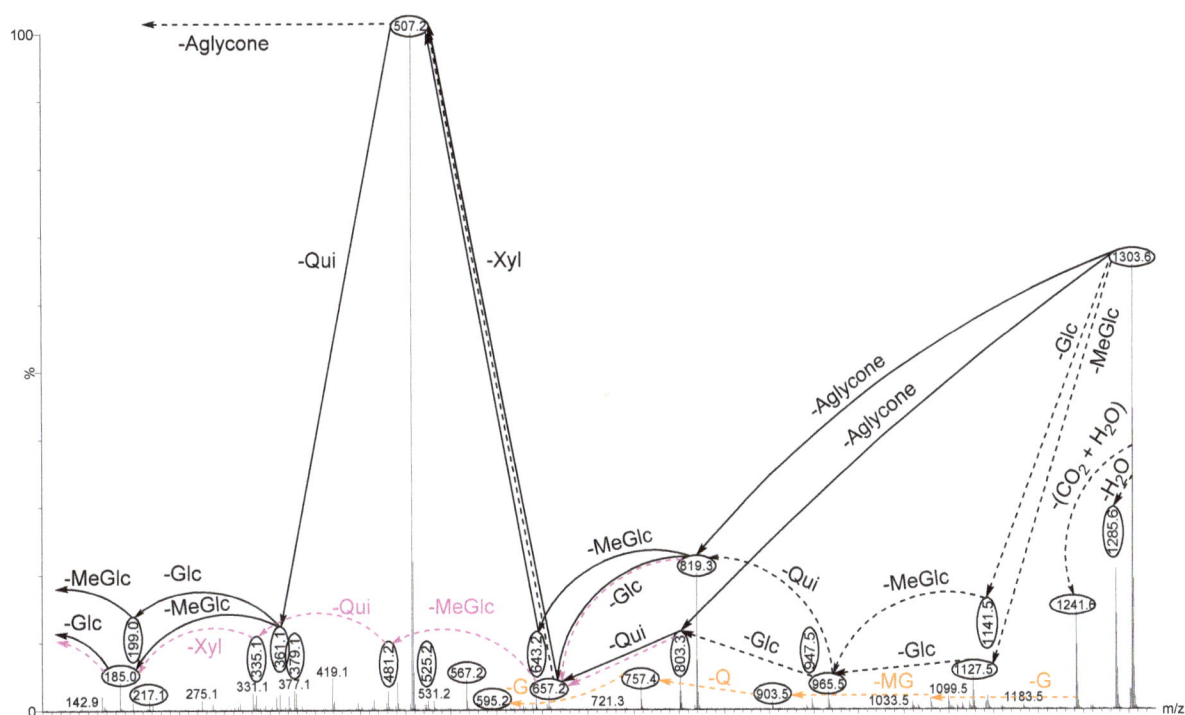

(B)

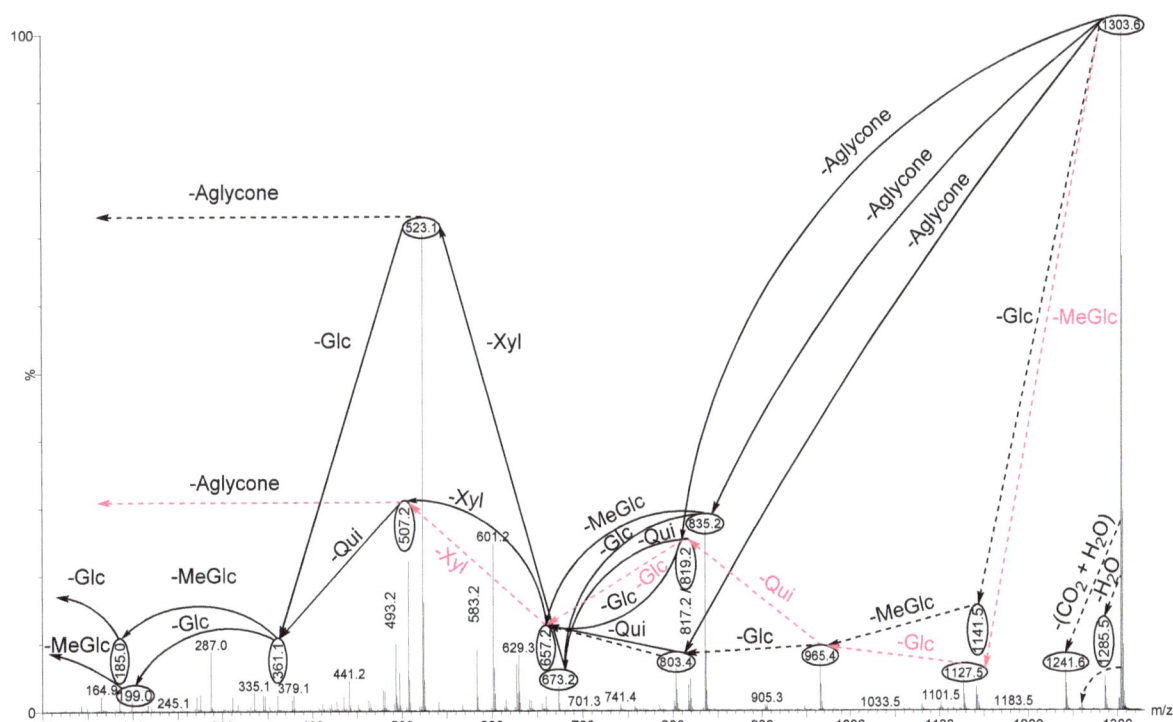

(C)

The full-scan MALDI mass spectrum of the isobutanol-enriched saponin extract obtained from the viscera of the *H. lessoni* is shown in Figure 4. A diverse range of saponins with various intensities was identified. This spectrum displays 13 intense peaks that could each correspond to at least one saponin congener. The most abundant ions observed under positive ion conditions were detected at *m/z* 1335, 1303, 1289, 1287, 1259, 1245, 1243, 1229, 1227, 1149, 1141, 1123 and 845. Further analysis revealed that some of these MS peaks represented more than one compound. For instance the peaks at *m/z* 1303 and 1287 were shown to contain at least six and five different congeners, respectively (Figures 3 and 5–7).

Figure 4. The full-scan MALDI mass spectrum of the isobutanol-enriched saponin extract from the viscera of the *H. lessoni*. A mass range of 600 to 1500 Da is shown here. It is noted that this spectrum is unique for this species.

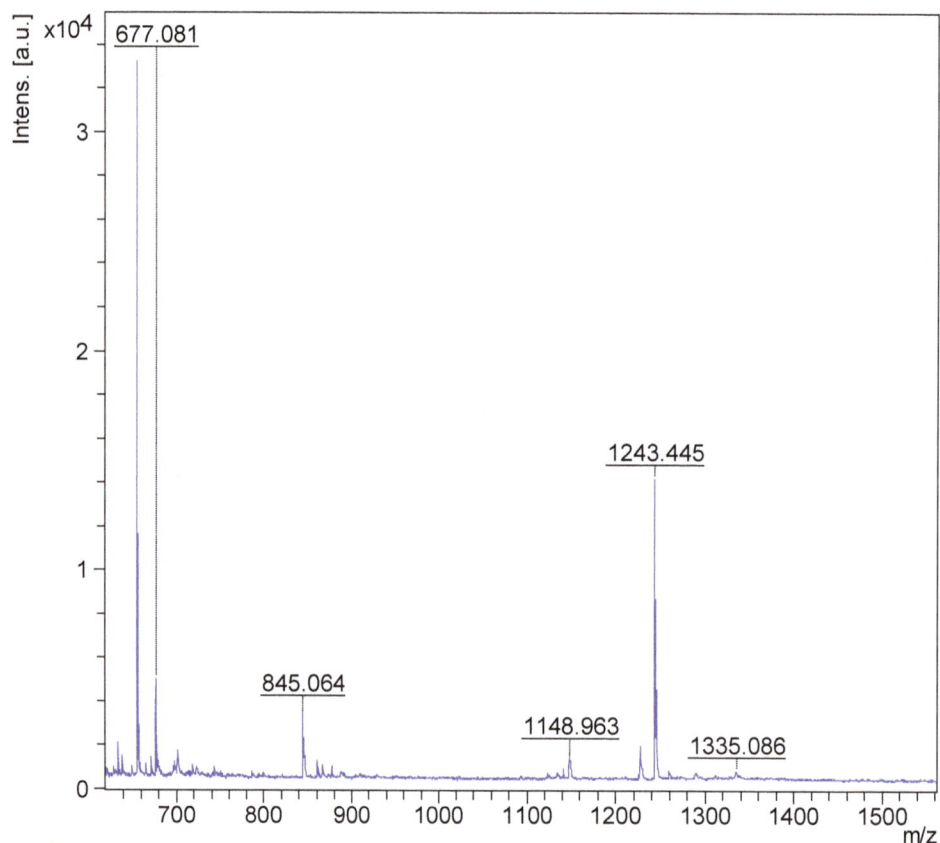

Figure 5. The schematic diagram of the proposed isomeric structures of ion at *m/z* 1303.6.

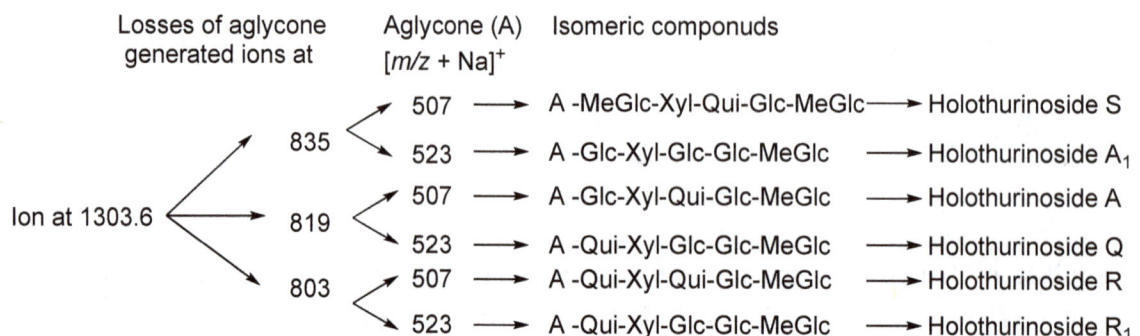

Figure 6. (+) ion mode ESI-MS/MS spectrum of saponins detected at *m/z* 1287.6. This spectrum shows the presence of two different aglycones, which led to the isomeric saponins. Full and dotted arrows illustrate the two main possible fragmentation pathways.

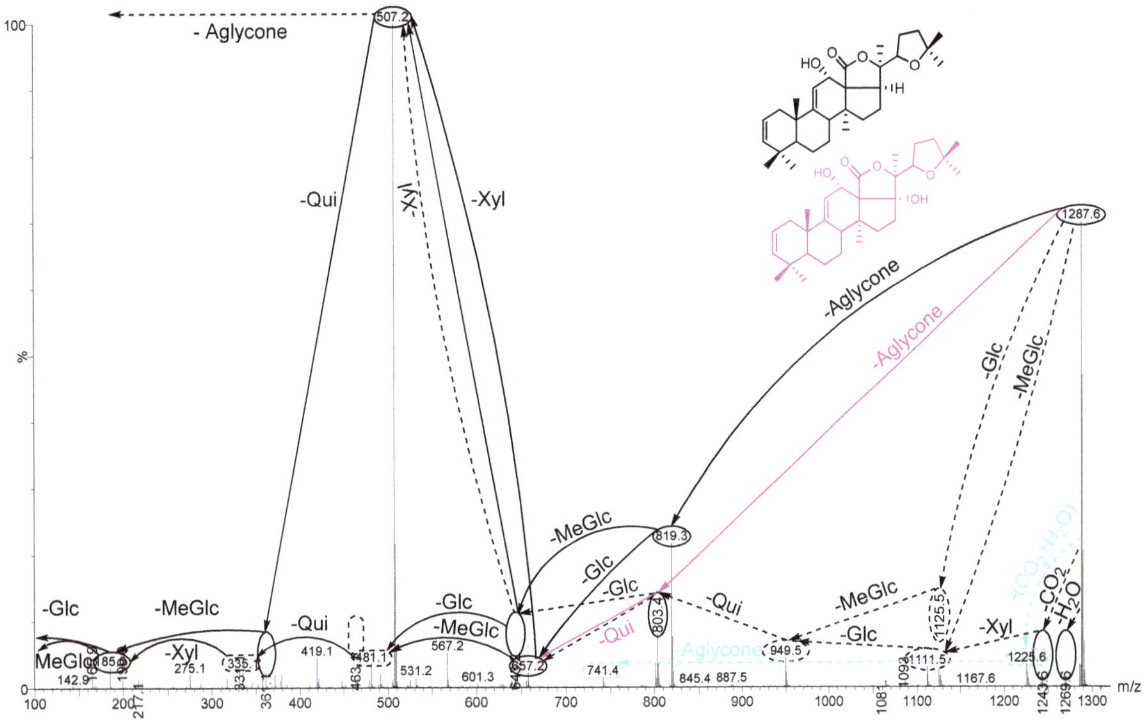

Figure 7. A schematic diagram of the proposed isomeric structures of ion at *m/z* 1287.6.

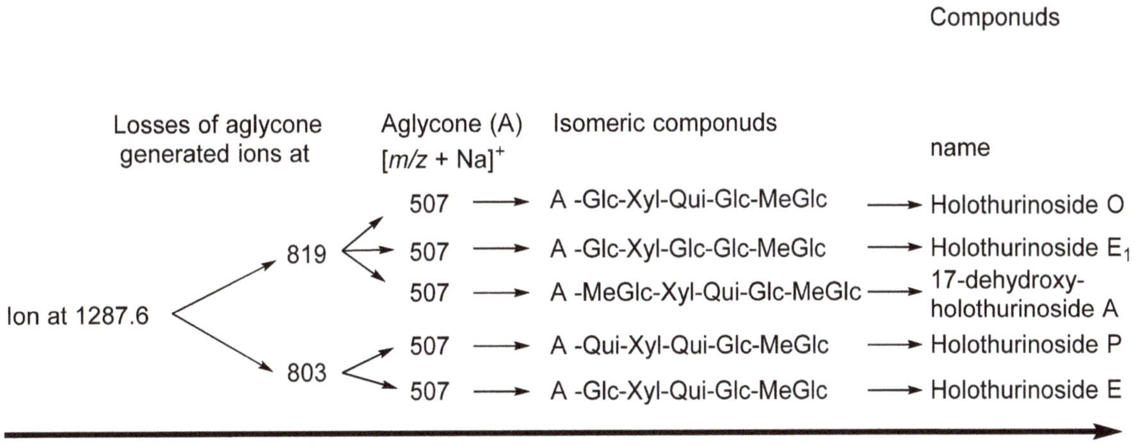

Componuds

The accurate mass measurements acquired by MALDI-MS detected the saponin peaks, and molecular formulae and elemental compositions were assigned by ESI-MS/MS as summarized in Table 1. Our results revealed that at least 75 saponins were detected in *H. lessoni*, including 39 new sulfated, non-sulfated and acetylated triterpene glycosides, containing a wide range of aglycone and sugar moieties.

Table 1. Summary of saponins identified from the viscera of *H. lessoni* by matrix-assisted laser desorption/ionization time-of-flight mass spectrometry (MALDI-ToF-MS) and electrospray-ionization mass spectrometry (ESI-MS). This table illustrates the 39 novel identified compounds (N) along with the 36 known compounds (P). This table also shows some identical saponins, which have been given different names by different researchers in which they might be isomeric congeners.

[M + Na]$^+$ *m/z*	MW	Formula	Compound's Name	Novel (N)/ Published (P)	References
889.4	866	$C_{41}H_{63}NaO_{16}S$	Holothurin B$_3$	P	[83]
		$C_{42}H_{67}NaO_{15}S$	Unidentified	N	–
905.4	882	$C_{41}H_{63}NaO_{17}S$	Holothurin B$_4$	P	[83]
			Holothurin B	P	[61,84–86]
			Nobiliside B	P	[87]
907.4	884	$C_{41}H_{65}NaO_{17}S$	Holothurin B$_2$	P	[83]
			Leucospilotaside B	P	[8]
911.6	888	$C_{45}H_{92}O_{16}$	Unidentified	N	–
917.4	994	$C_{44}H_{71}NaO_{15}S$	Unidentified	N	–
921.4	898	$C_{41}H_{63}NaO_{18}S$	Leucospilotaside A	P	[84]
1034.1	1011	a*	Unidentified	N	–
1065.5	1042	$C_{48}H_{82}O_{24}$	Unidentified	N	–
1071.5	1048	$C_{47}H_{93}NaO_{21}S$	Unidentified	N	–
1078.5	1055	a *	Unidentified	N	–
1083.3	1060	$C_{58}H_{64}O_{25}$	Unidentified	N	–
1087.6	1064	$C_{47}H_{93}NaO_{22}S$	Unidentified	N	–
1123.5	1100	$C_{54}H_{84}O_{23}$	Unidentified	N	–
1125.5	1102	$C_{54}H_{86}O_{23}$	Holothurinoside C	P	[62,69,88,89]
			Holothurinoside C$_1$		
1127.6	1104	$C_{54}H_{88}O_{23}$	Unidentified	N	–
			Unidentified	N	–
1141.6	1118	$C_{54}H_{86}O_{24}$	Desholothurin A (Nobiliside 2a),	P	[5,62,69,88–90]
			Desholothurin A$_1$ (Arguside E)		
1149.2	1126	a *	Unidentified	N	–
1157.5	1134	$C_{54}H_{109}O_{25}$	Holothurinoside J$_1$	P	[50]
		$C_{49}H_{91}NaO_{25}S$	Unidentified	N	–
1193.5	1170	$C_{55}H_{87}NaO_{23}S$	Unidentified	N	–
1199.4	1176	$C_{54}H_{64}O_{29}$	Unidentified	N	–
1221.5 **	1198	$C_{56}H_{78}O_{28}$	Unidentified	N	–
1225.5	1202	$C_{54}H_{83}NaO_{26}S$	Unidentified	N	–
1227.5	1204	$C_{54}H_{85}NaO_{26}S$	Fuscocineroside B/C,	P	[29,56,89,91,92]
			Scabraside A or		
			24-Dehydroechinoside A		
1229.5	1206	$C_{54}H_{87}NaO_{26}S$	Holothurin A$_2$,	P	[7,61,91,93–95]
			Echinoside A		

Table 1. *Cont.*

1243.5	1220	$C_{54}H_{85}NaO_{27}S$	Holothurin A	P	[29,58,61,95–97]
			Scabraside B		
			17-Hydroxy fuscocineroside B		
			25-Hydroxy fuscocinerosiden B		
1245.5	1222	$C_{54}H_{87}NaO_{27}S$	Holothurin A_1	P	[91]
			Holothurin A_4		[36]
			Scabraside D		[92]
1259.5	1236	$C_{54}H_{85}NaO_{28}S$	Holothurin A_3	P	[36]
			Unidentified	N	–
1265.5	1242	$C_{56}H_{83}NaO_{27}S$	Unidentified	N	–
1271.6	1248	$C_{60}H_{96}O_{27}$	Impatienside B	P	[5,98]
1287.6	1264	$C_{60}H_{96}O_{28}$	Holothurinoside E,	P	[62,69]
			Holothurinoside E_1		
			Unidentified	N	–
			Unidentified	N	–
			17-Dehydroxyholothurinoside A	P	[5,99]
1289.6	1266	$C_{60}H_{98}O_{28}$	Griseaside A	P	[99]
1301.6	1278	$C_{61}H_{98}O_{28}$	Holothurinoside M	P	[65]
		$C_{60}H_{94}O_{29}$	Unidentified	N	–
1303.6	1280	$C_{60}H_{96}O_{29}$	Holothurinoside A	P	[5,62,69,88]
			Holothurinoside A_1		
			Unidentified	N	–
			Unidentified	N	–
			Unidentified	N	–
			Unidentified	N	–
1305.6	1282	a *	Unidentified	N	–
1317.6	1294	$C_{61}H_{98}O_{29}$	Unidentified	N	–
1335.3	1312	a *	Unidentified	N	–
1356.4	1333	a *	Unidentified	N	–
1409.4	1386	$C_{61}H_{78}O_{36}$	Unidentified	N	–
1411.7	1388	$C_{62}H_{116}O_{33}$	Unidentified	N	–
1419.7	1396	$C_{66}H_{108}O_{31}$	Unidentified	N	–
1435.7	1412	$C_{66}H_{108}O_{32}$	Unidentified	N	–
1465.7	1442	$C_{67}H_{110}O_{33}$	Arguside B	P	[5,32]
			Arguside C		
1475.6	1452	$C_{65}H_{96}O_{36}$	Unidentified	N	–
1477.7 **	1454	$C_{61}H_{114}O_{38}$	Unidentified	N	–
1481.7	1458	$C_{66}H_{106}O_{35}$	Unidentified	N	–
1493.7	1470	$C_{65}H_{114}O_{36}$	Unidentified	N	–
1495.7	1472	$C_{61}H_{116}O_{39}$	Holothurinoside K_1	P	[50]
		$C_{72}H_{112}O_{31}$	Unidentified	N	–
1591.7	1568	$C_{66}H_{120}O_{41}$	Unidentified	N	–

a * The composition was not measured through the ESI analysis; ** acetylated compounds.

A number of studies have reported the presence of multiple saponins. Elbandy *et al.* [5] described the structures of 21 non-sulfated saponins from the body wall of *Bohadschia cousteaui*. These authors reported 10 new compounds together with 11 known triterpene glycosides including Holothurinoside I, Holothurinoside H, Holothurinoside A, Desholothurin A, 17-dehydroxyholothurinoside A, Arguside C, Arguside F, Impatienside B, Impatienside A, Marmoratoside A and Bivittoside. Bondoc *et al.* [64] investigated saponin congeners in three species from Holothuriidae (*H. scabra* Jaeger 1833, *H. fuscocinerea* Jaeger 1833, and *H. impatiens* Forskal 1775). This group reported 20 saponin ion peaks, with an even number of sulfated and non-sulfated types, in *H. scabra,* which contained the highest saponin diversity among the examined species, followed by *H. fuscocinerea* and *H. impatiens* with 17 and 16 saponin peaks, respectively. These authors also described a total of 32 compounds in *H. scabra* and *H. impatiens* and 33 compounds in *H. fuscocinerea*. The saponin content of five tropical sea cucumbers including *H. atra, H. leucospilota, P. graeffei, A. echinites* and *B. subrubra* was also studied by Van Dyck *et al.* [50]. These authors reported the presence of four, six, eight, ten and nineteen saponin congeners in these species, respectively. In addition, this group [69] also detected a higher number of saponins (26) in the cuvierian tubules of *H. forskali* compared to the body wall (12 saponins). These results further support the evidence, suggested by the present study, of greater saponin congeners in viscera.

2.1. MALDI-MS/MS Data of Compound Holothurin A in the Positive Ion Mode

The conventional procedures to differentiate between isomeric saponins, including chemical derivatization and stereoscopic analysis, are tedious and time-consuming [100]. Tandem mass spectrometry was conducted to obtain more structural information about the saccharide moiety and elucidate their structural features. In order to ascertain that ions (signals) detected in the full-scan MALDI MS spectrum indeed correspond to saponin ions, tandem mass spectrometry analyses were performed for each ion, and saponin ion peaks were further analyzed using MS/MS fingerprints generated with the aid of collision-induced dissociation (CID) from their respective glycan structures. CID can provide a wealth of structural information about the nature of the carbohydrate components, as it preferentially cleaves glycosides at glycosidic linkages, allowing a straightforward interpretation of data. Almost all of observed daughter ions originated from the cleavage of glycosidic bonds (Figure 8). Therefore, the reconstruction of their fingerprints (fragmentation patterns) created by the glycosidic bond cleavages was utilized to deduce the structure of sugar moieties. This technique was also able to distinguish the structural differences between the isomers following HPCPC separation. However, in some cases, the MS/MS spectra obtained from the CID could be essentially identical for isomeric precursor ions. As a typical example, the MALDI-MS/MS mass spectrum for the ion detected at *m/z* 1243.5 is shown in Figure 8. The fragmentation pattern of the sodiated compound at *m/z* 1243.5 [M + Na]$^+$ in successive MS experiments is discussed in detail below for stepwise elucidation of the molecular structure of these compounds.

Collisional induced-dissociation activates two feasible fragmentation pathways of cationized parent ions shown in full and dotted arrows. First, the loss of the sugar unit; the successive losses of 3-*O*-methylglucose (-MeGlc), glucose (-Glc), quinovose (-Qui), sulfate and xylose (-Xyl) units generate ion products detected at *m/z* 1067, 905, 759, 639 and 507, respectively. As this figure illustrates, the consecutive losses of the (MeGlc + Glc) simultaneously generated the ion at *m/z* 905.3, and Qui (−146 Da) resulted in the peak at *m/z* 759.1 which corresponds to [Aglycone + sulXyl-H + 2Na]$^+$.

Figure 8. Positive tandem MALDI spectrum analysis of the precursor ion (saponin) detected at m/z 1243.5. The figure shows the collision-induced fragmentation of parent ions at m/z 1243.5. The consecutive losses of sulfate group, aglycone, xylose (Xyl), quinovose (Qui) and 3-O-methylglucose (MeGlc) residues affords product ions detected at m/z 1123, 639, 507, 361and 185, respectively.

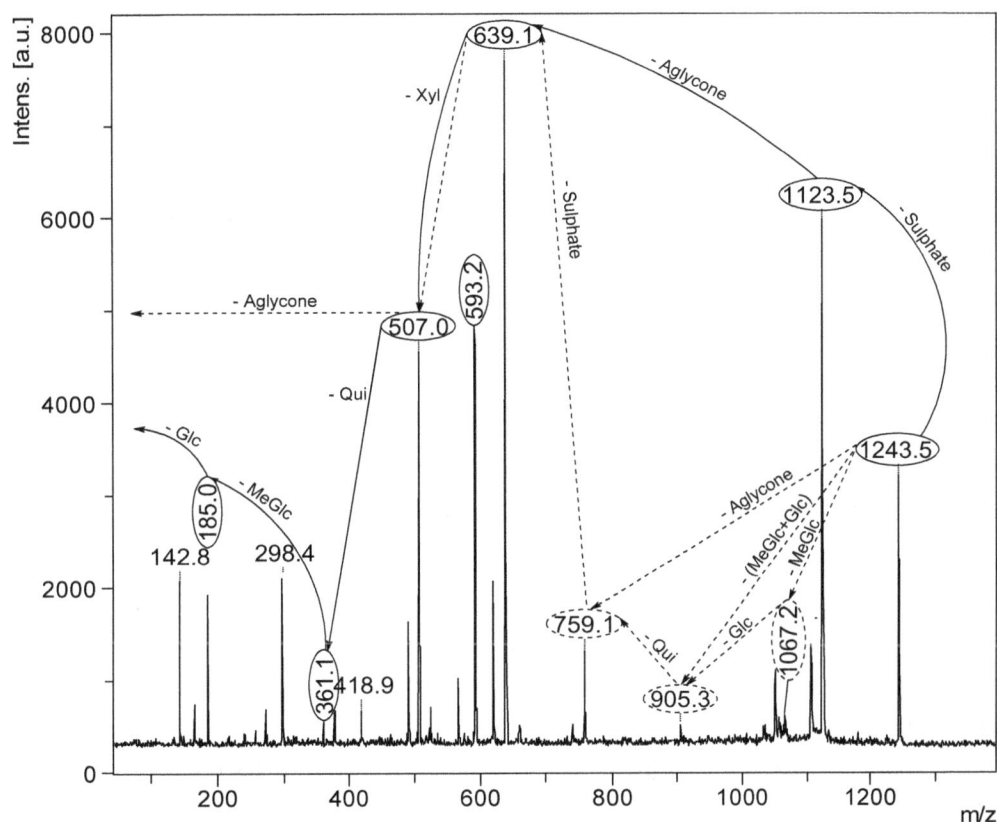

Secondly the decomposition of the precursor ions can also be triggered by the loss of the aglycone residue, creating peaks at m/z 759 (Figure 8) corresponding to the sugar moieties of 1243.5. The losses of the NaHSO$_4$ (ion generated at m/z 1123.5), aglycone residue (ion generated at m/z 639.1), and xylose (ion generated at m/z 507.0), respectively, were produced by glycone and aglycone fingerprint peaks from the precursor ion. Therefore, the consecutive losses of the sodium monohydrogen sulfate (NaHSO$_4$) from 1243.5 and aglycone unit produced signals observed at m/z 1123 and 639 (Figure 8); the latter peak corresponding to the total desulfated sugar moiety. Furthermore, the consecutive losses of Xyl, Qui and MeGlc presenting signals observed at m/z 507, 361 and 185, respectively, additionally proved that the decomposing ions were definitely generated from sodiated Holothurin A (m/z 1243.5). The implementations of these molecular techniques on all ions detected in the MALDI spectra allow us to identify the molecular structures of the saponins. All spectra were analyzed and fragmented, and some of them shared common fragmentation patterns. Key fragments from the tandem MS spectra of the positive ion mode of MALDI and ESI were reconstructed according to the example illustrated in order to propose the saponin structures. On the bases of these fragment signatures, 39 new saponins can be postulated. Some of these compounds, which share the common m/z 507 or/and m/z 523 key signals as a signature of the sodiated MeGlc-Glc-Qui and the sodiated MeGlc-Glc-Glc oligosaccharide residues, respectively, were easily identified. The identified saponins possess different aglycone structural elements.

The loss of 18 Da from the sodiated molecular ion, suggested the elimination of a neutral molecule (H_2O) from the sugar group [19]. The simultaneous loss of two sugar units indicated characteristics of a branched sugar chain. Other visible peaks correspond to saponin product ions produced by the losses of water and/or carbon dioxide from sodiated saponins upon MALDI ionization. Hereby the sugar sequence of saponins can be determined by applying CID. The MALDI MS/MS data for this m/z value were in complete agreement with those reported in a previous study [50,64]. The predominant fragment signal at m/z 593.2 results from a $^{1,5}A_4$ cross-ring cleavage of the sulXyl residue, which was consistent with previous findings for the MS/MS analyses of sea cucumber saponins [64]. However, this peak was only detected as an intense signal in the sulfated saponins such as Holothurin A, whereas it was not observed in the non-sulfated saponins such as Holothurinoside A. Therefore, this cross-ring cleavage seems to occur only with the sulfated Xyl. Analysis by MALDI resulted in an information rich tandem mass spectrum containing glycosidic bond and cross-ring cleavages that provided more structural information than previous studies on the same precursor ion. The sugar moiety of saponins developed from non-sulfated hexaosides to sulfated tetraosides [64]. The assignment of the sulfate group was determined by the mass difference between the parent ion at m/z 1243 and daughter ion m/z 1123 peaks based on knowing the molecular weight of the sulfate unit (120 Da). Complete glycosidic bond cleavage was observed, which enabled us to determine the locations of the sulfate (m/z 1123), the entire sugar moieties (m/z 639), and each component of sugar residue.

The losses of the aglycone and sugar residues are largely observed from glycosidic bond cleavages. Even though one cross-ring cleavage is assigned, the generation of glycosidic bond cleavages in combination with accurate mass is sufficient to assign the position of the sulfate group along the tetrasaccharide sequence for Holothurin A. The ion detected at m/z 1105 (Figure 8) is the water-loss ion derived from the ion at m/z 1123, whereas the ion observed at m/z 1061 corresponds to the neutral loss of CO_2 (44 Da).

As described by Song et al. [100], the cross-ring cleavages that occurred in the CID spectra of saccharides with α 1–2 linkage, such as the sugar residue for Holothurin A, are X and A types, whereas the glycoside bond cleavages are C and B types. The major peak at m/z 593.2 was attributed to cross-ring cleavage of the sugar unit.

This MS/MS spectrum allows us to reconstruct the collision-induced fragmentation pattern of the parent ion (Figure 4) and consequently to confirm that ions monitored at m/z 1243.5 correspond to the Holothurin A elucidated by Van Dyck et al. [50], Kitagawa et al. [96] and Rodriguez et al. [88].

The occurrence of a sulfate group ($NaHSO_4$) in saponin compounds, such as in the case of Holothurin A, was assigned by a loss of 120 Da during the MS/MS. By the combination of accurate mass and MS/MS information, saponins were categorized into seven distinct carbohydrate structural types: (A) MeGlc-Glc-Qui-Xyl-Aglycone; (B) MeGlc-Glc-Glc-Xyl-Aglycone; (C) (MeGlc-Glc)-Qui-sulXyl-Aglycone; (D) MeGlc-Glc-Qui-(Qui-Glc)-Xyl-Aglycone; (E) MeGlc-Glc-Qui-(MeGlc-Glc)-Xyl-Aglycone; (F) MeGlc-Glc-Glc- (MeGlc-Glc)-Xyl-Aglycone; and (G) MeGlc-Glc-Glc-(Qui-Glc)-Xyl-Aglycone. Non-sulfated saponins had one to six monosaccharide units and six distinct structural types. All sulfated saponins ranging from m/z 889 to 1259 had a structure (C), in which Xyl was sulfated. However, in some cases, the sulfation of Xyl, MeGlc and Glc was reported [13]. The MS analyses also indicated that this sea cucumber species produced a mixture of common and unique saponin types. Unique saponin types were also identified when the mass spectra of this species were compared with others. Saponin peaks with the ion signatures at m/z values of 1477, 1335, 1221, 1149 and 1123 were unique in *H. lessoni*. In the tandem MS, in general, the most abundant ions were attributed to the losses of aglycones and/or both key diagnostic sugar moieties (507 and 523). For 1243.5, the most abundant ions observed under positive ion conditions were at m/z 1123, 639 and 507, corresponding to the losses of sulfate, aglycone and Xyl moieties. The major ion at m/z 621.2 corresponded to the loss of water from ion at m/z 639. Some saponins were commonly found among species (e.g., Holothurins A and B), whereas others were unique to each species (e.g., 1221 in *H. lessoni*), as Bondoc *et al.* [64] and Caulier *et al.* [6] have also indicated. The saponin profile (peaks) of sea cucumbers indicated the different relative intensities of saponins in the viscera. The peaks observed (Figure 4) at m/z 1149.0, 1227.5, 1229.5, 1243.5, and 1259.5 in the positive ion mode corresponded to an unidentified saponin, Scabraside A or Fuscocineroside B/C (isomers), Holothurin A_2 (Echinoside A), Holothurin A, and Holothurin A_3, respectively [36,56,61,89,93,94]. Most of these sulfated saponins were also reported by Kitagawa *et al.* [89] and Bondoc *et al.* [64]. The ion peaks of the non-sulfated saponins at m/z 1125, 1141, 1287, 1289, 1301 and 1303 corresponded to Holothurinosides C/C_1 (isomers), Desholothurin A (synonymous with Nobiliside 2A) or Desholothurin A_1, Holothurinosides E/E_1, Griseaside A, Holothurinosides M and A, respectively [69]. *H. scabra*, *H. impatiens* and *H. fuscocinerea* were also reported to contain Holothurin A, Scabraside B and Holothurinoside C [64]. This group also detected 24-dehydroechinoside A and Scabraside A in *H. scabra*. The presence of Holothurinosides C/C_1 (isomers), Holothurinosides A/A_1 (isomers), Desholothurin A (synonymous with Nobiliside 2A), Desholothurin A_1 and Holothurinosides E/E1 were also described in *H. forskali* by several groups [65,69,88]. We were not able to identify all the saponin congeners detected in the semi-pure extract in the HPCPC-fractionated samples. Bondoc *et al.* [64] experienced a similar issue in that they observed some peaks in MALDI MS, which were not seen in the isomeric separation done in LC-ESI MS. For instance, we could not find ions at m/z 1149 and 1335 in the spectra of HPCPC fractions by ESI-MS. The MALDI mass spectra of the semi-pure and HPCPC fractionated samples of the *H. lessoni* revealed 75 ions (29 sulfated and 46 non-sulfated) in which a total of 13 isomers was found (Table 1), of which 36 congeners had previously been identified in other holothurians. It is the first time that the presence of these identified saponins has been reported in *H. lessoni,* apart from the saponins reported by Caulier *et al.* [58] that were found in the seawater surrounding *H. lessoni*. They reported saponins with m/z values of 1141, 1229, 1243 and 1463 namely Desholothurin A, Holothurin A_2, Scabraside B (synonymous with Holothurin A) and Holothurinoside H, respectively [58]. However, we could not detect the ion at m/z 1463 in our sample.

Most of the sulfated saponins that had previously been reported were detected in this species, including Holothurin B_3 (m/z 889), Holothurin B/B_4 (m/z 905), Holothurin B_2 (m/z 907), Fuscocinerosides B or C, which are functional group isomers (m/z 1227), Holothurin A_2 (m/z 1229), Holothurin A (m/z 1243), Holothurin A_1/A_4 (m/z 1245), and Holothurin A_3 (m/z 1259). The common sulfated congeners among this species and other sea cucumbers are Holothurin B (m/z 905) and Holothurin A (m/z 1243). Among these saponins, Holothurin A is the reported to be the major congener with the highest relative abundance in this species.

To illustrate the identification of a novel compound at m/z 1149.0, the parent ion at m/z 1149.0 was subjected to MS/MS fragmentation. The MALDI fingerprints revealed that the compound contained a novel aglycone at m/z 493 and a tetrasaccharide moiety with m/z value of 656 Da including -Xyl, -Qui, -Glc and -MeGlc in the ratio of 1:1:1:1. This saponin possessed the common m/z 507 key signal as a fingerprint of MeGlc-Glc-Qui + Na$^+$. We propose to name Holothurinoside T.

The isomers within one sample showed different MS^n spectra [101] allowing their structures to be elucidated based on the ion fingerprints. Here we indicate that the occurrence of many product ions in the spectrum of viscera extract is due to the presence of a mixture of saponins and isomeric saponins (Figures 3 and 5–7). This observation is consistent with the findings proposed by Van Dyck and associates [69] for the Cuvierian tubules of *H. forskali*. Mass spectrometry alone, however, is not powerful enough to obtain more structural information about the isomeric congeners. Nonetheless, it provides a quick and straightforward characterization of the element components and saponin distributions by the presence of ions at m/z 507 and 523 in the tandem spectra of the viscera extracts.

2.2. Key Fragments and Structure Elucidation of Novel Saponins

The common key fragments facilitated the structure elucidation of novel saponins. Tandem mass spectrometry analyses of saponins led to identification of several diagnostic key fragments corresponding to certain common structural element of saponins as summarized in Table 2.

Table 2. Key diagnostic ions in the MS/MS of the holothurians saponins.

Diagnostic ions in CID Spectra of [M + Na]$^+$		
m/z Signals (Da)		
507	**523**	**639**
Chemical signatures MeGlc-Glc-Qui + Na	MeGlc-Glc-Glc + Na	MeGlc-Glc-Qui-Xyl + Na

The structures of saponins were deduced by the identification and implementation of the key fragment ions generated by tandem mass spectrometry. The presence of these oligosaccharide residues (m/z 507 and/or 523) facilitated the determination of the saponin structure. However, some compounds with a m/z value of less than 1100 Da including 921, 907, 905 and 889 did not yield the peak m/z 523, which reflected the lack of this oligosaccharide unit in their structures. Unlike other compounds, the MS/MS spectrum of the ion at m/z 1477.7 illustrated the unique fingerprint profile, which contained ions at m/z 511 and 493 instead of an ion at m/z 507. The structure of compound was further confirmed by MS/MS analyses.

The MALDI analysis revealed that the ion with m/z 1243.5 was the prominent peak in the spectrum, which corresponded to Holothurin A, which was found in several species of sea cucumbers [6,29,50,58,61,64,89,95–97]. The MALDI data were confirmed by ESI-MS.

Table 1 summarizes data of all analyses performed on the saponin-enriched sample and HPCPC fractionated samples using MALDI and ESI on compounds from the viscera of *H. lessoni*. The identified saponin mixture contains a diverse range of molecular weights and structures. The chemical structures of the identified compounds are illustrated in Figure 9. The isobutanol and HPCPC fractionated samples indicated 29 sulfated and 46 non-sulfated saponin ions. The number of MS ion peaks was lower than the number of isomers identified by MS/MS following HPCPC separation (Figure 3).

Figure 9. The structure of identified saponins in the viscera of *H. lessoni*.

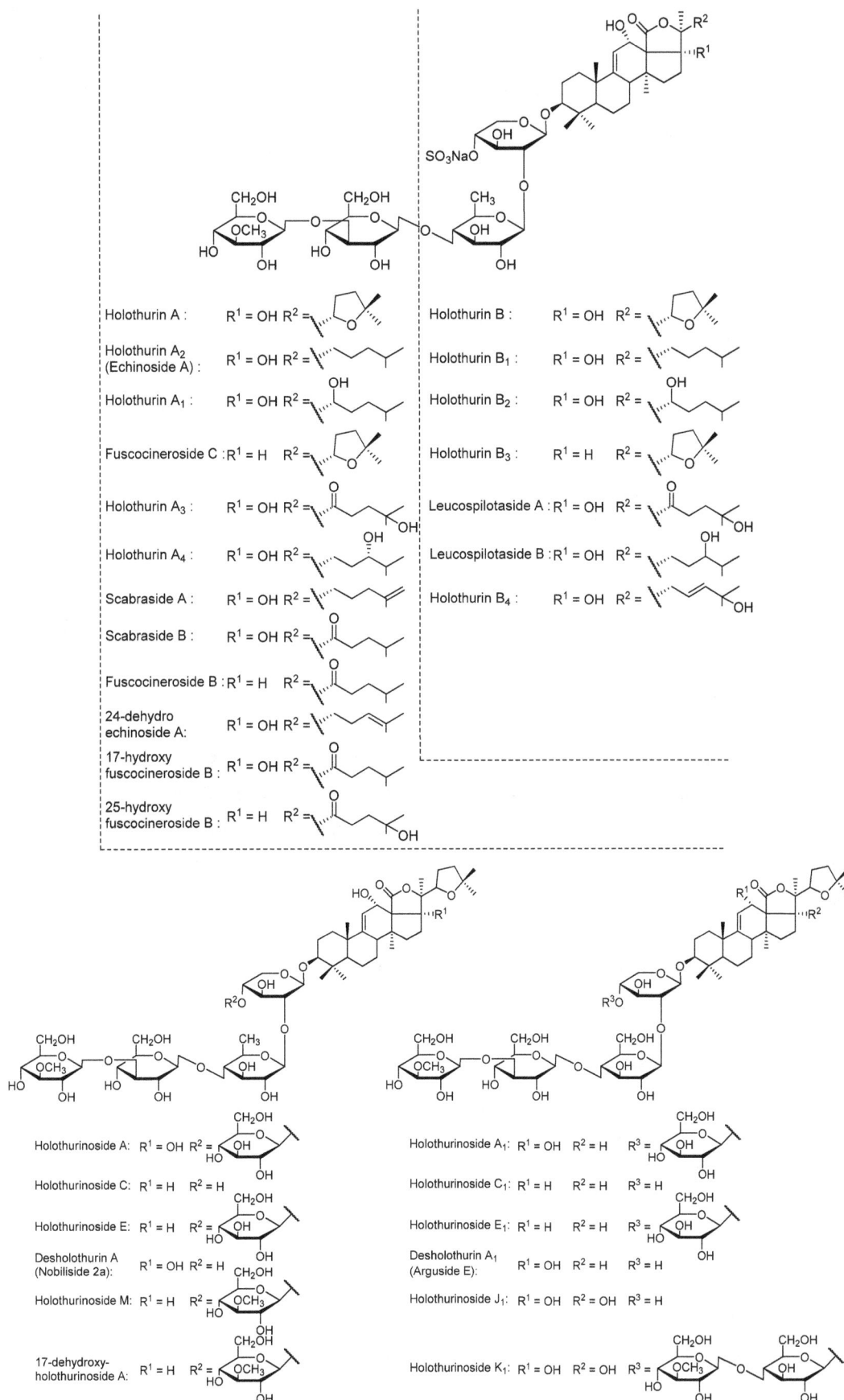

Figure 9. *Cont.*

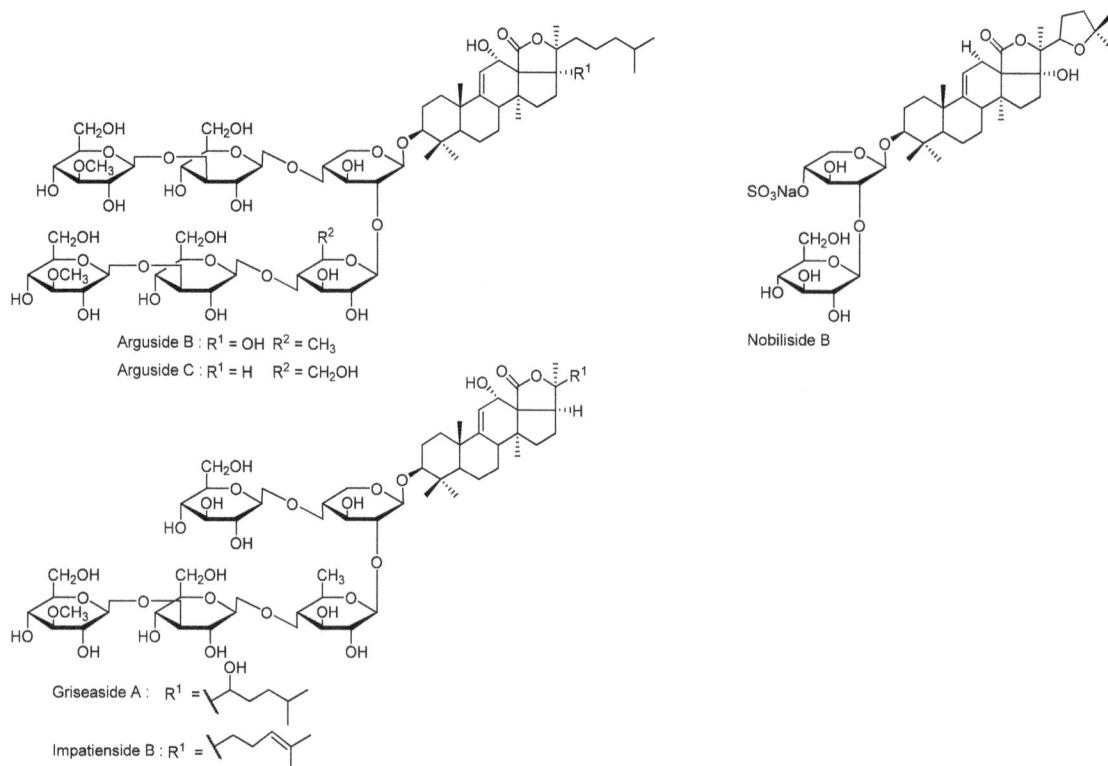

Arguside B : R¹ = OH R² = CH₃
Arguside C : R¹ = H   R² = CH₂OH

Nobiliside B

Griseaside A : R¹ =

Impatienside B : R¹ =

2.3. Analyses of Saponins by ESI-MS

The positive ion mode ESI-MS analyses were also conducted on the samples. ESI mass spectra of the saponins are dominated by $[M + Na]^+$. There were some instances where peaks observed in the MALDI-MS spectra were not monitored from the isomer separation done in the ESI-MS, such as the peak detected at m/z 1149 in the MALDI spectra. Other researchers had experienced the same issue [50,64].

ESI-MS^n is a very effective and powerful technique to differentiate isomeric saponins [100]. Tandem MS analyses on $[M + Na]^+$ ions provided abundant structural information about saponins. The positive ion mode ESI-MS/MS analyses were also performed on all compound ions detected in the ESI-MS spectrum of HPCPC fractions. This technique also confirmed the existence of saponins reported in the literature and allowed the discovery of new saponin congeners in the species examined. The molecular masses of the identified compounds are summarized in Table 1. The ESI-MS spectrum of the saponin extract from the viscera of *H. lessoni* is shown in the Figure 10.

Several major peaks were detected. The peaks at m/z 1123 and 1243 correspond to a novel compound and Holothurin A with the elemental compositions of $C_{54}H_{84}O_{23}$ and $C_{54}H_{85}NaO_{27}S$, respectively. The ESI-MS analyses were also carried out on all HPCPC fractions. As a typical example, Figure 11 shows the ESI-MS spectrum of Fraction 14.

Figure 10. (+) ESI-MS spectrum of saponins extract from the viscera of *H. lessoni*.

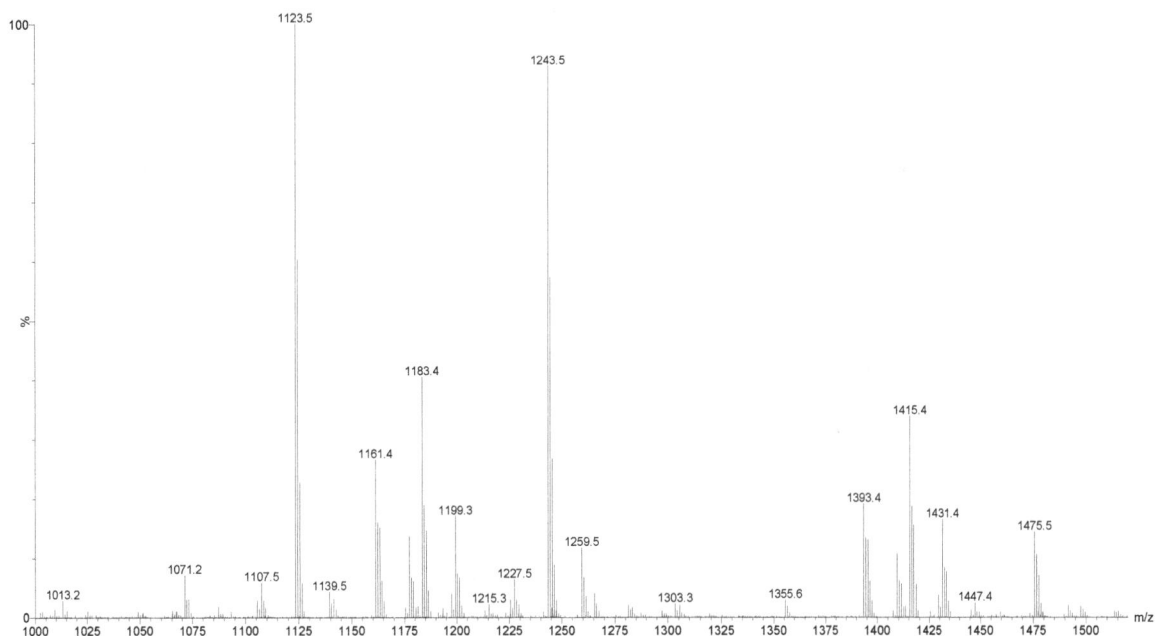

Figure 11. (+) ion mode ESI-MS spectrum of saponins extract from Fraction 14.

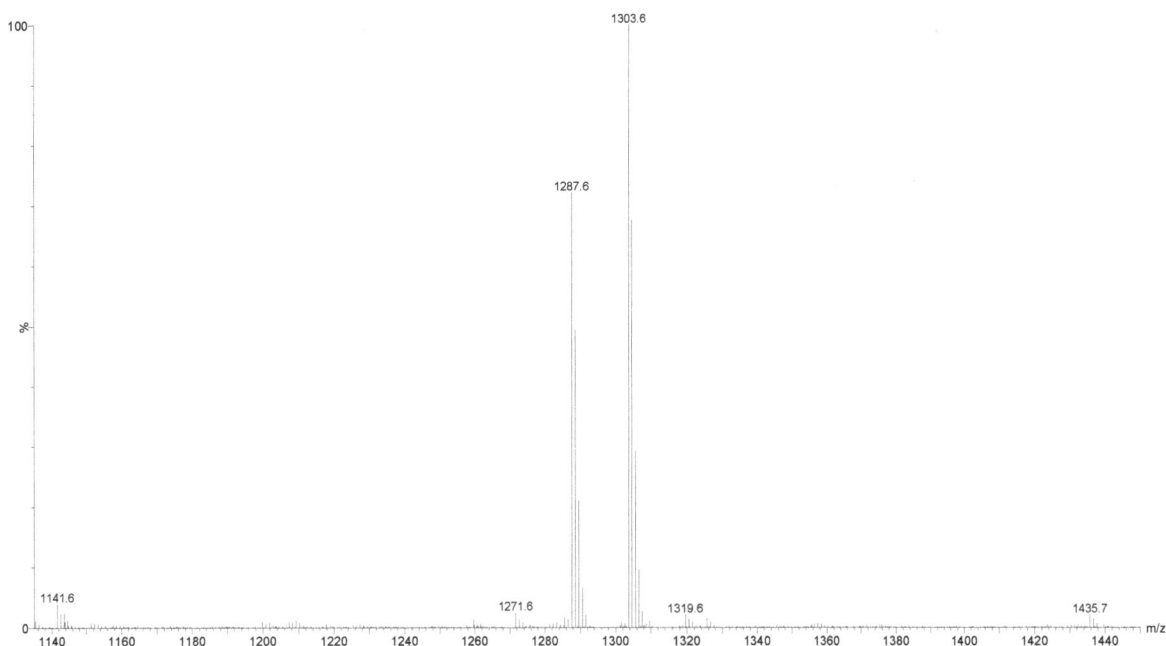

As can be seen in Figure 11, there are two major peaks at *m/z* 1287.6 and 1303.6, which correspond to Holothurinosides E/E_1 and Holothurinosides A/A_1, respectively. These two peaks, as the MS/MS analyses show which will be discussed later, were found to correspond to at least five and six isomers, respectively (Figures 3, 6 and Supplementary Figure S2). A comparison of the molecular weights of both saponins revealed some mass differences between them, such as a 16 Da (O) mass differences between Holothurinosides E/E_1 and Holothurinosides A/A_1, reflecting the small structural alterations and the intrinsic connections between them. Their MS/MS analyses indicated, as will be discussed later, the presence of some identical aglycones in both ions.

2.3.1. Molecular Mass of Saponins by ESI

ESI/MS provide considerable structural information with very high sensitivity for saponins [60,63]. Peaks corresponding to the sodium adduct of the complete sugar side chains were often quite intense in the product ion spectra of the sodiated saponin precursor. Tandem mass spectra of saponins reflected the different fingerprints with different relative intensities.

2.3.2. Structure Elucidation of the Saponins by ESI-MS/MS

Seventy-five different triterpene saponins purified from sea cucumber were investigated by MALDI and electrospray ionization tandem mass spectrometry (ESI-MS/MS) in the positive ion modes. All spectra were analyzed and fragmented, and some of them shared common fragmentation patterns. Key fragments from the positive ion mode MS/MS spectra of MALDI and ESI were reconstructed with an example illustrated that proposes the saponin structures. Peak intensities of fragment ions in MS/MS spectra were also correlated with structural features and fragmentation preferences of the investigated saponins. In general, the formation of fragments occurred predominantly by cleavages of glycosidic bonds in the positive mode (Figure 12), which was applied to identify the structure of saponins. Interpretation of fragment ions of MS/MS spectra provided the key information for the structural elucidation of saponins as exemplified in Figure 12.

Fragmentation of the ion at m/z 1243.5 (sulfated saponin) under collisionally activated dissociation (CAD) conditions is shown in Figure 12. Full and dotted arrows show the two main fragmentation pathways in this saponin. The peak at m/z 507 corresponds to both the aglycone and the key diagnostic fragment of sugar moiety.

The most abundant peaks were detected at m/z 1123 $[M + Na - 120 \text{ (sulfate)}]^+$, 639 $[M + Na - 120 - 484 \text{ (aglycone)}]^+$ and 507 $[M + Na - 120 - 484 - 132]^+$. In addition, the peaks observed at m/z 1225.5 and 1199.5 were generated by the losses of H_2O and CO_2 from their respective parent ion.

The most intensive peak was observed at m/z 593 stemming from a cross-ring cleavage. The observed fragments are consistent with the structure of the Holothurin A proposed by Van Dyck et al. [50]. This ESI-MS/MS analysis confirmed the MALDI data on the ion at m/z 1243.5. The full analysis can be seen in Supplementary Figure S1.

ESI-MS was applied to distinguish the isomeric saponins by Song et al. [100]. Isomers of saponins were also identified using tandem mass spectrometry combined with electrospray ionization (ESI-MS/MS) following HPCPC separation. MS/MS spectra of these ions gave detailed structural information and enabled differentiation of the isomeric saponins. The results are exemplified in the following figures. The analyses applied on the ion at m/z 1303.6 (non-sulfated saponins), which was obtained from Fractions 15, 14 and 12, are shown in Figure 3A–C. The main fragmentation patterns observed for this isomeric compound are shown with full and dotted arrows.

Figure 12. (+) Ion mode ESI-MS/MS spectrum of saponin detected at 1243.5 (Holothurin A). Full and dotted arrows show the two main feasible fragmentation pathways. The structure of saponin was elucidated on the base of tandem mass spectrometry.

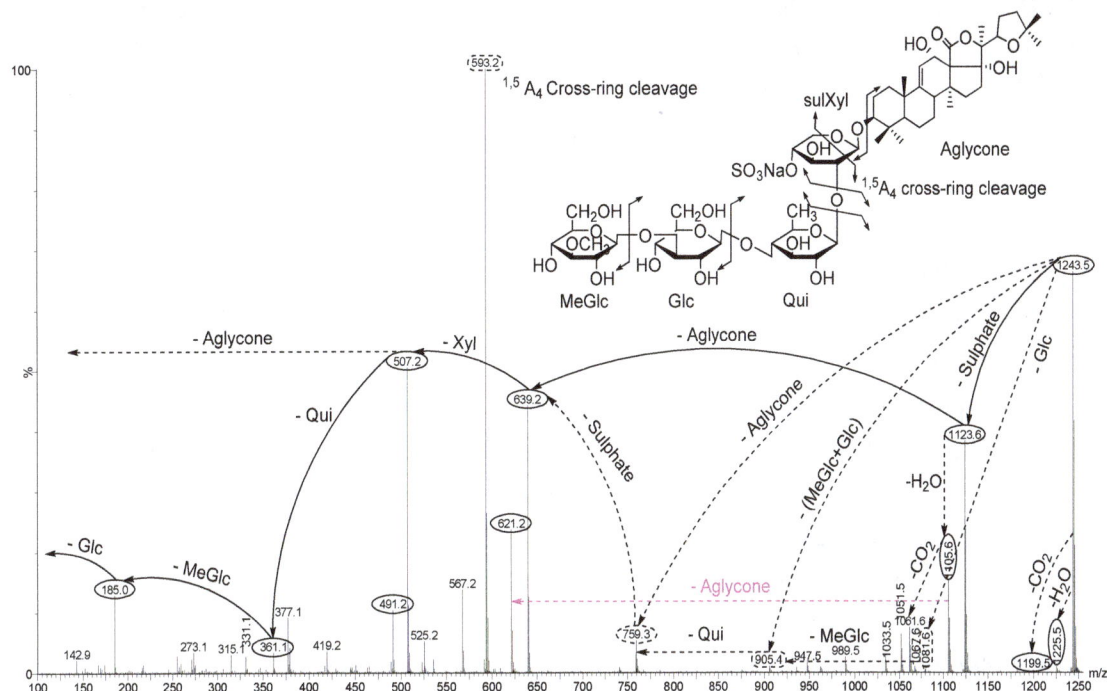

The structures of six isomeric saponins were ascribed to the ions detected at m/z 1303.6 (Figure 3A–C). These isomers have at least three different aglycone structures with m/z 468, 484 and 500 and contain five different monosaccharaide residues. These figures illustrate different isomers of ions detected at m/z 1303. For instance, Figure 3A (Fraction 15) shows the stepwise structure elucidation of Holothurinoside A. The consecutive losses of MeGlc, Glc, Qui, Glc, and Xyl units generate signals detected at m/z 1127.5, 965.5, 819.3, 657.2 and 507.2, respectively, which correspond to Holothurinoside A [5,62,69,88]. As can be seen in Figure 3A, this saponin fraction is quite pure.

In one of these isomers (Figure 3B), the consecutive losses of aglycone, Glc, Xyl, Qui and Glc units provided signals detected at m/z 819, 657, 507, 361 and 199, respectively, further confirming that the fragment ions unambiguously originate from sodium-cationized Holothurinoside A. In addition (m/z 1303.6), the precursor ion sequentially lost MeGlc (m/z 1127.5), Glc (m/z 965.5), Glc (m/z 803.4), Qui (m/z 657.2) and Xyl (m/z 507.2) (Figure 3B) thereby indicating the structure of another isomer of this molecule. The characteristic peak observed at m/z 507.2 generated by tandem MS was identified either as a sodiated MeGlc-Glc-Qui residue or sodiated aglycone residue. The ion at m/z 803 resulted from the loss of aglycone from the parent ion at m/z 1303.6, which is the fragment ion corresponding to the complete saccharide chain, which subsequently (Figure 3B) produces the ions at m/z 657 and ion at m/z 507 by the losses of Qui and Xyl residues. Moreover, the ions (m/z 507) further fragmented to form ions of the same m/z value at m/z 361 and m/z 199 or 185. The observation of ions at m/z 507 and 657 further supports the above conclusion. The ions detected at m/z 1285.5 and 1241.6 correspond to the losses of H_2O and $H_2O + CO_2$, respectively. These two fragments correspond to the sequential losses of water and carbon dioxide. It is notable that the configurations of all the sugars in all previously known sea cucumber triterpene glycosides are D-configurations.

A similar analysis was carried out (Figure 3C) on the ion at m/z 1303.6 of Fraction 12. As can be seen in the figure, the spectrum has a different fragmentation pattern compared to the spectra in Figure 3A,B even though they have the same m/z value. In one of the isomers, the consecutive losses of aglycone, Glc, Xyl, Glc and MeGlc units generated signals detected at m/z 835.2, 673.2, 523.1, 361.1 and 185, respectively, further confirming the structure of one of the isomeric compounds (Figures 3C and 5). The full analysis can be seen in the Supplementary Figure S2 (Fraction 12).

Further, the cleavage of the C_2 ion at m/z 673, 643, 629, 601, 583, 541 and 523 (Figure 3C) produced the ion at m/z 613, 583, 569, 541, 523, 481 and 463, respectively, through the loss of $C_2H_4O_2$ (60 Da) which indicates an α 1–4-linked glycosidic bond in the α-chain, which is in agreement with a previous study [100]. This observation is consistent with the fragmentation rules for ions of 1–4-linked disaccharides.

The MS/MS spectra show the presence of three different aglycone structures, namely ions detected at m/z 835.2, 819.2 and 803.4 by the losses of aglycone moieties. This analysis reveal the presence of at least six different isomers with different aglycones and sugar moieties, as the MS/MS spectra generate both key diagnostic fragments at m/z 507 and 523. These isomers are composed of five monosaccharaides including MeGlc-Glc-Qui (Glc)-Xyl-MeGlc (Glc or Qui). The proposed structures are shown in Figure 5 and correspond to Holothurinoside A_1 and Holothurinoside A, and four novel saponins [5,69,88]. We propose to call these molecules Holothurinosides S, Q, R and R_1, respectively.

It should be noted that both major fragment ions (507 and 523) can correspond to partial glycoside compositions or aglycone moieties, further supporting the presence of isomeric saponins. The predominant fragment ion at m/z 507 results from the sodium adduct ion of the [MeGlc-Glc-Qui + Na] side chain or the aglycone. Similarly, the abundant fragment ion at m/z 523 arises from the sodium adduct ion of the [MeGlc-Glc-Glc + Na]$^+$ side chain or the aglycone. Since the masses of sodiated aglycones are identical with their relative partial sugar residues, namely [MeGlc-Glc-Qui + Na]$^+$ and [MeGlc-Glc-Glc + Na]$^+$, the ions at m/z 507.2 and 523.2, respectively, correspond to both sugar residues and their aglycones. When the decomposition of the parent ion (m/z 1303.6) is triggered by the losses of sugar residues, as an exemplified by the black and pink dotted arrows in Figure 3A–C, the ions at m/z 507.2 and 523.2 correspond to the aglycone moieties. Alternatively, the fragmentation of the parent ion can proceed by the losses of all five sugar residues, which generates ions at m/z 507.2 and 523.2, which correspond to the aglycone moieties. Similar conclusions were drawn by Van Dyck *et al.* (2009) [69] for triterpene glycosides. Losses of H_2O and CO_2 or their combination result from cleavage at the glycosidic linkages as noted by Waller and Yamasaki [3].

Different fractions of the HPCPC separation were compared to show the presence of one aglycone (Figure 3A), the presence of two different aglycones (Figure 3B) and the presence of three different aglycones (Figure 3C) indicating that the HPCPC allowed the separation of the isomers.

On comparison of the MS/MS spectra of 1303.6 and 1243.5 (Figures 3 and 12), it is notable that the m/z 523 fragment (aglycone loss) of the [M + Na]$^+$ ions was only observed with 1303.6, which corresponds to the presence of a new aglycone unit at m/z 500 (sodiated 523). Individual patterns were detected from sulfated and non-sulfated saponins as indicated in Holothurin A and Holothurinoside A as representative examples. This sequential decomposition confirms the proposed Holothurin A and Holothurinoside A structures.

Another typical chemical structure elucidation of isomeric saponins by tandem MS is exemplified in Figure 6. This spectrum shows the ion signature of the sample under tandem MS from the ion detected at m/z 1287.6. Tandem MS analyses revealed the presence of two different aglycones with m/z values of 484 and 468, confirming the presence of chemical isomeric structures. The same fragmentation behaviors have been observed from the positive ESI-MS/MS spectra of saponins with m/z 1303. The structures of aglycones are identical with those reported for the ion at m/z 1303. The possible fragmentation pathways were shown using full and dotted arrows. The losses of aglycone moieties (Figure 6) generated ions at m/z 819.3 and 803.4, which correspond to the complete sugar components. The successive losses of aglycone, Glc or MeGlc, Xyl, Qui and MeGlc yielded to ion fragments at m/z 819, 657 or 643, 507, 361 and 185, respectively.

The decomposition of the parent ion can also be triggered by the loss of a sugar moiety, namely MeGlc, Glc, Qui, Qui or Glc and Xyl, followed by the aglycone, which generates daughter ions at m/z 1111.5, 949.5, 803.4, 657.2 or 643.2 and 507.2. It is clear that the ion at m/z 507 is the most abundant fragment ion and is the signature of the sodiated aglycone and/or the key sugar component. The losses of water (-18 Da) and/or carbon dioxide (-44 Da) are observed from the spectrum, and some of the peaks are also designated to those molecules.

This analysis revealed the presence of at least five different isomers with different aglycones and sugar moieties. These isomers contain some identical aglycone structures with those identified in the ion at 1303 (Figure 5). These isomers are also pentaglycosidic saponins. The proposed structures are shown in Figure 7, which correspond to Holothurinoside E_1, Holothurinoside E, 17-dehydroxy-holothurinoside A and two novel saponins (the first and fourth compounds). We propose to name these molecules Holothurinosides O and P, respectively.

The data indicate that the terminal sugar is preferentially lost first in glycosidic bond cleavages. Since Holothurinosides A and E contain the same terminal sugar units in their sugar residue, they yield the ions with the same m/z value (m/z 507).

3. Experimental Section

3.1. Sea Cucumber Sample

Twenty sea cucumber samples of *Holothuria lessoni* Massin *et al.* 2009, commonly known as Golden sandfish were collected off Lizard Island (latitude 14°41′29.46″ S; longitude 145°26′23.33″ E), Queensland, Australia in September 2010. The viscera (all internal organs) were separated from the body wall and kept separately in zip-lock plastic bags which were snap-frozen, then transferred to the laboratory and kept at -20 °C until use.

3.2. Extraction Protocol

The debris and sand particles were separated from the viscera (all internal organs) manually and the visceral mass was freeze-dried (VirTis, BenchTop K, New York, NY, USA). The dried specimens were then pulverized to a fine powder using liquid nitrogen and a mortar and pestle.

All aqueous solutions were prepared with ultrapure water generated by a Milli-Q systems (18.2 MΩ, Millipore, Bedford, MA, USA). All organic solvents were purchased from Merck (Darmstadt, Germany) except when the supplier was mentioned, and were either of HPLC grade or the highest degree of purity.

Extraction of Saponins

The extraction and purification procedures were adapted from Campagnuolo et al. [34], Van Dyck et al. [69], Garneau et al. [102] and Grassia et al. [103]. The pulverized viscera sample (40 g) was extracted four times with 70% ethanol (EtOH) (400 mL) followed by filtration through Whatman filter paper (No.1, Millipore, Bedford, MA, USA) at room temperature. The extract was concentrated under reduced pressure at 30 °C using a rotary evaporator (Büchi AG, Flawil, Switzerland) to remove the ethanol, and the residual sample was freeze-dried to remove water (VirTis, BenchTop K, New York, NY, USA). The dried residue was successively extracted using a modified Kupchan partition procedure [104]: The dried extract (15 g) was dissolved in 90% aqueous methanol (MeOH) (any remaining solid residue was removed by filtration), and partitioned against 400 mL of n-hexane (v/v) twice. The water content of the hydromethanolic phase was then adjusted to 20% (v/v) and then to 40% (v/v) and the solutions partitioned against CH_2Cl_2 (450 mL) and $CHCl_3$ (350 mL), respectively. In the next step, the hydromethanolic phase was concentrated to dryness using a rotary evaporator and freeze-drier. The dry powder was solubilized in 10 mL of MilliQ water (the aqueous extract) in order to undergo chromatographic purification.

3.3. Purification of the Extract

A solution of the aqueous extract was then subjected to a prewashed Amberlite® XAD-4 column (250 g XAD-4 resin 20–60 mesh; Sigma-Aldrich, MO, USA; 4 × 30 cm column) chromatography. After washing the column extensively with water (1 L), the saponins were eluted sequentially with MeOH (450 mL) and acetone (350 mL) and water (250 mL). The eluates (methanolic, acetone and water fractions) were then concentrated, dried, and redissolved in 5 mL of MilliQ water. Finally, the aqueous extract was partitioned with 5 mL isobutanol (v/v). The isobutanolic saponin-enriched fraction was either stored for subsequent mass spectrometry analyses or concentrated to dryness and the components of the extract were further examined by HPCPC and RP-HPLC. The profile of fractions was also monitored by Thin Layer Chromatography (TLC) using the lower phase of $CHCl_3/MeOH/H_2O$ (7:13:8 v/v/v) solvent system.

3.4. Thin Layer Chromatography (TLC)

Samples were dissolved in 90% or 50% aqueous MeOH and 10 microliters were loaded onto silica gel 60 F_{254} aluminum sheets (Merck #1.05554.0001) and developed with the lower phase of $CHCl_3/MeOH/H_2O$ (7:13:8) biphasic solvent system. The profile of separated compounds on the TLC plate was visualized by UV light and by spraying with a 15% sulfuric acid in EtOH solution and heating for 15 min at 110 °C until maroon-dark purple spots developed.

3.5. High Performance Centrifugal Partition Chromatography (HPCPC or CPC)

The solvent system containing $CHCl_3/MeOH/H_2O$–0.1% HCO_2H (7:13:8) was mixed vigorously using a separating funnel and allowed to reach hydrostatic equilibration. Following the separation of the two-immiscible phase solvent systems, both phases were degassed using a sonicator-degasser (Soniclean Pty Ltd., Adelaide, SA, Australia). Then the rotor column of HPCPC™, CPC240 (Ever Seiko Corporation, Tokyo, Japan) was filled with the liquid stationary phase at a flow rate of 5 mL/min by Dual Pump model 214 (Tokyo, Japan).

The CPC was loaded with the aqueous upper phase of the solvent system in the descending mode at a flow rate of 5 mL/min with a revolution speed of 300 rpm. The lower mobile phase was pumped in the descending mode at a flow rate of 1.2 mL/min with a rotation speed of 900 rpm within 2 h. One hundred and twenty milligrams of isobutanol-enriched saponin mixture was dissolved in 10 mL of the upper phase and lower phase in a ratio of 1:1 and injected to the machine from the head-end direction (descending mode) following hydrostatic equilibration of the two phases indicated by a clear mobile phase eluting at the tail outlet. This indicated that elution of the stationary phase had stopped and the back pressure was constant. The chromatogram was developed at 254 nm for 3.0 h at 1.2 mL/min and 900 rpm using the Variable Wavelength UV-VIS Detector S-3702 (Soma optics Ltd., Tokyo, Japan) and chart recorder (Ross Recorders, Model 202, Topac Inc., Cohasset, MA, USA). The fractions were collected in 3 mL/tubes using a Fraction collector. The elution of the sample with the lower organic phase proceeded to remove the compounds with low polarity from the sample, within 200 mL of which several peaks were eluted. At this point (Fraction 54), the elution mode was switched to ascending mode and the aqueous upper phase was pumped at the same flow rate for 3.0 h. Recovery of saponins was achieved by changing the elution mode to the aqueous phase which allowed the elution of the remaining compounds with high polarity in the stationary phase. A few minor peaks were also monitored. Fractions were analyzed by TLC using the lower phase of $CHCl_3/MeOH/H_2O$ (7:13:8) as the developing system. The monitoring of the fractions is necessary, as most of the saponins were not detected by UV due to the lack of a chromophore structure. Fractions were concentrated with nitrogen gas.

3.6. Mass Spectrometry

The isobutanol saponin-enriched fractions and the resultant HPCPC purified polar samples were further analyzed by MALDI and ESI MS to elucidate and characterize the molecular structures of compounds.

3.6.1. MALDI-MS

MALDI analysis was performed on a Bruker Autoflex III Smartbeam (Bruker Daltonik, Bremen, Germany). All MALDI MS equipment, software and consumables were from Bruker Daltonics (Bremen, Germany). The laser (355 nm) had a repetition rate of 200 Hz and operated in the positive reflectron ion mode for MS data over the mass range of 400 to 2200 Da under the control of the FlexControl and FlexAnalysis software (V 3.3 build 108, Bruker Daltonik, Bremen, Germany). External calibration was performed using PEG. MS spectra were processed in FlexAnalysis (version 3.3, Bruker Daltonik, Bremen, Germany). MALDI MS/MS spectra were obtained using the

LIFT mode of the Bruker Autoflex III with the aid of CID. The isolated ions were submitted to collision against argon in the collision cell to collisionally activate and fragment, and afford intense product ion signals. For MALDI, a laser energy was used that provided both good signal levels and mass resolution, the laser energy for MS/MS analysis was generally 25% higher than for MS analysis.

The samples were placed onto a MALDI stainless steel MPT AnchorChip TM 600/384 target plate. Alpha-cyano-4-hydroxycinnamic acid (CHCA) in acetone/ iso-propanol in ratio of 2:1 (15 mg/mL) was used as a matrix to produce gas-phase ions. The matrix solution (1 μL) was spotted onto the MALDI target plate and air-dried. Subsequently 1μL of sample was added to the matrix crystals and air-dried. Finally, 1 μL of NaI (Sigma-Aldrich #383112, St Louis, MO, USA) solution (2 mg/mL in acetonitrile) was applied onto the sample spots. The samples were mixed on the probe surface and dried prior to analysis.

3.6.2. ESI-MS

The ESI mass spectra were obtained with a Waters Synapt HDMS (Waters, Manchester, UK). Mass spectra were obtained in the positive ion mode with a capillary voltage of 3.0 kV and a sampling cone voltage of 100 V.

The other conditions were as follows: extraction cone voltage, 4.0 V; ion source temperature, 80 °C; desolvation temperature, 350 °C; desolvation gas flow rate, 500 L/h. Data acquisition was carried out using Waters MassLynx (V4.1, Waters Corporation, Milford, CT, USA). Positive ion mass spectra were acquired in the V resolution mode over a mass range of 100–2000 m/z using continuum mode acquisition. Mass calibration was performed by infusing sodium iodide solution (2 μg/μL, 1:1 (v/v) water/isopropanol). For accurate mass analysis a lock mass signal from the sodium attached molecular ion of Raffinose (m/z 527.1588) was used.

MS/MS spectra were obtained by mass selection of the ion of interest using the quadrupole, fragmentation in the trap cell where argon was used as collision gas. Typical collision energy (Trap) was 50.0 V. Samples were infused at a flow rate of 5 μL/min, if dilution of the sample was required then acetonitrile was used [100]. Chemical structures were determined from fragmentation schemes calculated on tandem mass spectra and from the literature.

4. Conclusions

The extract of the viscera of sea cucumber *H. lessoni* was processed by applying HPCPC to purify the saponin mixture and to isolate saponin congeners and isomeric saponins. The tandem MS approach enabled us to determine the structure of a range of saponins. The purity of HPCPC fractions allowed mass spectrometry analyses to reveal the structure of isomeric compounds containing different aglycones and/or sugar residues. Several novel saponins, along with known compounds, were identified from the viscera of sea cucumber.

This study is the first on saponins from the viscera of sea cucumbers. Our results to date highlight that there are a larger number of novel saponins in the viscera compared to the body wall (data not shown) indicating the viscera as a major source of these compounds. This paper is the first not only to report the presence of several novel saponins in the viscera of *H. lessoni* but also to indicate the highest number of saponin congeners detected in the viscera of any sea cucumber species. The mass of reported saponins for this species ranged from 460 Da to 1600 Da. So far we have identified more than ten aglycone structures in this species. Evidence from MALDI-MS suggested that the most intensive saponin ion was m/z 1243.5, a major component which seemed to correspond to Holothurin A. However, in the tandem MS, the most abundant ions are generally attributed to the loss of aglycones and/or both key diagnostic sugar moieties (507 and 523). Our results also showed that the incidence of the cross-ring cleavages was higher in the sulfated compounds compared to non-sulfated glycosides. It can be concluded that the presence of a sulfate group in the sugar moiety of saponins made them more vulnerable to cross-ring cleavages.

At the moment, MS is one of the most sensitive techniques of molecular analysis to determine saponin structures. This methodology of molecular structure identification using fragmentation patterns acquired from MS/MS measurements helps to propose and identify the structure of saponins. It was found that under CID some of the identified saponins had the same ion fingerprints for their aglycone units, yielding the same m/z daughter ions. Some of these saponins were easily characterized based on MS/MS measurement since their CID spectra contained the key diagnostic signals at m/z 507 and 523, corresponding to the oligosaccharide chains [MeGlc-Glc-Qui + Na$^+$] and [MeGlc-Glc-Glc + Na$^+$], respectively. The simultaneous loss of two sugar units indicated characteristics of a branched sugar chain. This methodology also permitted the structural elucidation of isomers.

Sea cucumbers have developed a chemical defense against potential predators based upon saponins. Our finding indicates that the viscera are rich in saponins, in both diversity and quantity, and that these saponins are apparently more localized in the viscera than in the body wall.

The chromatography techniques used in this study were able to for the first time, separate high purity saponins from sea cucumber, highlight the diversity of saponin congeners, and stress the unique profile of saponins for this species. MALDI and ESI-MS proved to be sensitive, ultra-high-throughput methodologies to identify these secondary metabolites in a complex mixture. Therefore, mass spectrometry has become the preferred techniques for analysis of saponins, as both ESI-MS and the MALDI-MS spectra provide remarkable structural information. However, the MALDI data is simpler to interpret compared to ESI-MS data due to the singly charged ions. This ancient creature with a long evolutionary history is a unique source of high-value novel compounds.

This manuscript describes the structure elucidation of seven novel compounds; Holothurinoside O, Holothurinoside P, Holothurinoside Q, Holothurinoside R, Holothurinoside R$_1$, Holothurinoside S and Holothurinoside T in addition to six known compounds, including Holothurin A, Holothurinoside A, Holothurinoside A$_1$, Holothurinoside E, Holothurinoside E$_1$ and 17-dehydroxy-holothurinoside A.

In conclusion, our findings show that the viscera of *H. lessoni* contain numerous unique and novel saponins with a high range of structural diversity, including both sulfated and non-sulfated congeners, and with different aglycone and sugar moieties. Furthermore, the tremendous range of structural biodiversity of this class of natural metabolites, which enables them to present in a remarkable functional diversity, is potentially an important source for the discovery of high-value compounds for biotechnological applications.

Acknowledgments

We would like to express our gratitude to the Australian SeaFood CRC for financially supporting this project, Ben Leahy for supplying the sea cucumber samples. The authors gratefully acknowledge the technical assistance provided by Daniel Jardine at Flinders Analytical Laboratory and Tim Chataway at Flinders Proteomics Facility.

Author Contributions

Y.B., C.F. and W.Z. designed the experiments. Y.B. carried out the experiments with guidance of C.F. and W.Z., who assisted in setting up the HCPCP analysis. Y.B., C.F. and W.Z. worked together on chemical structure elucidation, and all three authors contributed in writing the manuscript.

References

1. Lovatelli, A.; Conand, C. *Advances in Sea Cucumber Aquaculture and Management*; FAO: Rome, Italy, 2004.

2. Purcell, S.W.; Samyn, Y.; Conand, C. *Commercially Important Sea Cucumbers of the World*; FAO Species Catalogue for Fishery Purposes No. 6; FAO: Rome, Italy, 2012; p. 150.

3. Waller, G.R.; Yamasaki, K. *Saponins Used in Food and Agriculture*; Plenum Press: New York, NY, USA, 1996; Volume 405.

4. Hostettmann, K.; Marston, A. *Saponins*; Cambridge University Press: Cambridge, MA, USA, 1995.

5. Elbandy, M.; Rho, J.; Afifi, R. Analysis of saponins as bioactive zoochemicals from the marine functional food sea cucumber *Bohadschia cousteaui*. *Eur. Food Res. Technol.* **2014**, doi:10.1007/s00217-014-2171-6.

6. Caulier, G.; van Dyck, S.; Gerbaux, P.; Eeckhaut, I.; Flammang, P. Review of saponin diversity in sea cucumbers belonging to the family Holothuriidae. *SPC Beche-de-mer Inf. Bull.* **2011**, *31*, 48–54.

7. Dong, P.; Xue, C.; Du, Q. Separation of two main triterpene glycosides from sea cucumber *Pearsonothuria graeffei* by high-speed countercurrent chromatography. *Acta Chromatogr.* **2008**, *20*, 269–276.

8. Han, H.; Zhang, W.; Yi, Y.H.; Liu, B.S.; Pan, M.X.; Wang, X.H. A novel sulfated holostane glycoside from sea cucumber *Holothuria leucospilota*. *Chem. Biodivers.* **2010**, *7*, 1764–1769.

9. Naidu, A.S. *Natural Food Antimicrobial Systems*; CRC Press: New York, NY, USA, 2000.

10. Zhang, S.L.; Li, L.; Yi, Y.H.; Sun, P. Philinopsides E and F, two new sulfated triterpene glycosides from the sea cucumber *Pentacta quadrangularis*. *Nat. Prod. Res.* **2006**, *20*, 399–407.

11. Zhang, S.L.; Li, L.; Yi, Y.H.; Zou, Z.R.; Sun, P. Philinopgenin A, B, and C, three new triterpenoid aglycones from the sea cucumber *Pentacta quadrangulasis*. *Mar. Drugs* **2004**, *2*, 185–191.

12. Zhang, S.Y.; Yi, Y.H.; Tang, H.F.; Li, L.; Sun, P.; Wu, J. Two new bioactive triterpene glycosides from the sea cucumber *Pseudocolochirus violaceus*. *J. Asian Nat. Prod. Res.* **2006**, *8*, 1–8.

13. Chludil, H.D.; Muniain, C.C.; Seldes, A.M.; Maier, M.S. Cytotoxic and antifungal triterpene glycosides from the Patagonian sea cucumber *Hemoiedema spectabilis*. *J. Nat. Prod.* **2002**, *65*, 860–865.

14. Francis, G.; Kerem, Z.; Makkar, H.P.; Becker, K. The biological action of saponins in animal systems: A review. *Br. J. Nutr.* **2002**, *88*, 587–605.

15. Maier, M.S.; Roccatagliata, A.J.; Kuriss, A.; Chludil, H.; Seldes, A.M.; Pujol, C.A.; Damonte, E.B. Two new cytotoxic and virucidal trisulfated triterpene glycosides from the Antarctic sea cucumber *Staurocucumis liouvillei*. *J. Nat. Prod.* **2001**, *64*, 732–736.

16. Osbourn, A.; Goss, R.J.M.; Field, R.A. The saponins-polar isoprenoids with important and diverse biological activities. *Nat. Prod. Rep.* **2011**, *28*, 1261–1268.

17. Jha, R.K.; Zi-rong, X. Biomedical Compounds from Marine organisms. *Mar. Drugs* **2004**, *2*, 123–146.

18. Kalinin, V.I.; Aminin, D.L.; Avilov, S.A.; Silchenko, A.S.; Stonik, V.A. Triterpene glycosides from sea cucucmbers (Holothurioidea, Echinodermata). Biological activities and functions. In *Studies in Natural Products Chemistry*; Atta-ur, R., Ed.; Elsevier: Amsterdam, The Netherlands, 2008; Volume 35, pp. 135–196.

19. Liu, J.; Yang, X.; He, J.; Xia, M.; Xu, L.; Yang, S. Structure analysis of triterpene saponins in *Polygala tenuifolia* by electrospray ionization ion trap multiple-stage mass spectrometry. *J. Mass Spectrom.* **2007**, *42*, 861–873.

20. Kim, S.K.; Himaya, S.W.; Kang, K.H. Sea Cucumber Saponins Realization of Their Anticancer Effects. In *Marine Pharmacognosy: Trends and Applications*; Kim, S.K., Ed.; CRC Press: New York, NY, USA, 2012; pp. 119–128.

21. Mohammadizadeh, F.; Ehsanpor, M.; Afkhami, M.; Mokhlesi, A.; Khazaali, A.; Montazeri, S. Antibacterial, antifungal and cytotoxic effects of a sea cucumber *Holothuria leucospilota*, from the north coast of the Persian Gulf. *J. Mar. Biol. Assoc. UK* **2013**, *93*, 1401–1405.

22. Mohammadizadeh, F.; Ehsanpor, M.; Afkhami, M.; Mokhlesi, A.; Khazaali, A.; Montazeri, S. Evaluation of antibacterial, antifungal and cytotoxic effects of *Holothuria scabra* from the north coast of the Persian Gulf. *J. Med. Mycol.* **2013**, *23*, 225–229.

23. Mokhlesi, A.; Saeidnia, S.; Gohari, A.R.; Shahverdi, A.R.; Nasrolahi, A.; Farahani, F.; Khoshnood, R.; Es' haghi, N. Biological activities of the sea cucumber *Holothuria leucospilota*. *Asian J. Anim. Vet. Adv.* **2012**, *7*, 243–249.

24. Sarhadizadeh, N.; Afkhami, M.; Ehsanpour, M. Evaluation bioactivity of a sea cucumber, *Stichopus hermanni* from Persian Gulf. *Eur. J. Exp. Biol.* **2014**, *4*, 254–258.

25. Kim, S.K.; Himaya, S.W. Triterpene glycosides from sea cucumbers and their biological activities. *Adv. Food Nutr. Res.* **2012**, *65*, 297–319.

26. Yamanouchi, T. On the poisonous substance contained in holothurians. *Publ. Seto Mar. Biol. Lab.* **1955**, *4*, 183–203.

27. Avilov, S.A.; Drozdova, O.A.; Kalinin, V.I.; Kalinovsky, A.I.; Stonik, V.A.; Gudimova, E.N.; Riguera, R.; Jimenez, C. Frondoside C, a new nonholostane triterpene glycoside from the sea cucumber *Cucumaria frondosa*: Structure and cytotoxicity of its desulfated derivative. *Can. J. Chem.* **1998**, *76*, 137–141.

28. Girard, M.; Bélanger, J.; ApSimon, J.W.; Garneau, F.X.; Harvey, C.; Brisson, J.R.; Frondoside, A. A novel triterpene glycoside from the holothurian *Cucumaria frondosa*. *Can. J. Chem.* **1990**, *68*, 11–18.

29. Han, H.; Yi, Y.; Xu, Q.; La, M.; Zhang, H. Two new cytotoxic triterpene glycosides from the sea cucumber *Holothuria scabra*. *Planta Med.* **2009**, *75*, 1608–1612.

30. Kalinin, V.I.; Avilov, S.A.; Kalinina, E.Y.; Korolkova, O.G.; Kalinovsky, A.I.; Stonik, V.A.; Riguera, R.; Jimenez, C. Structure of eximisoside A, a novel triterpene glycoside from the Far-Eastern sea cucumber *Psolus eximius*. *J. Nat. Prod.* **1997**, *60*, 817–819.

31. Kitagawa, I.; Yamanaka, H.; Kobayashi, M.; Nishino, T.; Yosioka, I.; Sugawara, T. Saponin and sapogenol. XXVII. Revised structures of holotoxin A and holotoxin B, two antifungal oligoglycosides from the sea cucumber *Stichopus japonicus* Selenka. *Chem. Pharm. Bull. (Tokyo)* **1978**, *26*, 3722–3731.

32. Liu, B.S.; Yi, Y.H.; Li, L.; Sun, P.; Yuan, W.H.; Sun, G.Q.; Han, H.; Xue, M. Argusides B and C, two new cytotoxic triterpene glycosides from the sea cucumber *Bohadschia argus* Jaeger. *Chem. Biodivers.* **2008**, *5*, 1288–1297.

33. Miyamoto, T.; Togawa, K.; Higuchi, R.; Komori, T.; Sasaki, T. Structures of four new triterpenoid oligoglycosides: DS-penaustrosides A, B, C, and D from the sea cucumber *Pentacta australis*. *J. Nat. Prod.* **1992**, *55*, 940–946.

34. Campagnuolo, C.; Fattorusso, E.; Taglialatela-Scafati, O. Feroxosides A–B, two norlanostane tetraglycosides from the Caribbean sponge *Ectyoplasia ferox*. *Tetrahedron* **2001**, *57*, 4049–4055.

35. Thompson, J.; Walker, R.; Faulkner, D. Screening and bioassays for biologically-active substances from forty marine sponge species from San Diego, California, USA. *Mar. Biol.* **1985**, *88*, 11–21.

36. Dang, N.H.; Thanh, N.V.; Kiem, P.V.; Huong le, M.; Minh, C.V.; Kim, Y.H. Two new triterpene glycosides from the Vietnamese sea cucumber *Holothuria scabra*. *Arch. Pharm. Res.* **2007**, *30*, 1387–1391.

37. Kerr, R.G.; Chen, Z. *In vivo* and *in vitro* biosynthesis of saponins in sea cucumbers. *J. Nat. Prod.* **1995**, *58*, 172–176.

38. Chludil, H.D.; Murray, A.P.; Seldes, A.M.; Maier, M.S. Biologically active triterpene Glycosides from sea cucumbers (Holothuroidea, Echinodermata). In *Studies in Natural Products Chemistry*; Atta-ur, R., Ed.; Elsevier: Amsterdam, The Netherlands, 2003; Volume 28, Part I, pp. 587–615.

39. Habermehl, G.; Volkwein, G. Aglycones of the toxins from the Cuvierian organs of *Holothuria forskali* and a new nomenclature for the aglycones from Holothurioideae. *Toxicon* **1971**, *9*, 319–326.

40. Kalinin, V.I.; Silchenko, A.S.; Avilov, S.A.; Stonik, V.A.; Smirnov, A.V. Sea cucumbers triterpene glycosides, the recent progress in structural elucidation and chemotaxonomy. *Phytochem. Rev.* **2005**, *4*, 221–236.

41. Stonik, V.A.; Kalinin, V.I.; Avilov, S.A. Toxins from sea cucumbers (holothuroids): Chemical structures, properties, taxonomic distribution, biosynthesis and evolution. *J. Nat. Toxins* **1999**, *8*, 235–248.

42. Zhang, S.Y.; Tang, H.F.; Yi, Y.H. Cytotoxic triterpene glycosides from the sea cucumber *Pseudocolochirus violaceus*. *Fitoterapia* **2007**, *78*, 283–287.

43. Aminin, D.L.; Chaykina, E.L.; Agafonova, I.G.; Avilov, S.A.; Kalinin, V.I.; Stonik, V.A. Antitumor activity of the immunomodulatory lead Cumaside. *Int. Immunopharmacol.* **2010**, *10*, 648–654.

44. Antonov, A.S.; Avilov, S.A.; Kalinovsky, A.I.; Anastyuk, S.D.; Dmitrenok, P.S.; Evtushenko, E.V.; Kalinin, V.I.; Smirnov, A.V.; Taboada, S.; Ballesteros, M.; *et al.* Triterpene glycosides from antarctic sea cucumbers. 1. Structure of liouvillosides A_1, A_2, A_3, B_1, and B_2 from the sea cucumber *Staurocucumis liouvillei*: New procedure for separation of highly polar glycoside fractions and taxonomic revision. *J. Nat. Prod.* **2008**, *71*, 1677–1685.

45. Antonov, A.S.; Avilov, S.A.; Kalinovsky, A.I.; Dmitrenok, P.S.; Kalinin, V.I.; Taboada, S.; Ballesteros, M.; Avila, C. Triterpene glycosides from Antarctic sea cucumbers III. Structures of liouvillosides A_4 and A_5, two minor disulphated tetraosides containing 3-*O*-methylquinovose as terminal monosaccharide units from the sea cucumber *Staurocucumis liouvillei* (Vaney). *Nat. Prod. Res.* **2011**, *25*, 1324–1333.

46. Avilov, S.A.; Silchenko, A.S.; Antonov, A.S.; Kalinin, V.I.; Kalinovsky, A.I.; Smirnov, A.V.; Dmitrenok, P.S.; Evtushenko, E.V.; Fedorov, S.N.; Savina, A.S.; *et al.* Synaptosides A and A_1, triterpene glycosides from the sea cucumber *Synapta maculata* containing 3-*O*-methylglucuronic acid and their cytotoxic activity against tumor cells. *J. Nat. Prod.* **2008**, *71*, 525–531.

47. Iniguez-Martinez, A.M.D.M.; Guerra-Rivas, G.; Rios, T.; Quijano, L. Triterpenoid oligoglycosides from the sea cucumber *Stichopus parvimensis*. *J. Nat. Prod.* **2005**, *68*, 1669–1673.

48. Stonik, V.A.; Elyakov, G.B. Secondary metabolites from echinoderms as chemotaxonomic markers. *Bioorg. Mar. Chem.* **1988**, *2*, 43–86.

49. Bordbar, S.; Anwar, F.; Saari, N. High-value components and bioactives from sea cucumbers for functional foods–A review. *Mar. Drugs* **2011**, *9*, 1761–1805.

50. Van Dyck, S.; Gerbaux, P.; Flammang, P. Qualitative and quantitative saponin contents in five sea cucumbers from the Indian ocean. *Mar. Drugs* **2010**, *8*, 173–189.

51. Avilov, S.A.; Antonov, A.S.; Drozdova, O.A.; Kalinin, V.I.; Kalinovsky, A.I.; Stonik, V.A.; Riguera, R.; Lenis, L.A.; Jiménez, C. Triterpene glycosides from the far-eastern sea cucumber *Pentamera calcigera*. 1. Monosulfated glycosides and cytotoxicity of their unsulfated derivatives. *J. Nat. Prod.* **2000**, *63*, 65–71.

52. Avilov, S.A.; Kalinovsky, A.I.; Kalinin, V.I.; Stonik, V.A.; Riguera, R.; Jiménez, C. Koreoside A, a new nonholostane triterpene glycoside from the sea cucumber *Cucumaria koraiensis*. *J. Nat. Prod.* **1997**, *60*, 808–810.

53. Avilov, S.A.; Antonov, A.S.; Drozdova, O.A.; Kalinin, V.I.; Kalinovsky, A.I.; Riguera, R.; Lenis, L.A.; Jimenez, C. Triterpene glycosides from the far eastern sea cucumber *Pentamera calcigera* II: Disulfated glycosides. *J. Nat. Prod.* **2000**, *63*, 1349–1355.

54. Avilov, S.A.; Antonov, A.S.; Silchenko, A.S.; Kalinin, V.I.; Kalinovsky, A.I.; Dmitrenok, P.S.; Stonik, V.A.; Riguera, R.; Jimenez, C. Triterpene glycosides from the far eastern sea cucumber *Cucumaria conicospermium*. *J. Nat. Prod.* **2003**, *66*, 910–916.

55. Jia, L.; Qian, K. An Evidence-Based Perspective of *Panax Ginseng* (Asian Ginseng) and *Panax Quinquefolius* (American Ginseng) as a Preventing or Supplementary Therapy for Cancer Patients. In *Evidence-Based Anticancer Materia Medica*; Springer Verlag: New York, NY, USA, 2011; pp. 85–96.

56. Zhang, S.Y.; Yi, Y.H.; Tang, H.F. Bioactive triterpene glycosides from the sea cucumber *Holothuria fuscocinerea. J. Nat. Prod.* **2006**, *69*, 1492–1495.

57. Zhang, S.Y.; Yi, Y.H.; Tang, H.F. Cytotoxic sulfated triterpene glycosides from the sea cucumber *Pseudocolochirus violaceus. Chem. Biodivers.* **2006**, *3*, 807–817.

58. Caulier, G.; Flammang, P.; Gerbaux, P.; Eeckhaut, I. When a repellent becomes an attractant: Harmful saponins are kairomones attracting the symbiotic *Harlequin crab. Sci. Rep.* **2013**, *3*, doi:10.1038/srep02639.

59. Mercier, A.; Sims, D.W.; Hamel, J.F. Advances in Marine Biology: Endogenous and Exogenous Control of Gametogenesis and Spawning in Echinoderms; Academic Press: New York, NY, USA, 2009; Volume 55.

60. Matsuno, T.; Ishida, T. Distribution and seasonal variation of toxic principles of sea-cucumber (*Holothuria leucospilota*; Brandt). *Cell. Mol. Life Sci.* **1969**, *25*, doi:10.1007/BF01897485.

61. Kobayashi, M.; Hori, M.; Kan, K.; Yasuzawa, T.; Matsui, M.; Suzuki, S.; Kitagawa, I. Marine natural products. XXVII: Distribution of lanostane-type triterpene oligoglycosides in ten kinds of Okinawan Sea cucumbers. *Chem. Pharm. Bull. (Tokyo)* **1991**, *39*, 2282–2287.

62. Van Dyck, S.; Flammang, P.; Meriaux, C.; Bonnel, D.; Salzet, M.; Fournier, I.; Wisztorski, M. Localization of secondary metabolites in marine invertebrates: contribution of MALDI MSI for the study of saponins in Cuvierian tubules of *H. forskali. PLoS One* **2010**, *5*, e13923.

63. Bakus, G.J. Defensive mechanisms and ecology of some tropical holothurians. *Mar. Biol.* **1968**, *2*, 23–32.

64. Bondoc, K.G.V.; Lee, H.; Cruz, L.J.; Lebrilla, C.B.; Juinio-Meñez, M.A. Chemical fingerprinting and phylogenetic mapping of saponin congeners from three tropical holothurian sea cucumbers. *Comp. Biochem. Physiol. B Biochem. Mol. Biol.* **2013**, *166*, 182–193.

65. Van Dyck, S.; Caulier, G.; Todesco, M.; Gerbaux, P.; Fournier, I.; Wisztorski, M.; Flammang, P. The triterpene glycosides of *Holothuria forskali*: Usefulness and efficiency as a chemical defense mechanism against predatory fish. *J. Exp. Biol.* **2011**, *214*, 1347–1356.

66. Kalyani, G.A.; Kakrani, H.K.N.; Hukkeri, V.I. Holothurin—A Review. *Indian J. Nat. Prod.* **1988**, *4*, 3–8.

67. Kalinin, V.; Anisimov, M.; Prokofieva, N.; Avilov, S.; Afiyatullov, S.S.; Stonik, V. Biological activities and biological role of triterpene glycosides from holothuroids (Echinodermata). *Echinoderm Stud.* **1996**, *5*, 139–181.

68. Kalinin, V.I.; Prokofieva, N.G.; Likhatskaya, G.N.; Schentsova, E.B.; Agafonova, I.G.; Avilov, S.A.; Drozdova, O.A. Hemolytic activities of triterpene glycosides from the holothurian order Dendrochirotida: Some trends in the evolution of this group of toxins. *Toxicon* **1996**, *34*, 475–483.

69. Van Dyck, S.; Gerbaux, P.; Flammang, P. Elucidation of molecular diversity and body distribution of saponins in the sea cucumber *Holothuria forskali* (Echinodermata) by mass spectrometry. *Comp. Biochem. Physiol. B Biochem. Mol. Biol.* **2009**, *152*, 124–134.

70. Elyakov, G.B.; Stonik, V.A.; Levina, E.V.; Slanke, V.P.; Kuznetsova, T.A.; Levin, V.S. Glycosides of marine invertebrates—I. A comparative study of the glycoside fractions of pacific sea cucumbers. *Comp. Biochem. Physiol. B Comp. Biochem.* **1973**, *44*, 325–336.

71. Cai, Z.; Liu, S.; Asakawa, D. *Applications of MALDI-TOF Spectroscopy*; Springer: Berlin, Germany, 2013.

72. Du, Q.; Jerz, G.; Waibel, R.; Winterhalter, P. Isolation of dammarane saponins from *Panax notoginseng* by high-speed counter-current chromatography. *J. Chromatogr.* **2003**, *1008*, 173–180.

73. Cui, M.; Song, F.; Zhou, Y.; Liu, Z.; Liu, S. Rapid identification of saponins in plant extracts by electrospray ionization multi-stage tandem mass spectrometry and liquid chromatography/tandem mass spectrometry. *Rapid Commun. Mass Spectrom.* **2000**, *14*, 1280–1286.

74. Schöpke, T.; Thiele, H.; Wray, V.; Nimtz, M.; Hiller, K. Structure elucidation of a glycoside of 2β, 3β, t23-trihydroxy-16-oxoolean-12-en-28-oic acid from *Bellis bernardii* using mass spectrometry for the sugar sequence determination. *J. Nat. Prod.* **1995**, *58*, 152–155.

75. Bankefors, J.; Broberg, S.; Nord, L.I.; Kenne, L. Electrospray ionization ion-trap multiple-stage mass spectrometry of Quillaja saponins. *J. Mass Spectrom.* **2011**, *46*, 658–665.

76. Liu, S.; Cui, M.; Liu, Z.; Song, F.; Mo, W. Structural analysis of saponins from medicinal herbs using electrospray ionization tandem mass spectrometry. *J. Am. Soc. Mass Spectrom.* **2004**, *15*, 133–141.

77. Wang, X.; Sakuma, T.; Asafu-Adjaye, E.; Shiu, G.K. Determination of ginsenosides in plant extracts from *Panax ginseng* and *Panax quinquefolius* L. by LC/MS/MS. *Anal. Chem.* **1999**, *71*, 1579–1584.

78. Wolfender, J.L.; Rodriguez, S.; Hostettmann, K. Liquid chromatography coupled to mass spectrometry and nuclear magnetic resonance spectroscopy for the screening of plant constituents. *J. Chromatogr.* **1998**, *794*, 299–316.

79. Zheng, Z.; Zhang, W.; Kong, L.; Liang, M.; Li, H.; Lin, M.; Liu, R.; Zhang, C. Rapid identification of C_{21} steroidal saponins in *Cynanchum versicolor* Bunge by electrospray ionization multi-stage tandem mass spectrometry and liquid chromatography/tandem mass spectrometry. *Rapid Commun. Mass Spectrom.* **2007**, *21*, 279–285.

80. Cui, M.; Song, F.; Liu, Z.; Liu, S. Metal ion adducts in the structural analysis of ginsenosides by electrospray ionization with multi-stage mass spectrometry. *Rapid Commun. Mass Spectrom.* **2001**, *15*, 586–595.

81. Fang, S.; Hao, C.; Sun, W.; Liu, Z.; Liu, S. Rapid analysis of steroidal saponin mixture using electrospray ionization mass spectrometry combined with sequential tandem mass spectrometry. *Rapid Commun. Mass Spectrom.* **1998**, *12*, 589–594.

82. Li, L. *MALDI Mass Spectrometry for Synthetic Polymer Analysis*; Wiley & Sons: Hoboken, NJ, USA, 2009; Volume 175.

83. Silchenko, A.S.; Stonik, V.A.; Avilov, S.A.; Kalinin, V.I.; Kalinovsky, A.I.; Zaharenko, A.M.; Smirnov, A.V.; Mollo, E.; Cimino, G. Holothurins B₂, B₃, and B₄, new triterpene glycosides from mediterranean sea cucumbers of the genus *holothuria*. *J. Nat. Prod.* **2005**, *68*, 564–567.

84. Han, H.; Yi, Y.H.; Li, L.; Wang, X.H.; Liu, B.S.; Sun, P.; Pan, M.X. A new triterpene glycoside from sea cucumber *Holothuria leucospilota*. *Chin. Chem. Lett.* **2007**, *18*, 161–164.

85. Kitagawa, I.; Nishino, T.; Matsuno, T.; Akutsu, H.; Kyogoku, Y. Structure of holothurin B a pharmacologically active triterpene-oligoglycoside from the sea cucumber *Holothuria leucospilota* Brandt. *Tetrahedron Lett.* **1978**, *19*, 985–988.

86. Han, H.; Yi, Y.H.; Liu, B.S.; Wang, X.H.; Pan, M.X. Leucospilotaside C, a new sulfated triterpene glycoside from sea cucumber *Holothuria leucospilota*. *Chin. Chem. Lett.* **2008**, *19*, 1462–1464.

87. Wu, J.; Yi, Y.H.; Tang, H.F.; Wu, H.M.; Zou, Z.R.; Lin, H.W. Nobilisides A–C, three new triterpene glycosides from the sea cucumber *Holothuria nobilis*. *Planta Med.* **2006**, *72*, 932–935.

88. Rodriguez, J.; Castro, R.; Riguera, R. Holothurinosides: New antitumour non sulphated triterpenoid glycosides from the sea cucumber *Holothuria forskalii*. *Tetrahedron* **1991**, *47*, 4753–4762.

89. Kitagawa, I.; Kobayashi, M.; Kyogoku, Y. Marine natural products. IX. Structural elucidation of triterpenoidal oligoglycosides from the Bahamean sea cucumber *Actinopyga agassizi* Selenka. *Chem. Pharm. Bull. (Tokyo)* **1982**, *30*, 2045–2050.

90. Liu, B.S.; Yi, Y.H.; Li, L.; Sun, P.; Han, H.; Sun, G.Q.; Wang, X.H.; Wang, Z.L. Argusides D and E, two new cytotoxic triterpene glycosides from the sea cucumber *Bohadschia argus* Jaeger. *Chem. Biodivers.* **2008**, *5*, 1425–1433.

91. Han, H.; Yi, Y.H.; Li, L.; Liu, B.S.; La, M.P.; Zhang, H.W. Antifungal active triterpene glycosides from sea cucumber *Holothuria scabra*. *Acta Pharm. Sin.* **2009**, *44*, 620–624.

92. Han, H.; Li, L.; Yi, Y.-H.; Wang, X.-H.; Pan, M.-X. Triterpene glycosides from sea cucumber *Holothuria scabra* with cytotoxic activity. *Chin. Herbal Med.* **2012**, *4*, 183–188.

93. Kalinin, V.; Stonik, V. Glycosides of marine invertebrates. Structure of Holothurin A₂ from the holothurian *Holothuria edulis*. *Chem. Nat. Compd.* **1982**, *18*, 196–200.

94. Kitagawa, I.; Kobayashi, M.; Inamoto, T.; Fuchida, M.; Kyogoku, Y. Marine natural products. XIV. Structures of echinosides A and B, antifungal lanostane-oligosides from the sea cucumber *Actinopyga echinites* (Jaeger). *Chem. Pharm. Bull. (Tokyo)* **1985**, *33*, 5214–5224.

95. Thanh, N.V.; Dang, N.H.; Kiem, P.V.; Cuong, N.X.; Huong, H.T.; Minh, C.V. A new triterpene glycoside from the sea cucumber *Holothuria scabra* collected in Vietnam. *ASEAN J. Sci. Technol. Dev.* **2006**, *23*, 253–259.

96. Kitagawa, I.; Nishino, T.; Kyogoku, Y. Structure of holothurin A a biologically active triterpene-oligoglycoside from the sea cucumber *Holothuria leucospilota* Brandt. *Tetrahedron Lett.* **1979**, *20*, 1419–1422.

97. Yuan, W.; Yi, Y.; Tang, H.; Xue, M.; Wang, Z.; Sun, G.; Zhang, W.; Liu, B.; Li, L.; Sun, P. Two new holostan-type triterpene glycosides from the sea cucumber *Bohadschia marmorata* JAEGER. *Chem. Pharm. Bull. (Tokyo)* **2008**, *56*, 1207–1211.

98. Yuan, W.H.; Yi, Y.H.; Tan, R.X.; Wang, Z.L.; Sun, G.Q.; Xue, M.; Zhang, H.W.; Tang, H.F. Antifungal triterpene glycosides from the sea cucumber *Holothuria (Microthele) axiloga*. *Planta Med.* **2009**, *75*, 647–653.

99. Sun, G.Q.; Li, L.; Yi, Y.H.; Yuan, W.H.; Liu, B.S.; Weng, Y.Y.; Zhang, S.L.; Sun, P.; Wang, Z.L. Two new cytotoxic nonsulfated pentasaccharide holostane (=20-hydroxylanostan-18-oic acid γ-lactone) glycosides from the sea cucumber *Holothuria grisea*. *Helv. Chim. Acta* **2008**, *91*, 1453–1460.

100. Song, F.; Cui, M.; Liu, Z.; Yu, B.; Liu, S. Multiple-stage tandem mass spectrometry for differentiation of isomeric saponins. *Rapid Commun. Mass Spectrom.* **2004**, *18*, 2241–2248.

101. Van Setten, D.C.; Jan ten Hove, G.; Wiertz, E.J.H.J.; Kamerling, J.P.; van de Werken, G. Multiple-stage tandem mass spectrometry for structural characterization of saponins. *Anal. Chem.* **1998**, *70*, 4401–4409.

102. Garneau, F.X.; Simard, J.; Harvey, O.; ApSimon, J.; Girard, M. The structure of psoluthurin A, the major triterpene glycoside of the sea cucumber *Psolus fabricii*. *Can. J. Chem.* **1983**, *61*, 1465–1471.

103. Grassia, A.; Bruno, I.; Debitus, C.; Marzocco, S.; Pinto, A.; Gomez-Paloma, L.; Riccio, R. Spongidepsin, a new cytotoxic macrolide from Spongia sp. *Tetrahedron* **2001**, *57*, 6257–6260.

104. Kupchan, S.M.; Britton, R.W.; Ziegler, M.F.; Sigel, C.W. Bruceantin, a new potent antileukemic simaroubolide from *Brucea antidysenterica*. *J. Org. Chem.* **1973**, *38*, 178–179.

11

Lipids and Fatty Acids of Nudibranch Mollusks: Potential Sources of Bioactive Compounds

Natalia V. Zhukova [1,2]

[1] Institute of Marine Biology, Far East Branch, Russian Academy of Sciences,
Vladivostok 690041, Russia;
E-Mail: nzhukova35@list.ru

[2] School of Biomedicine, Far Eastern Federal University, Vladivostok 690950, Russia

Abstract: The molecular diversity of chemical compounds found in marine animals offers a good chance for the discovery of novel bioactive compounds of unique structures and diverse biological activities. Nudibranch mollusks, which are not protected by a shell and produce chemicals for various ecological uses, including defense against predators, have attracted great interest for their lipid composition. Lipid analysis of eight nudibranch species revealed dominant phospholipids, sterols and monoalkyldiacylglycerols. Among polar lipids, 1-alkenyl-2-acyl glycerophospholipids (plasmalogens) and ceramide-aminoethyl phosphonates were found in the mollusks. The fatty acid compositions of the nudibranchs differed greatly from those of other marine gastropods and exhibited a wide diversity: very long chain fatty acids known as demospongic acids, a series of non-methylene-interrupted fatty acids, including unusual $21:2\Delta7,13$, and an abundance of various odd and branched fatty acids typical of bacteria. Symbiotic bacteria revealed in some species of nudibranchs participate presumably in the production of some compounds serving as a chemical defense for the mollusks. The unique fatty acid composition of the nudibranchs is determined by food supply, inherent biosynthetic activities and intracellular symbiotic microorganisms. The potential of nudibranchs as a source of biologically active lipids and fatty acids is also discussed.

Keywords: mollusks; symbiotic bacteria; fatty acids; phospholipids

1. Introduction

The molecular diversity of chemical compounds found in marine animals is the result of the evolution of the organisms and their unique physiological and biochemical adaptations and offers a good chance for the discovery of novel bioactive compounds with a variety of unique structures and diverse biological activities [1]. Marine mollusks have become the focus of many chemical studies aimed at isolating and identifying novel natural products [2]. Phylum Mollusca is the second largest phylum of animals. Nudibranch mollusks, which often are very colorful, are not protected by a shell and are named sea slugs, have attracted strong interest for their secondary metabolites, which are active in chemical defenses against predators [3]. These compounds exhibit a large variety of chemical structures [4,5] and have been shown to possess ichthyotoxic, feeding-deterrent and cytotoxic properties, to have antibacterial activity, to act as sexual pheromones [6] and are responsible for various bioactivities, such as antitumor, anti-inflammatory and antioxidant activities. Clearly, dietary sources contribute significantly to the chemical diversity of metabolites found in some mollusks [6]. However, their *de novo* biosynthesis has been reported for several mollusk species [7]. The secondary metabolites isolated from mollusks fall into a wide range of structural classes, with some compounds predominating in certain taxa. In the Gastropoda, terpenes dominate, whereas fatty acid derivatives are relatively uncommon [2].

Mollusks, as well as the invertebrates, in general, constitute a source of lipid bioactive compounds offering a variety of nutraceutical and pharmaceutical applications [2]. Among them, the omega-3 polyunsaturated fatty acids (PUFA), such as eicosapentaenoic acid, 20:5n-3, and docosahexaenoic acid, 22:6n-3, are known for their beneficial effects on human health [8]. These PUFA n-3 fatty acids are widely known for their capacities for cardioprotection; they reduce triacylglycerol and cholesterol levels and have anti-inflammatory and anticancer effects [9]. Numerous experiments on animals confirmed the cancer preventive properties of PUFA n-3 fatty acids from marine sources [9,10].

Some other marine lipids also show many potential bioactive properties. Monogalactosyldiacylglycerols and digalactosyldiacylglycerols from the marine microalga, *Nannochloropsis granulata*, have been reported to have a nitric oxide inhibitory activity [11]. The betaine lipid from microalgae *N. granulata*, diacylglyceryltrimethylhomoserine, shows a nitric oxide inhibitory activity, indicating a possible value as an anti-inflammatory agent [12]. The glycolipid, sulfoquinovosyl diacylglycerol, from red alga *Osmundaria obtusiloba* [13] and from brown alga *Sargassum vulgare* [14] exhibits a potent antiviral activity against herpes simplex virus type 1 and 2. This glycolipid from a brown alga, *Lobophora variegata*, possess a pronounced antiprotozoal activity [15]. Studies on glycosphingolipids from marine sponge *Axinyssa djiferi* proved their good antiplasmodial activity [16].

Although interest in the fatty acid composition of mollusks has not been abated, it has become increasingly obvious that phyla of marine invertebrates may be a source of unusual marine lipids, such as plasmalogens, phospholipids, glycolipids and diverse fatty acids.

The aim of the work was to fill a gap in the knowledge of the lipid biochemistry of mollusks. In particular, we consider data on the lipid of the nudibranchs (Mollusca, Gastropoda, Opisthobranchia, Nudibranchia). Herein, we report the investigation of the eight common species of nudibranchs with the use of the high-performance thin-layer chromatography (HPTLC), gas chromatography coupled with flame ionization detection (GC-FID) and gas chromatography coupled with mass spectrometry

(GC-MS) methods to elucidate their lipid, phospholipid and fatty acid composition. A suggestion on the origin of the fatty acid variety in nudibranchs and their potential as bioactive compounds is also given.

2. Results and Discussion

2.1. Lipids and Phospholipids

Lipids exert important biological functions as energy storage compounds, structural components of the cell membranes and as signaling molecules. The lipid content of the nudibranchs accounts for 14.2–21.4 mg·g^{-1} wet weight. The eight studied species of the nudibranchs appeared to have similar lipid compositions. According to this similarity in the lipid classes of these nudibranchs, the amounts of the lipid classes insignificantly vary depending on species and environmental conditions. Statistical analysis confirmed that lipid class values differed insignificantly among species. Hence, Figure 1 gives the average results for all studied species. The lipid composition of nudibranchs revealed that the major lipid class was phospholipids (PLs) and, to a lesser extent, sterols (STs) (13.5%–16.1% of total lipids). The PL concentration varied within a range from 73.8% in *Chromodoris geometrica* to 81.7% in *Glossodoris cincta*; this was much more than was found in other mollusks and invertebrates in total [17]. Triacylglycerols (TAGs), monoalkyldiacylglycerols (MADAGs) and free fatty acids (FFAs), which are the storage compounds of the cells, were minor components (2.6%, 3.4% and 2.6%, respectively). The detected distribution was similar to that found in two other tropical species of nudibranchs [18] and confirmed the high membrane phospholipids and low storage lipids in the tissues of these mollusks. The level of the neutral storage lipids is known to be species specific and depends mainly on the life history strategy and food availability [19].

Figure 1. Lipid classes (% of total lipids) of nudibranchs. Results are expressed as the mean of eighth species (*n* = 8). PL, phospholipid; ST, sterol; FFA, free fatty acid; TAG, triacylglycerol; MADAG, monoalkyldiacylglycerol.

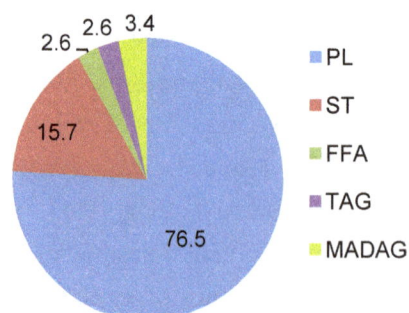

The PL composition of the studied species was similar, with the dominance of phosphatidylcholine (PC) (up to 62.8% of total PL in *Chromodoris tinctoria*) and further, in descending order: phosphatidylethanolamine (PE), phosphatidylserine (PS), ceramide-aminoethylphosphonate (CAEP), phosphatidylinositol (PI) and diphosphatidylglycerol (DPG) (Table 1). The PL data obtained for the nudibranchs were different from those of other mollusks species by the elevated concentration of PC.

The phosphonolipid, CAEP, is relatively abundant in some invertebrates, and it has been detected previously in freshwater and marine mollusks [20].

Table 1. Composition of phospholipids in the nudibranchs (mol%). Results are expressed as the mean ± SD of four replicates ($n = 4$). PC, phosphatidylcholine; PE, phosphatidylethanolamine; PS, phosphatidylserine; CAEP, ceramide-aminoethylphosphonate; PI, phosphatidylinositol; DPG, diphosphatidylglycerol.

	PC	PE	PS	CAEP	PI	DPG
Chromodoris tinctoria	60.9 ± 2.1	11.7 ± 0.9	12.5 ± 1.1	5.1 ± 0.4	6.5 ± 0.8	2.0 ± 0.6
C. michaeli	53.1 ± 2.1	21.4 ± 1.1	13.4 ± 0.8	5.6 ± 0.5	4.9 ± 0.6	1.7 ± 0.4
C. geometrica	53.8 ± 1.5	15.4 ± 0.9	12.6 ± 0.7	12.2 ± 1.1	3.7 ± 1.1	1.2 ± 0.3
Chromodoris sp.	51.2 ± 1.1	17.4 ± 1.3	14.5 ± 1.2	9.1 ± 1.9	5.0 ± 0.6	1.8 ± 0.4
Glossodoris cincta	56.1 ± 0.6	16.4 ± 1.4	15.4 ± 1.3	5.1 ± 0.7	4.2 ± 0.6	1.8 ± 0.3
G. atromarginata	53.5 ± 2.3	18.2 ± 1.9	12.2 ± 1.4	9.9 ± 1.7	5.1 ± 0.6	1.1 ± 0.2
Risbecia tryoni	49.6 ± 0.4	18.2 ± 1.1	13.8 ± 0.8	10.6 ± 1.1	4.6 ± 1.0	3.2 ± 0.8
Platydoris sp.	50.9 ± 2.0	21.2 ± 1.5	10.2 ± 0.8	8.0 ± 1.7	5.4 ± 0.5	2.7 ± 0.5

Marine invertebrates are known as a rich source of 1-alkenyl-2-acyl glycerophospholipids, commonly called plasmalogens [17,21]. Plasmalogens are particular phospholipids characterized by the presence of a vinyl ether bond at the C1 position of the glycerol skeleton. Plasmalogens are also ubiquitously found in animal cells. In mammals, the brain, heart, lymphocytes, spleen, macrophages and polymorphonuclear leukocytes contain the highest amount of plasmalogen-ethanolamine [22]. Two PLs, PE and PS, were represented as diacyl- and alkenyl-forms, and more than half of these aminophospholipids were plasmalogens (Table 2). 1-Alkenyl-2-acyl-PE made up 50.3%–65.1% of total PE; and 1-alkenyl-2-acyl-PS reached 47.1%–61.3% of total PS. The highest percentage of PE plasmalogens was found in *Risbecia tryoni*, accounting for 65.1% of total PE, and the PS plasmalogen contribution reached 61.3% of total PS in *Platydoris* sp. In contrast to many marine and freshwater mollusks, the nudibranchs contained PC only as a diacyl-form. Earlier, plasmalogens have been detected in PE, PS and PC in common edible mollusk species; the PE fraction is very often composed predominantly of the plasmalogens [23,24].

Table 2. Content of plasmalogens (1-alkenyl-2-acyl glycerophospholipids) in the nudibranchs. Expressed as the proportion of the plasmalogen forms relative to the whole of the same class forms; mean ± SD of four replicates ($n = 4$).

	1-Alkenyl-2-acyl-PE	1-Alkenyl-2-acyl-PS
Chromodoris tinctoria	60.4 ± 1.6	47.1 ± 2.1
Chromodoris michaeli	52.9 ± 2.4	51.1 ± 2.4
Chromodoris geometrica	58.8 ± 2.1	50.6 ± 1.7
Chromodoris sp.	59.2 ± 1.8	48.7 ± 1.1
Glossodoris cincta	61.1 ± 1.1	47.8 ± 1.5
Glossodoris atromarginata	60.7 ± 2.1	48.6 ± 1.2
Risbecia tryoni	65.1 ± 1.6	56.5 ± 1.0
Platydoris sp.	50.3 ± 1.8	61.3 ± 1.5

Serving as a structural component of the mammalian and invertebrate cell membrane, plasmalogens are widely distributed in excitable tissues, like heart and brain. Plasmalogens mediate the dynamics of the cell membrane. They provide storage for polyunsaturated fatty acids and can contribute to endogenous antioxidant activity, thus protecting cells from oxidative stress [25]. Plasmalogen phospholipids are suggested to be involved in signal transduction [26]. Plasmalogens are not only components of the plasma membrane and of lung surfactant, they serve as a reservoir for secondary messengers and may be also involved in membrane fusion, ion transport and cholesterol efflux. Low levels of these metabolites have trophic effects, but at a high concentration, they are cytotoxic and may be involved in allergic response, inflammation and trauma. Decreased levels of plasmalogens are associated with several neurological disorders, including Alzheimer's disease, ischemia and spinal cord trauma [27].

2.2. Fatty Acids

The fatty acid profiles of the studied species were rather similar and differed only in their qualitative proportions of the fatty acids. Figure 2 shows the GC-MS chromatogram of the 4,4-dimethyloxazoline (DMOX) derivatives from the sea slug, *Chromodoris michaeli*. Table 3 reports the qualitative and quantitative data obtained, respectively, from GC-MS and GC-FID analyses. The components were eluted according their chain length and the degree of unsaturation in the chain on the MDN-5S capillary column. The chromatographic analyses allowed us to detect and identify about 50 individual fatty acids. The nudibranchs exhibited a wide diversity of fatty acids, including common saturated fatty acids (SFA) (8.6%–16.5% of total fatty acids), monounsaturated fatty acids (MUFA) (22.7%–31.2%) and polyunsaturated fatty acids (PUFA) (15.1%–31.4%), as well as non-methylene-interrupted dienoic fatty acids (NMID FA) (8.0%–21.5%), very long chain fatty acids (VLCFAs) (7.7%–16.6%) and odd-chain and branched fatty acids (5.0%–17.4%) (Figure 3).

Figure 2. GC-MS chromatogram of 4,4-dimethyloxazoline derivatives from *Chromodoris michaeli*.

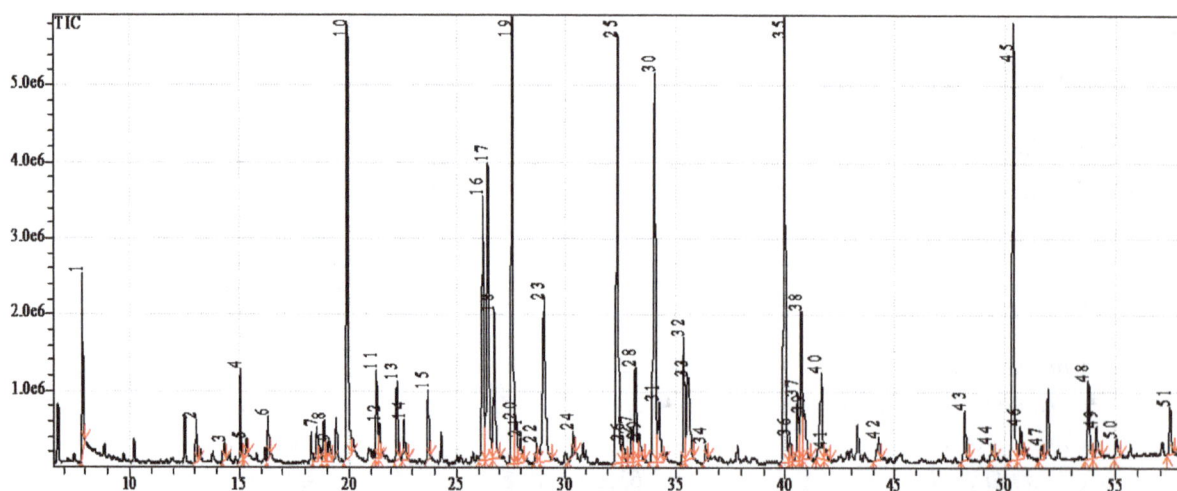

Marine mollusks are generally characterized by the predominance of essential *n*-3 PUFA, mainly 20:5*n*-3 and 22:6*n*-3, which constitute usually almost half of the total fatty acids [17,28]. In contrast, the nudibranchs did not show this property; these two marine PUFA were minor components and constituted in sum about 2% of the total fatty acids (range 1.4%–7.3%) (Figure 4). Nevertheless, sea slugs exhibited some unique features in their fatty acid composition. Their fatty acid profiles were distinguished drastically from those of other mollusks. The differences seem to be more obvious compared with fatty acids of a common marine snail, *Nucella heyseana*, and limpet, *Acmea pallida* [28] (Figure 4).

Table 3. Identification of the 4,4-dimethyloxazoline derivatives and fatty acid composition (wt%) of *Chromodoris michaeli*. Results are expressed as the mean ± SD of four replicates (*n* = 4).

FA	Molecular Ion (*m/z*)	% of Total FA	FA	Molecular Ion (*m/z*)	% of Total FA
12:0	253	0.4 ± 0.1	20:5*n*-3	355	0.2 ± 0.1
14:0	281	0.9 ± 0.3	20:2Δ5,11	355	1.8 ± 0.5
iso-15:0	295	1.9 ± 0.5	20:2Δ5,13	355	1.3 ± 0.4
anteiso-15:0	295	0.3 ± 0.1	20:3*n*-6	359	0.7 ± 0.3
15:0	295	1.1 ± 0.2	20:1*n*-11	363	5.6 ± 0.7
iso-16:0	309	0.5 ± 0.1	20:1*n*-9	363	0.2 ±0.1
anteiso-16:0	309	0.7 ± 0.1	20:1*n*-7	363	2.5 ± 0.6
16:1*n*-7	307	1.7 ± 0.6	21:2Δ7,13	375	0.1 ± 0.0
16:0	309	5.9 ± 0.8	*iso*-21:1	377	0.4 ± 0.1
iso-17:0	323	1.4 ± 0.3	21:1*n*-7	377	2.0 ± 0.1
anteiso-17:0	323	0.9 ± 0.1	21:1*n*-5	377	1.3 ± 0.5
17:1*n*-8	321	1.4 ± 0.4	22:5*n*-6	383	0.4 ± 0.1
17:1*n*-6	321	0.4 ± 0.1	22:6*n*-3	381	0.7 ± 0.2
17:0	323	1.4 ±0.1	22:4*n*-6	385	10.2 ± 1.3
18:3*n*-6	331	0.2 ± 0.1	22:5*n*-3	383	0.2 ± 0.1
iso-18:0	337	0.4 ± 0.2	22:3*n*-6	387	0.2 ± 0.1
anteiso-18:0	337	0.4 ± 0.2	22:2Δ7,13	389	3.6 ± 0.7
18:2*n*-6	333	7.0 ± 0.9	22:2Δ7,15	289	1.2 ± 0.3
18:1*n*-9	335	5.4 ± 0.4	22:1*n*-9	391	0.4 ± 0.2
18:1*n*-7	335	3.4 ± 1.0	22:1*n*-7	391	0.1 ± 0.1
18:0	337	7.4 ± 1.3	*iso*-24:2Δ5,9	417	0.2 ± 0.1
iso-19:1	349	0.1 ± 0.0	24:2Δ5,9	417	3.1 ± 0.5
anteiso-19:1	349	0.2 ± 0.1	*iso*-25:2Δ5,9	431	4.0 ± 1.0
iso-19:0	351	0.2 ± 0.1	*anteiso*-25:2Δ5,9	431	0.6 ± 0.2
anteiso-19:0	351	0.2 ± 0.1	25:2Δ5,9	431	0.1 ± 0.1
19:1*n*-8	349	0.1 ± 0.0	*iso*-26:2Δ5,9	445	0.8 ± 0.3
19:1*n*-12	349	2.6 ± 0.8	*anteiso*-26:2Δ5,9	445	0.1 ± 0.0
19:0	351	0.3 ± 0.1	26:2Δ5,9	445	0.8 ± 0.3
20:4*n*-6	357	10.5 ± 1.2	*anteiso*-27:2Δ5,9	459	0.2 ± 0.1

Figure 3. Distribution of fatty acids in nudibranch species. Results are expressed as the mean ± SD of eight studied species. OBFA, odd-chain and branched; NMID, non-methylene-interrupted dienoic; VLCFA, very long chain fatty acids.

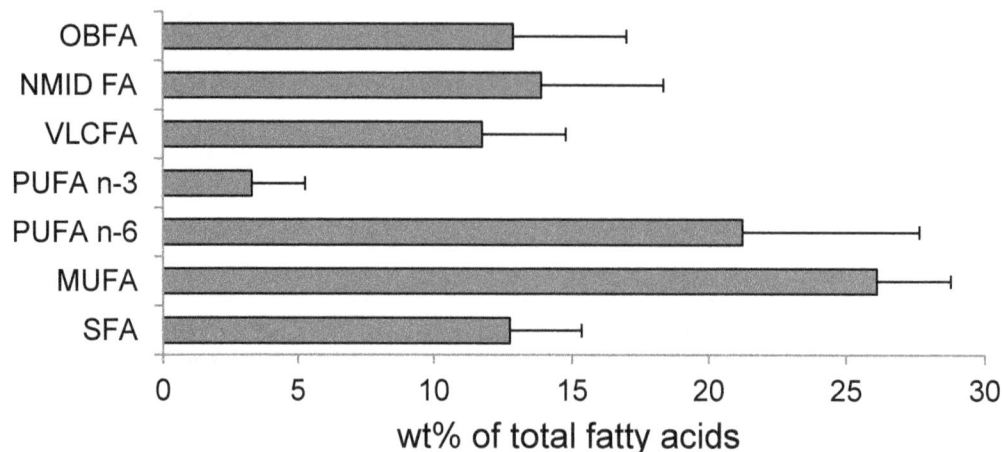

Figure 4. Fatty acid composition (wt%) of nudibranchs. Results are expressed as the mean ± SD of eight studied species.

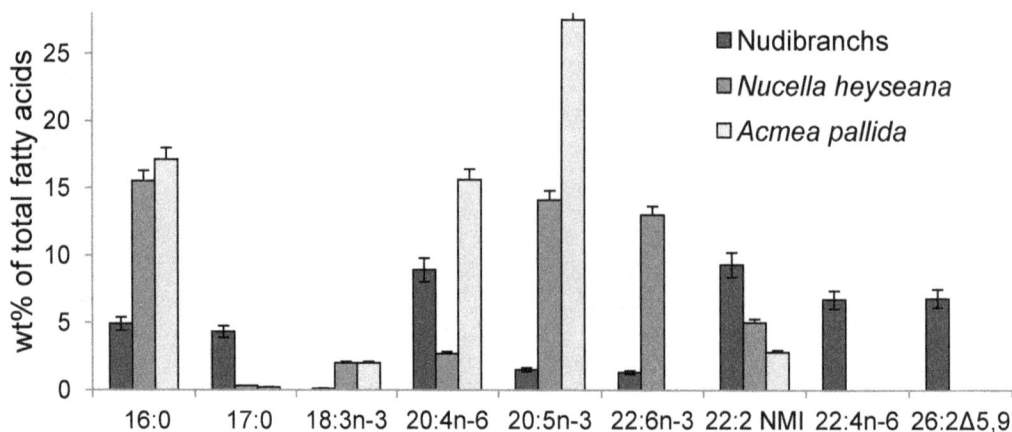

In the nudibranchs, a significant amount of VLCFA specific to sponges, so-called demospongic acids, was found. These nudibranchs are carnivorous and specialized feeders on sponges. Utilization of this food probably is responsible for the high level of the demospongic acids in these mollusks. It is suspected that the majority of sea slugs feed on certain sponge species, which are known to be distinguished in their fatty acid composition [29]. Indeed, a series of VLCFA with double bonds at $\Delta 5,9$ positions in the chain was identified in the tropical nudibranchs (Table 4). Concentrations of these components differed among the species. Among VLCFA of *Platydoris* sp., only hexacosadienoic acid $26:2\Delta 5,9$ was identified, whereas in *Chromodoris michaeli*, tetracosadienoic $24:2\Delta 5,9$ and branched *iso*-$25:2\Delta 5,9$ were dominant, with some minor VLCFA. Moreover, branched hexacosatrienoic acids, *iso*-$26:3\Delta 5,9,19$ and *anteiso*-$26:3\Delta 5,9,19$, were found only in *Glossodoris cincta*, and *iso*-$27:2\Delta 5,9$ was identified only in *C. michaeli*. The specific distribution of the VLCFA suggests that these nudibranchs may feed on different sponge species.

Table 4. The distribution of very long-chain fatty acids of nudibrans according to the degree of unsaturation and chain length (% of total fatty acids). Results are expressed as the mean ± SD of four replicates ($n = 4$).

VLCFA	*Chromodoris* sp.	*C. geometrica*	*C. tinctoria*	*C. michaeli*	*Glossodoris atromarginata*	*G. cincta*	*Risbecia tryoni*	*Platydoris* sp.
i-24:2Δ5,9	-	-	-	0.2	0.2	-	-	-
24:2Δ5,9	1.0	4.7	0.4	3.1	1.1	1.2	1.4	-
24:1	-	0.3	0.1	-	-	-	-	-
i-25:2Δ5,9	-	-	-	4.0	-	-	-	-
25:2Δ5,9	4.0	1.6	0.7	0.2	2.3	2.6	0.5	-
25:3Δ5,9	0.5	0.1	-	-	-	1.3	-	-
26:0	-	0.1	-	-	-	0.3	-	-
i-26:2Δ5,9	0.8	0.1	0.8	0.8	-	0.7	-	-
26:2Δ5,9	6.0	3.0	13.6	0.8	6.3	8.4	5.8	10.5
i-26:3Δ5,9,19	-	0.2	-	-	1.2	2.3	-	-
ai-26:3Δ5,9,19	-	-	-	-	-	1.1	-	-
ai-27:2Δ5,9	-	-	-	0.2	-	-	-	-

Thus, this study demonstrated that sponges are not the only source of these Δ5,9 dienoic acids, since they were found also in other marine organisms, such as nudibranch mollusks. Various biological activities have been reported to date for the most encountered VLCFA double bonds at Δ5,9 positions in the chain. The use of an antiplasmodial bioassay revealed that fatty acids with 23 to 26 carbon atoms and double bonds in the position Δ5,9 displayed considerable antiprotozoal activity [30]. Thus, these demospongic fatty acids may be the source of very potent antimalarial drugs. These fatty acids are also potent inhibitors of the enzyme, topoisomerase I; this property could lead to the development of effective anti-cancer drugs. The 14-methyl-5,9-pentadecadienoic acid from phospholipids of the gorgonian *Eunicea succinea* was active against Gram-positive bacteria, such as *Staphylococcus aureus* and *Streptococcus faecalis* [31]. The natural compound, 30:3Δ5,9,23, was isolated from the sponge, *Chondrilla nucula*, and was found to be an elastase inhibitor, which is known to be a potential therapeutic agent in various diseases, such as pulmonary emphysema, chronic bronchitis and several inflammatory disorders. In addition, the C23–C26Δ5,9 fatty acids had almost no cytotoxicity on mammalian L6 cells. Therefore, the Δ5,9 FA may be of use against the parasites without damage to the host [32]. Mixtures of the branched 22-methyl-5,9-tetracosadienoic and 23-methyl-5,9-tetracosadienoic acids showed cytotoxic activity against mouse Ehrlich carcinoma cells and a hemolytic effect on mouse erythrocytes [33].

As referenced above, marine mollusks are probably a unique source of unusual unsaturated fatty acids, the non-methylene-interrupted dienoic fatty acids (NMID FA), as opposed to the common methylene-interrupted PUFA, in that their double bonds are separated by more than one methylene group. These ubiquitous and, in some species, major components of the mollusk lipids [28] have been, to date, extensively studied [34]. The NMID FAs were found in all studied species (Figures 3 and 4); among them, common 20:2Δ5,11, 20:2Δ5,13, 22:2Δ7,13 and 22:2Δ7,15 and a novel isomer, 21:2Δ7,13, were identified. FA 21:2Δ7,13 has been reported earlier in other nudibranch species, *Phyllidia coelestis* [18], in the edible bivalve, *Megangulus zyonoensis* [35], and the gonads of the limpets, *Cellana grata* and *Collisella dorsuosa* [36]. The largest concentration of NMID FA was detected in *Risbecia tryoni* (21.5%). The mollusks are able to synthesize the C20 and C22 NMID FA by a $\Delta5$ desaturase acting upon the appropriate precursor, such as 18:1n-7 and 18:1n-9, and further chain elongation. The potential precursors of 21:2$\Delta7,13$ are 17:1n-8 and 19:1n-8, which are of bacterial origin and abundant in the nudibranchs (Table 3). It has been recently shown that mollusks expressed a Fad-like gene that encodes an enzyme with $\Delta5$-desaturation activity, which participates in the biosynthesis of NMID FA [37]. Although their biological role and function is not fully understood, it has been suggested that NMID FAs play structural and protective roles in cell membranes, since they are esterified phospholipids and occur in amounts that are often in a reverse relation to 20:5n-5 and 22:6n-3 [18,34]. The unusual double bond positions in NMID FAs are considered to confer to cell membranes a higher resistance to oxidative processes and microbial lipases than the common PUFA [34]. The introduction of a *trans*-5 double bond into the linoleic acid, 18:2n-6, molecule provides a fatty acid, columbinic acid, 18:3$\Delta5$*trans*,9*cis*,12*cis*, with fascinating biological properties [38]. In experiments with essential fatty acid-deficient rats, it has been shown that columbinic acid is effective in maintaining the proper epidermal layer and improves the fertility of the rats, while the inhibition of prostaglandin synthesis has a beneficial effect, since inflammation and the thrombotic tendency are reduced.

Another unique feature of the nudibranchs is associated with an aberrant level of the odd-chain and branched fatty acids (OBFA) that are specific for bacteria and usually named "bacterial fatty acids" (Figure 3). They are normally minor metabolites in most animals, but a high abundance of bacterial acids found in the nudibranchs was extraordinary. The sum of OBFA, predominantly 15:0, 17:0, 17:1n-8 and *iso*- and *anteiso*-C15, C16, C17, C18 and C19 fatty acids, reached up to 15.8% in the *Chromodoris geometrica* and 17.4% in *Glossodoris atromarginata*, whereas their concentration in *Platydoris* sp. was the lowest (5.0%). A high level of bacterial fatty acids in the nudibranchs may serve as an indicator that the symbiotic bacteria provide the host with nutrients. Earlier, an abundance of OBFA discovered in *Dendrodoris nigra* allowed us to suggest that symbiotic bacteria may be their source in this nudibranch; transmission electron microscopy (TEM) confirmed the presence of symbiotic bacteria in the cytoplasm of the epithelial cells and the glycocalyx layer covering the epithelium of the notum and the mantle of *D. nigra* [39]. It was found that the bacteria in the glycocalyx sometimes undergo destructive lysis, with their components being utilized by the epithelial cells. The high concentration of typical bacterial fatty acids in the lipids of the nudibranch *D. nigra* agrees well with the results of TEM and confirms that the lysed bacterial cells are utilized by the mollusk tissues [39]. Moreover, some bacterial OBFA, such as 17:1n-8 and 19:1n-8, evidently serve as potential precursors for the biosynthesis of odd-chain PUFA identified in the nudibranchs, such as 21:2$\Delta7,13$, as well as 21:4n-7 isolated from other marine opisthobranch mollusk, *Scaphander lignarius*, and possessed activity against a range of human cancer cell lines (melanoma, colon carcinoma and breast carcinoma) [40].

There is increasing evidence that microbial symbionts are the true source of biologically-active compounds isolated from some species of chemically-rich invertebrates, mainly sponges, bryozoans, isopods and tunicates [41,42]. The symbionts are reported to be producers of the host's secondary metabolites that have defensive and protective functions for their hosts [43,44]. Many biologically active compounds, including toxic and deterrent secretions, have been isolated from nudibranchs [45].

Interestingly, endobacterial morphotypes have been recently described for twelve of thirteen species of nudibranchs tested [46]. Moreover, the epithelium of the temperate nudibranch, *Rostanga alisae*, rich in OBFA, demonstrated high numbers of symbiont-containing cells (*i.e.*, bacteriocytes) [47]. Taken all together, these results suggest that symbiotic bacteria might be involved in the defense against predators and, so, in production of the bioactive compounds.

3. Experimental Section

3.1. Site and Samples

Eight species of nudibranchs, *Chromodoris* sp., *C. geometrica*, *C. tinctoria*, *C. michaeli*, *Glossodoris atromarginata*, *G. cincta*, *Risbecia tryoni* and *Platydoris* sp. (phyla: Mollusca, Class: Opisthobranchia; orders: Nudibranchia, Suborder: Doridina) were collected from the Research Vessel Akademik Oparin by SCUBA divers in Nha Trang Bay of the South China Sea, Vietnam, in January 2005, October 2006, June 2007, and April–May 2013. The nudibranchs collected were placed immediately in tanks under water at the site of collection and transported to the laboratory. Three to five specimens of each species were used for lipid analysis.

3.2. Lipid Analysis

Tissues of mollusks were crushed, and total lipids were extracted by homogenization in a chloroform/methanol mixture (1:2, v/v) [48]. Lipid classes were separated by one-dimensional silica gel thin-layer chromatography (TLC). The Merk Kieselgel 60 G plates (6 cm × 6 cm) were first developed in hexane/diethyl ether/acetic acid (80:20:1, v/v) to resolve nonpolar compounds. After development, the TLC plates were dried under air flow and developed to 20% length in a polar solvent system of chloroform/acetone/methanol/acetic acid/water (50:20:10:10:1, v/v). Lipids were detected on the TLC plates using 10% H_2SO_4/methanol with subsequent heating to 180 °C. The TLC plates were scanned using an image scanner (Epson Perfection 2400 Photo) in grayscale mode. Lipid class concentrations were based on band intensity using an image analysis program (Sorbifil TLC Videodensitometer). Units were calibrated using standards for each lipid class.

Polar lipids were separated by two-dimensional silica gel TLC in the solvent systems: chloroform/methanol/28% NH_4OH, 65:25:4, v/v, for the first direction; chloroform/acetone/methanol/ acetic acid/water, 50:20:10:10:1, v/v, for the second one. Lipids were detected on TLC pales using 10% H_2SO_4/methanol with heating to 180 °C and by specific reagents for phospholipids [49], amino-containing lipids (0.5% ninhydrin in waterlogged butanol) and choline lipids (Dragendorff's reagent). Phospholipids were quantified with the molybdenum reagent [49].

3.3. Fatty Acid Analysis

Fatty acid methyl esters (FAME) were prepared by a sequential treatment of the total lipids with 1% sodium methylate/methanol and 5% HCl/methanol in a screw-capped vial [50] and purified by preparative silica gel TLC using benzene as a solvent. 4,4-Dimethloxazoline (DMOX) derivatives were prepared from FAME [51]. The GC analysis of FAME was carried out on a Shimadzu GC-2010 chromatograph (Kyoto, Japan) with a flame ionization detector on a SUPELCOWAX (Supelco,

Bellefonte, PA, USA) capillary column (30 m × 0.25-mm internal diameter, 0.25-μm film thickness) at 210 °C. Helium was used as a carrier gas at a linear velocity of 30 cm s^{-1} (the split ratio was 1:30). Injector and detector temperatures were 250 °C. Fatty acids were identified by a comparison with authentic standards and equivalent chain length values (ECL) [52]. Identification of fatty acids was confirmed by gas chromatography-mass spectrometry (GC-MS) of their methyl esters and DMOX derivatives. The GC-MS analysis of FAME was performed on a model Shimadzu GCMS-QP5050A (Kyoto, Japan) fitted with a Supelco MDN-5S capillary column (30 m × 0.25 mm i.d. Supelco, Bellefonte, PA, USA). Ionization of the samples was performed by an electron impact at 70 eV. The column temperature was programmed from 170 °C, held for 1 min, followed by an increase to 240 °C at a rate of 2 °C min^{-1} and then held for 20 min. The temperature of the injector and detector was 250 °C. GC-MS of DMOX derivatives was performed using the same instrument at a column temperature of 210 °C with a 3 °C min^{-1} increase to 270 °C, which was held for 40 min. The injector and detector temperatures were 300 °C. Spectra were compared with the NIST library and fatty acid mass spectra archive [53].

3.4. Statistical Analysis

Difference in the mean of lipid concentrations was examined with a one-way ANOVA. In all cases, statistical significance was indicated by $p < 0.05$. All data were expressed as mean ± SD.

4. Conclusions

Mollusks, as well as the invertebrates in general, constitute a source of lipid bioactive compounds offering a variety of activities. This study has demonstrated for the first time that nudibranchs exhibit a wide diversity of lipids that differed greatly from that of other marine gastropods. Lipids of nudibranchs were composed mainly of phospholipids rich in plasmalogen PE and plasmalogen PS. The nudibranchs exhibited some unique features in their fatty acid composition. They displayed large amounts of VLCFA, various NMID FAs and a high abundance of OBFA. Many of these fatty acids originate in nudibranchs from unusual biosynthetic pathways, specific dietary sources and symbiotic partnerships with bacteria. The results of this study and of previous research suggest that symbiotic bacteria may play an important role in producing bioactive chemicals or their precursors within the host. The current study has shown that these mollusks may be an important resource of a wide range of bioactive compounds.

Acknowledgments

I thank Alexey V. Chernyshev for the identification of the nudibranch species. This work was supported by the Government of the Russian Federation (Grant 11.G34.31.0010).

Abbreviations

CAEP: ceramide-aminoethylphosphonate; DPG: diphosphatidylglycerol; FA: fatty acid; FAME: fatty acid methyl ester; FFA: free fatty acids; GC-MS: gas chromatography-mass spectrometry; MADAG: monoalkyldiacylglycerol; MUFA: monounsaturated fatty acid; NMID FA:

non-methylene-interrupted dienoic fatty acids; OBFA: odd-chain and branched fatty acids; PC: phosphatidylcholine; PE: phosphatidylethanolamine; PI: phosphatidylinositol; PL: phospholipids; PS: phosphatidylserine; PUFA: polyunsaturated fatty acid; SFA: saturated fatty acid; ST: sterols; TLC: thin-layer chromatography; VLCFA: very long chain fatty acid.

References

1. Evans-Illidge, E.A.; Logan, M.; Doyle, J.; Fromont, J.; Battershill, C.N.; Ericson, G.; Wolff, C.W.; Muirhead, A.; Kearns, P.; Abdo, D.; *et al.* Phylogeny drives large scale patterns in australian marine bioactivity and provides a new chemical ecology rationale for future biodiscovery. *PLoS One* **2013**, *8*, e73800.

2. Benkendorff, K. Molluscan biological and chemical diversity: Secondary metabolites and medicinal resources produced by marine molluscs. *Biol. Rev.* **2010**, *85*, 757–775.

3. Cimino, G.; Gavagnin, M. *Molluscs: Progress in Molecular and Subcellular Biology Subseries Marine Molecular Biochemistry*; Springer-Verlag: Berlin, Heidelburg, Germany, 2006; p. 387.

4. Faulkner, D.J. Marine natural products. *Nat. Prod. Rep.* **2000**, *17*, 7–55.

5. Gavagnin, M.; Fontana, A. Diterpenes from marine opisthobranch mollusks. *Curr. Org. Chem.* **2000**, *4*, 1201–1248.

6. Avila, C. Natural products of opisthobranch molluscs: A biological review. *Oceanogr. Mar. Biol.* **1995**, *33*, 487–559.

7. Fontana, A. Biogenetical proposals and biosynthetic studies on secondary metabolites of opisthobranch molluscs. In *Molluscs: Progress in Molecular and Subcellular Biology Subseries Marine Molecular Biochemistry*; Cimino, G., Gavagnin, M., Eds.; Springer-Verlag: Berlin, Heidelburg, Germany, 2006; pp. 303–328.

8. Simopoulos, A.P. The importance of the omega-6/omega-3 fatty acid ratio in cardiovascular disease and other chronic diseases. *Exp. Biol. Med.* **2008**, *233*, 674–688.

9. Wendel, M.; Heller, A.R. Anticancer actions of omega-3 fatty acids—Current state and future perspectives. *Anticancer Agents Med. Chem.* **2009**, *9*, 457–470.

10. Candela, C.G.; Lopez, L.; Kohen, V. Importance of a balanced omega 6/omega 3 ratio for the maintenance of health: Nutritional recommendations. *Nutr. Hosp.* **2011**, *26*, 323–329.

11. Banskota, A.H.; Stefanova, R.; Gallant, P.; McGinn, P. Mono- and digalactosyldiacylglycerols: Potent nitric oxide inhibitors from the marine microalga *Nannochloropsis granulata*. *J. Appl. Phycol.* **2013**, *25*, 349–357.

12. Banskota, A.H.; Stefanova, R.; Sperker, S.; McGinn, P.J. New diacylglyceryltrimethyl-homoserines from the marine microalga *Nannochloropsis granulata* and their nitric oxide inhibitory activity. *J. Appl. Phycol.* **2013**, *25*, 1513–1521.

13. De Souza, L.M.; Sassaki, G.L.; Villela, R.; Maria, T.; Barreto-Bergter, E. Structural Characterization and anti-HSV-1 and HSV-2 activity of glycolipids from the marine algae *Osmundaria obtusiloba* isolated from Southeastern Brazilian coast. *Mar. Drugs* **2012**, *10*, 918–931.

14. Plouguerne, E.; de Souza, L.M.; Sassaki, G.L.; Cavalcanti, J.F.; Romanos, M.T.V.; da Gama, B.A.P.; Pereira, R.C.; Barreto-Bergter, E. Antiviral Sulfoquinovosyldiacylglycerols (SQDGs) from the Brazilian brown seaweed *Sargassum vulgare*. *Mar. Drugs* **2013**, *11*, 4628–4640.

15. Cantillo-Ciau, Z.; Moo-Puc, R.; Quijano, L.; Freile-Pelegrin, Y. The tropical brown alga *Lobophora variegata*: A source of antiprotozoal compounds. *Mar. Drugs* **2013**, *8*, 1292–1304.

16. Farokhi, F.; Grellier, P.; Clement, M.; Roussakis, C.; Loiseau, P.M.; Genin-Seward, E.; Kornprobst, J.-M.; Barnathan, G.; Wielgosz-Collin, G. Antimalarial activity of axidjiferosides, new beta-galactosylceramides from the African sponge *Axinyssa djiferi*. *Mar. Drugs* **2013**, *11*, 1304–1315.

17. Joseph, J.D. Lipid composition of marine and estuarine invertebrates: Mollusca. *Prog. Lipid. Res.* **1982**, *21*, 109–153.

18. Zhukova, N.V. Lipid classes and fatty acid composition of the tropical nudibranch mollusks *Chomodoris* sp. and *Phyllidia coelestis*. *Lipids* **2007**, *42*, 1169–1175.

19. Gannefors, C.; Boer, M.; Kattner, G.; Graeve, M.; Eiane, K.; Gulliksen, B.; Hop, H.; Falk-Petersen, S. The Arctic sea butterfly *Limacina helicina*: Lipids and life strategy. *Mar. Biol.* **2005**, *147*, 169–177.

20. Hanuš, L.O.; Levitsky, D.O.; Shkrob, I.; Dembitsky, V.M. Plasmalogens, fatty acids and alkyl glyceryl ethers of marine and freshwater clams and mussels. *Food Chem.* **2009**, *116*, 491–498.

21. Sargent, J.R. Ether-Linked Glycerides in Marine Animals. In *Marine Biogenic Lipids, Fats, and Oils*; Ackman, R.G., Ed.; CRC Press: Boca Raton, FL, USA, 1987; pp. 176–193.

22. Magnusson, C.D.; Haraldsson, G.G. Ether lipids. *Chem. Phys. Lipids* **2011**, *164*, 315–340.

23. Dembitsky, V.M.; Rezanka, T.; Kashin, A.G. Fatty acid and phospholipid composition of freshwater molluscs *Anodonta piscinalis* and *Limnaea fragilis* from the river Volga. *Comp. Biochem. Physiol.* **1993**, *105B*, 597–601.

24. Kraffe, E.; Sounant, P.; Marty, A. Fatty acids of Serine, ethanolamine, and choline plasmalogens in some marine bivalves. *Lipids* **2004**, *39*, 59–66.

25. Brosche, T.; Brueckmann, M.; Haase, K.K.; Sieber, C.; Bertsch, T. Decreased plasmalogen concentration as a surrogate marker of oxidative stress in patients presenting with acute coronary syndromes or supraventricular tachycardias. *Clin. Chem. Lab. Med.* **2007**, *45*, 689–691.

26. Latorre, E.; Collado, M.P.; Fernandez, I.; Aragones, M.D.; Catalan, R.E. Signaling events mediating activation of brain ethanolamine plasmalogen hydrolysis by ceramide. *Eur. J. Biochem.* **2003**, *270*, 36–46.

27. Hartmann, T.; Kuchenbecker, J.; Grimm, M.O. Alzheimer's disease: The lipid connection. *J. Neurochem.* **2007**, *103*, 159–170.

28. Zhukova, N.V.; Svetashev, V.I. Non-methylene-interrupted dienoic fatty acids in mollusks from the Sea of Japan. *Comp. Biochem. Physiol.* **1986**, *83B*, 643–646.

29. Bergquist, P.R.; Lawson, M.P.; Lavis, A.; Cambie, R.C. Fatty acid composition and the classification of the Porifera. *Biochem. Syst. Ecol.* **1984**, *12*, 63–84.

30. Carballeira, N.M. New advances in fatty acids as antimalarial, antimycobacterial and antifungal agents. *Prog. Lipid Res.* **2008**, *47*, 50–61.

31. Carballeira, N.M.; Reyes, E.D.; Sostre, A.; Rodriguez, A.D.; Rodriguez, J.L.; Gonzales, F.A. Identification of the novel antimicrobial fatty acid (5Z,9Z)-14-methyl-5,9-pentadecadienoic acid in *Eunicea succinea. J. Nat. Prod.* **1997**, *60*, 502–504.

32. Tasdemir, D.; Topaloglu, B.; Perozzo, R.; Brun, R.; O'Neill, R.; Carballeira, N.M.; Zhang, X.J.; Tonge, P.J.; Linden, A.; Ruedi, P. Marine natural products from the Turkish sponge *Agelas oroides* that inhibit the enoyl reductase from *Plasmodium falciparum, Mycobacterium tuberculosis* and *Escherichia coli. Bioorg. Med. Chem.* **2007**, *15*, 6834–6845.

33. Makarieva, T.N.; Santalova, E.A.; Gorshkova, I.A.; Dmitrenok, A.S.; Guzi, A.G.; Gorbach, V.I.; Svetashev, V.I.; Stonok, V.A. A new cytotoxic fatty acid (5Z,9Z)-22-methyl-5,9-tetrecosadienoic acid and the sterols from the far eastern sponge *Geodinella robusta. Lipids* **2002**, *37*, 75–80.

34. Barnathan, G. Non-methylene-interrupted fatty acids from marine invertebrates: Occurrence, characterization and biological properties. *Biochimie* **2009**, *91*, 671–678.

35. Kawashima, H.; Ohnishi, M. Identification of minor fatty acids and various nonmethylene-interrupted diene isomers in mantle, muscle, and viscera of the marine bivalve *Megangulus zyonoensis. Lipids* **2004**, *39*, 265–271.

36. Kawashima, H. Unusual minor nonmethylene-interrupted di-, tri-, and tetraenoic fatty acids in limpet gonads. *Lipids* **2005**, *40*, 627–630.

37. Monroig, O.; Navarro, J.C.; Dick, J.R.; Alemany, F.; Tocher, D.R. Identification of a Δ5-like fatty acyl desaturase from the cephalopod *Octopus vulgaris* (Cuvier 1797) involved in the biosynthesis of essential fatty acids. *Mar. Biotechnol.* **2012**, *14*, 411–422.

38. Houtsmuller, U.M.T. Columbinic acid, a new type of essential fatty acid. *Prog. Lipid Res.* **1981**, *20*, 889–896.

39. Zhukova, N.V.; Eliseikina, M.G. Symbiotic bacteria in the nudibranch mollusk *Dendrodoris nigra*: Fatty acid composition and ultrastructure analysis. *Mar. Biol.* **2012**, *159*, 1783–1794.

40. Vasskog, T.; Andersen, J.H.; Hansen, E.; Svenson, J. Characterization and cytotoxicity studies of the rare 21:4n-7 acid and other polyunsaturated fatty acids from the marine opisthobranch *Scaphander lignaris*, isolated using bioassay guided fractionation. *Mar. Drugs* **2012**, *10*, 2676–2690.

41. Proksch, P.; Edrada, R.A.; Ebel, R. Drugs from the seas—Current status and microbiological implications. *Appl. Microbiol. Biot.* **2002**, *59*, 125–134.

42. Schmidt, E.W.; Donia, M.S. Life in cellulose houses: Symbiotic bacterial biosynthesis of ascidian drugs and drug leads. *Curr. Opin. Biotechnol.* **2010**, *21*, 827–833.

43. Lindquist, N.; Barber, P.H.; Weisz, J.B. Episymbiotic microbes as food and defence for marine isopods, unique symbioses in a hostile environment. *Proc. Biol. Sci.* **2005**, *272*, 1209–1216.

44. Haine, E.R. Symbiont-mediated protection. *Proc. Biol. Sci.* **2008**, *275*, 353–361.

45. Wagele, H.; Ballesteros, M.; Avila, C. Defensive glandular structures in Opisthobranch mollusks—From histology to ecology. *Oceanogr. Mar. Biol.* **2006**, *44*, 197–276.

46. Schuett, C.; Doepke, H. Endobacterial morphotypes in nudibranch cerata tips: A SEM analysis. *Helgol. Mar. Res.* **2013**, *67*, 219–227.

47. Zhukova, N.V.; Eliseikina, M.G.; Balakirev, E.S.; Ayala, F.J. A novel symbiotic association of the nudibranch mollusk *Rostanga alisae* with bacteria. Unpublished work, 2014.

48. Bligh, E.G.; Dyer, W.J. A rapid method of total lipid extraction and purification. *Can. J. Biochem. Physiol.* **1959**, *37*, 911–918.

49. Vaskovsky, V.E.; Kostetsky, E.Y.; Vasendin, I.M. A universal reagent for phospholipids analysis. *J. Chromatogr.* **1975**, *114*, 129–142.

50. Carreau, J.P.; Dubacq, J.P. Adaptation of the macroscale method to the micro-scale for fatty acid methyl transesterification of biological lipid extracts. *J. Chromatogr.* **1979**, *151*, 384–390.

51. Svetashev, V.I. Mild method for preparation of 4,4-dimethyloxazoline derivatives of polyunsaturated fatty acids for GC-MS. *Lipids* **2011**, *46*, 463–467.

52. Christie, W.W. Equivalent chain lengths of methyl ester derivatives of fatty acids on gas chromatography—A reappraisal. *J. Chromatogr.* **1988**, *447*, 305–314.

53. Christie, W.W. 2013, Mass Spectrometry of Fatty Acid Derivatives. Available online: http://lipidlibrary.aocs.org/ms/masspec.html (accessed on 21 January 2014).

Permissions

All chapters in this book were first published in MDPI; hereby published with permission under the Creative Commons Attribution License or equivalent. Every chapter published in this book has been scrutinized by our experts. Their significance has been extensively debated. The topics covered herein carry significant findings which will fuel the growth of the discipline. They may even be implemented as practical applications or may be referred to as a beginning point for another development.

The contributors of this book come from diverse backgrounds, making this book a truly international effort. This book will bring forth new frontiers with its revolutionizing research information and detailed analysis of the nascent developments around the world.

We would like to thank all the contributing authors for lending their expertise to make the book truly unique. They have played a crucial role in the development of this book. Without their invaluable contributions this book wouldn't have been possible. They have made vital efforts to compile up to date information on the varied aspects of this subject to make this book a valuable addition to the collection of many professionals and students.

This book was conceptualized with the vision of imparting up-to-date information and advanced data in this field. To ensure the same, a matchless editorial board was set up. Every individual on the board went through rigorous rounds of assessment to prove their worth. After which they invested a large part of their time researching and compiling the most relevant data for our readers.

The editorial board has been involved in producing this book since its inception. They have spent rigorous hours researching and exploring the diverse topics which have resulted in the successful publishing of this book. They have passed on their knowledge of decades through this book. To expedite this challenging task, the publisher supported the team at every step. A small team of assistant editors was also appointed to further simplify the editing procedure and attain best results for the readers.

Apart from the editorial board, the designing team has also invested a significant amount of their time in understanding the subject and creating the most relevant covers. They scrutinized every image to scout for the most suitable representation of the subject and create an appropriate cover for the book.

The publishing team has been an ardent support to the editorial, designing and production team. Their endless efforts to recruit the best for this project, has resulted in the accomplishment of this book. They are a veteran in the field of academics and their pool of knowledge is as vast as their experience in printing. Their expertise and guidance has proved useful at every step. Their uncompromising quality standards have made this book an exceptional effort. Their encouragement from time to time has been an inspiration for everyone.

The publisher and the editorial board hope that this book will prove to be a valuable piece of knowledge for researchers, students, practitioners and scholars across the globe.

List of Contributors

Alane Beatriz Vermelho
BIOINOVAR—Biotechnology laboratories: Biocatalysis, Bioproducts and Bioenergy, Institute of Microbiology Paulo de Góes, Federal University of Rio de Janeiro, Av. Carlos Chagas Filho, 373, 21941-902 Rio de Janeiro, Brazil

Gabriel Zamith Leal Dalmaso
BIOINOVAR—Biotechnology laboratories: Biocatalysis, Bioproducts and Bioenergy, Institute of Microbiology Paulo de Góes, Federal University of Rio de Janeiro, Av. Carlos Chagas Filho, 373, 21941-902 Rio de Janeiro, Brazil
Graduate Program in Plant Biotechnology, Health and Science Centre, Federal University of Rio de Janeiro, Av. Carlos Chagas Filho, 373, 21941-902 Rio de Janeiro, Brazil

Davis Ferreira
BIOINOVAR—Biotechnology Laboratories: Virus-Cell Interaction, Institute of Microbiology Paulo de Góes, Federal University of Rio de Janeiro, Av. Carlos Chagas Filho, 373, 21941-902 Rio de Janeiro, Brazil

Nathan J. Kenny
School of Life Sciences, The Chinese University of Hong Kong, Shatin, Hong Kong, China
Department of Zoology, University of Oxford, Oxford OX1 3PS, UK

Yung Wa Sin, Xin Shen, Qu Zhe, Wei Wang, Ting Fung Chan, Ka Hou Chu and Jerome H. L. Hui
School of Life Sciences, The Chinese University of Hong Kong, Shatin, Hong Kong, China

Sebastian M. Shimeld
Department of Zoology, University of Oxford, Oxford OX1 3PS, UK

Stephen S. Tobe
Department of Cell and Systems Biology, University of Toronto, Toronto M5S 3G5, Canada

Evelyne Benoit and Jordi Molgó
Neurobiology and Development Laboratory, Research Unit # 3294, Institute of Neurobiology Alfred Fessard # 2118, National Center for Scientific Research, Gif sur Yvette Cedex 91198, France

Bao Nguyen and Hung Lamthanh
Neurobiology and Development Laboratory, Research Unit # 3294, Institute of Neurobiology Alfred Fessard # 2118, National Center for Scientific Research, Gif sur Yvette Cedex 91198, France

Institute of Biotechnology and Environment, University of Nha Trang, Nha Trang, Khanh Hoa 57000, Vietnam

Jean-Pierre Le Caer
Research Unit # 2301, Natural Product Chemistry Institute, National Center for Scientific Research, Gif sur Yvette Cedex 91198, France

Gilles Mourier, Robert Thai and Denis Servent
Molecular Engineering of Proteins, Institute of Biology and Technology Saclay, Atomic Energy Commission, Gif sur Yvette Cedex 91191, France

Qiang Liu, Di Wu, Na Ni, Huixia Ren, Chuanming Luo, Chengwei He, Jian-Bo Wan and Huanxing Su
State Key Laboratory of Quality Research in Chinese Medicine, Institute of Chinese Medical Sciences, University of Macau, Macao 999078, China

Jing-Xuan Kang
Laboratory for Lipid Medicine and Technology, Massachusetts General Hospital and Harvard Medical School, Boston, MA 02114, USA

Veronica Piazza, Marco Faimali and Francesca Garaventa
ISMAR—CNR Institute of Marine Science, U.O.S. Genova, Via De Marini 6, 16149 Genova, Italy

Ivanka Dragić, Kristina Sepčić, Tom Turk and Sabina Berne
Department of Biology, Biotechnical Faculty, University of Ljubljana, Večna pot 111, Ljubljana 1000, Slovenia

Elango Jeevithan, Bin Bao, Yongshi Bu, Yu Zhou, Qingbo Zhao and Wenhui Wu
Department of Marine Pharmacology, College of Food Science and Technology, Shanghai Ocean University, Shanghai 201306, China

Xiaohong Wang, Heinz C. Schröder, Vladislav Grebenjuk, Renate Steffen, Ute Schloßmacher and Werner E. G. Müller
ERC Advanced Investigator Grant Research Group, Institute for Physiological Chemistry, University Medical Center, Johannes Gutenberg University, Duesbergweg 6, D-55128 Mainz, Germany

Bärbel Diehl-Seifert
NanotecMARIN GmbH, 55128 Mainz, Germany

Volker Mailänder
Max Planck Institute for Polymer Research, Ackermannweg 10, 55129 Mainz, Germany
Medical Clinic, University Medical Center, Johannes Gutenberg University, Langenbeckstr. 1, D-55131 Mainz, Germany

Hsiao-Che Kuo, Hao-Hsuan Hsu and Young-Mao Chen
Laboratory of Molecular Genetics, Institute of Biotechnology, College of Bioscience and Biotechnology, National Cheng Kung University, Tainan 70101, Taiwan
Translational Center for Marine Biotechnology, National Cheng Kung University, Tainan 70101, Taiwan
Agriculture Biotechnology Research Center, National Cheng Kung University, Tainan 70101, Taiwan

Chee Shin Chua
Laboratory of Molecular Genetics, Institute of Biotechnology, College of Bioscience and Biotechnology, National Cheng Kung University, Tainan 70101, Taiwan
Agriculture Biotechnology Research Center, National Cheng Kung University, Tainan 70101, Taiwan

Ting-Yu Wang
Laboratory of Molecular Genetics, Institute of Biotechnology, College of Bioscience and Biotechnology, National Cheng Kung University, Tainan 70101, Taiwan
Translational Center for Marine Biotechnology, National Cheng Kung University, Tainan 70101, Taiwan

Tzong-Yueh Chen
Laboratory of Molecular Genetics, Institute of Biotechnology, College of Bioscience and Biotechnology, National Cheng Kung University, Tainan 70101, Taiwan
Translational Center for Marine Biotechnology, National Cheng Kung University, Tainan 70101, Taiwan
Agriculture Biotechnology Research Center, National Cheng Kung University, Tainan 70101, Taiwan
University Center of Bioscience and Biotechnology, National Cheng Kung University, Tainan 70101, Taiwan
Research Center of Ocean Environment and Technology, National Cheng Kung University, Tainan 70101, Taiwan

Ingrid Richter
Environmental Technology Group, Cawthron Institute, Private Bag 2, Nelson 7012, New Zealand
Environmental Technology Group, Victoria University of Wellington, P.O. Box 600, Wellington 6140, New Zealand

Andrew E. Fidler
Environmental Technology Group, Cawthron Institute, Private Bag 2, Nelson 7012, New Zealand
Maurice Wilkins Centre for Molecular Biodiscovery, University of Auckland, Auckland 1142, New Zealand
Institute of Marine Science, University of Auckland, Auckland 1142, New Zealand

Wei Zhang and Chris Franco
Department of Medical Biotechnology, School of Medicine, Flinders University, Adelaide 5001, SA 5042, Australia
Centre for Marine Bioproducts Development, Flinders University, Adelaide 5001, SA 5042, Australia
Australian Seafood Cooperative Research Centre, Mark Oliphant Building, Science Park, Adelaide 5001, SA 5042, Australia

Yadollah Bahrami
Department of Medical Biotechnology, School of Medicine, Flinders University, Adelaide 5001, SA 5042, Australia
Centre for Marine Bioproducts Development, Flinders University, Adelaide 5001, SA 5042, Australia
Australian Seafood Cooperative Research Centre, Mark Oliphant Building, Science Park, Adelaide 5001, SA 5042, Australia
Medical Biology Research Center, Kermanshah University of Medical Sciences, Kermanshah 6714415185, Iran

Natalia V. Zhukova
Institute of Marine Biology, Far East Branch, Russian Academy of Sciences, Vladivostok 690041, Russia
School of Biomedicine, Far Eastern Federal University, Vladivostok 690950, Russia

Index